Lecture Notes in Civil Engineering

Volume 1

Lecture Notes in Civil Engineering (LNCE) publishes the latest developments in Civil Engineering - quickly, informally and in top quality. Though original research reported in proceedings and post-proceedings represents the core of LNCE, edited volumes of exceptionally high quality and interest may also be considered for publication. Volumes published in LNCE embrace all aspects and subfields of, as well as new challenges in, Civil Engineering. Topics in the series include:

- Construction and Structural Mechanics
- Building Materials
- Concrete, Steel and Timber Structures
- Geotechnical Engineering
- Earthquake Engineering
- Coastal Engineering
- Hydraulics, Hydrology and Water Resources Engineering
- Environmental Engineering and Sustainability
- Structural Health and Monitoring
- Surveying and Geographical Information Systems
- Heating, Ventilation and Air Conditioning (HVAC)
- Transportation and Traffic
- Risk Analysis
- Safety and Security

More information about this series at http://www.springer.com/series/15087

Helena Cruz · José Saporiti Machado
Alfredo Campos Costa · Paulo Xavier Candeias
Nicola Ruggieri · José Manuel Catarino
Editors

Historical Earthquake-Resistant Timber Framing in the Mediterranean Area

HEaRT 2015

Springer

Editors
Helena Cruz
LNEC
Lisbon
Portugal

José Saporiti Machado
LNEC
Lisbon
Portugal

Alfredo Campos Costa
LNEC
Lisbon
Portugal

Paulo Xavier Candeias
LNEC
Lisbon
Portugal

Nicola Ruggieri
University of Calabria
Rende
Italy

José Manuel Catarino
LNEC
Lisbon
Portugal

ISSN 2366-2557 ISSN 2366-2565 (electronic)
Lecture Notes in Civil Engineering
ISBN 978-3-319-39491-6 ISBN 978-3-319-39492-3 (eBook)
DOI 10.1007/978-3-319-39492-3

Library of Congress Control Number: 2016945964

Printed on acid-free paper

This Springer imprint is published by Springer Nature
The registered company is Springer International Publishing AG Switzerland

Contents

Part I Historical and Constructive Aspects

Cooperating Timber and Stone Antiseismic Frames in Historic Structures of Greece 3
Panos Touliatos

The Role of a Post-Byzantine Timber Roof Structure in the Seismic Behavior of a Masonry Building—The Case of a Unique Type of Timber-Roofed Basilicas in Cyprus (15th–19th Century) 17
Marios Pelekanos

Historic Earthquake-Resistant Constructive Techniques Reinforced by Wooden Logs in Algeria 33
Amina Abdessemed-Foufa

A Comparative Evaluation of the Results of Two Earthquakes: Istanbul and Lisbon Earthquake in 18th Century 47
Şevket Erşan

Some Examples of Turkish Houses with Wooden Frame in the Seismic Zone Anatolia 55
Ayten Erdem and Uzay Yergün

Traditional Timber Housing Structure in Zeyrek 75
Aylin Akcabozan

Possible Precursors of Pombalino Cage 87
Alice Tavares, Aníbal Costa and Dina D'Ayala

Local Seismic Cultures: The Use of Timber Frame Structures in the South of Portugal 101
Javier Ortega, Graça Vasconcelos, Hugo Rodrigues and Mariana Correia

The Use of Wood with an Anti-seismic Function in the Architecture of Palermo During the 18th Century 113
Tiziana Campisi and Federica Scibilia

Urban Planning and Building Reconstruction of Southern Italy after the 1783 Earthquake: The Case of Mileto 125
Stefania Stellacci, Francesca Geremia, Enrico Pagano, Nicola Ruggieri, Ginevra Salerno and Michele Zampilli

The Masonry Timber Framed Load Bearing Structure of the *Palazzo Vescovile* in Mileto (Italy) 135
Nicola Ruggieri, Francesca Geremia, Enrico Pagano, Ginevra Salerno, Stefania Stellacci and Michele Zampilli

A Multidisciplinary Approach to the Analysis of the Traditional Lefkada Houses 147
Sandra Tonna and Claudio Chesi

Part II Numerical Analysis

The "Gothic House" in Arquata Scrivia (Italy): Constructive Details and Seismic Assessment of a Timber Framed Masonry Structure 161
S. Cattari, S. Lagomarsino, R. Vecchiattini and G. Stagno

Seismic Assessment of *Pombalino* Buildings 171
Rita Bento, Ana Paula Falcão, Ricardo Catulo and Jelena Milosevic

Numerical Approaches for the Analysis of Timber Frame Walls 183
Elisa Poletti, Graça Vasconcelos, Paulo B. Lourenço and Maria Pia Ciocci

Seismic Vulnerability of Borbone Masonry Reinforced with Timber Frames 193
Nicola Ruggieri, Carmen Sandhaas and Ario Ceccotti

The Masonry Timber Framed Load Bearing Structure of "Baraccato" System: A Numerical Model 205
Ginevra Salerno, Francesca Geremia, Enrico Pagano, Michele Zampilli, Nicola Ruggieri and Stefania Stellacci

Load-Bearing Capacity of Traditional Dovetail Carpentry Joints with and Without Dowels: Comparison of Experimental and Analytical Results 215
Karel Šobra, Coralie Avez, Yasemin D. Aktaş, Richard de Rijk, Isabela Burawska and Jorge M. Branco

Part III Codes Analysis and Safety Assessment

Reliability of Timber Framed Constructions in Seismic Prone Areas in the Practice Codes Issued from 18th to 19th Century 229
Nicola Ruggieri

Assessment and Safety Requirements for Existing Structures 241
J.M. Catarino, Teresa O. Santos, Ana S. Louro and Helena Cruz

Building Regulations and Existing Buildings: Setting the Requirements on Structural Safety in Portugal. 253
João Branco Pedro and Vitor Campos

Seismic Assessment Procedures for Flexible Timber Diaphragms 263
Ivan Giongo, Gianni Schiro, Roberto Tomasi, Dmytro Dizhur and Jason Ingham

Part IV Diagnosis

The Role of Terrestrial LiDAR for Mechanical and Safety Evaluation of Historic Timber Floors. 277
Clara Bertolini-Cestari, Antonia Spanò, Stefano Invernizzi, Elisabetta Donadio, Tanja Marzi and Giulia Sammartano

Traditional Timber-Framed Infill Structure Experimentation with Four Scales Analysis (To Connection from a House Scale) 287
Yannick Sieffert, Florent Vieux-Champagne, Stéphane Grange, Philippe Garnier, Jean-Charles Duccini and Laurent Daudeville

Characterization of Timber Masonry Walls with Dynamic Tests. 299
Ana Gonçalves, Paulo Candeias, Luís Guerreiro, João Ferreira and Alfredo Costa

Assessment and Diagnosis of Two Collar Timber Trusses by Means of Visual Grading and Non-destructive Tests 311
Isabel D. Gomes, Fotis Kondis, Hélder S. Sousa, Jorge M. Branco and Paulo B. Lourenço

Retrofitting Assessment and Numerical Models for the Historical Timber Roof Structures of the Towers of Valentino Castle in Turin 321
Clara Bertolini-Cestari, Antonia Spanò, Stefano Invernizzi and Tanja Marzi

Part V Repair/Reinforcement

Development of Dissipation Panels for Seismic Retrofitting of Historic Buildings. 337
Raquel Paula, Vítor Cóias and Luís Guerreiro

Pombaline Buildings—Consequences of Some Rehabilitation Interventions 349
Jorge Mascarenhas, Lurdes Belgas and Fernando G. Branco

Experimental and Analytical Assessment of the Capacity of Traditional Single Notch Joints and Impact of Retrofitting by Self-tapping Screws 359
Karel Šobra, Richard de Rijk, Yasemin D. Aktaş, Coralie Avez, Isabela Burawska and Jorge M. Branco

Repairing of a Timber Truss Through Two Different Techniques Using Timber Elements and Screwed Metal Plates 371
Daniele Cibecchini, Niclas Bjorngrim, Hélder S. Sousa, Marco Jorge and Jorge M. Branco

Stiffness of Prosthetic Repairs for Historic Timber Beams 383
Coralie Avez and Thierry Descamps

Consolidation of a Collar Truss with a Steel Cable System 395
Giacomo Crivellari and Jorge M. Branco

A Timber Roof Structure for the Milreu Rural House Rehabilitation 405
Jorge N. Bastos

Design and Maintenance of a Long-Span Timber Roof 415
Jorge N. Bastos

Part VI Timber Structures

Assessment of the Mechanical Characteristics of Timber Floors Through Simple Dynamic Identification 427
José Coelho, João M. Guedes, Bruno Silva and Tiago Ilharco

Assessment of the Mechanical Behaviour of Tabique Walls Through Experimental Tests 437
Andreia Rebelo, João M. Guedes, Bruno Quelhas and Tiago Ilharco

Robustness Assessment of an Ancient Timber Roof Structure 447
Leonardo G. Rodrigues, Chrysl A. Aranha and Hélder S. Sousa

FP1101 and RILEM TC 245 RTE Training School on Assessment and Reinforcement of Timber Elements and Structures: The Role of the Academic Community in Disseminating Knowledge 459
Jorge M. Branco, Hélder S. Sousa, Eleftheria Tsakanika, Marco Jorge and Paulo B. Lourenço

Part I
Historical and Constructive Aspects

Cooperating Timber and Stone Antiseismic Frames in Historic Structures of Greece

Panos Touliatos

Abstract On the Kefalonia Island a Historical Building, dated to 1670, is located. The structure withstood successfully three catastrophic earthquakes in the past, until the severe earthquakes of 2014, during which it partially collapsed. Apart from the building's historic value, it actually constitutes a rare example of double load bearing structure. The main concept behind this double anti seismic system is the following: Masonry walls transfer the loads of the wooden structure of the floor and the roof, which was also simultaneously supported by a row of wooden posts placed in parallel position and close to the masonry, by means of a wooden beam which was based on them. When, during earthquakes, parts of the masonry fail to function properly, then it is this second parallel system of wooden skeleton frame which takes over the load transfer, until masonry walls are repaired. Moreover, this system demonstrates very advanced and detailed connections of the wooden load bearing structures, which have pre-scheduled absorption capacity of the dynamic energy released by the earthquake. Therefore, this enables the structure to absorb the seismic energy during the severe earthquakes in the region of the Ionian Islands, avoiding in this way the building's failure.

Keywords Anti-seismic behavior · Combined · Timber · Stone · Frames

Introduction

On the western part of the Kefalonia Island, at the eastern part of Lixouri Peninsula, a Historical Building, called "Samoli Mansion", dated back approximately to 1670, is located.

P. Touliatos (✉)
Department of Architecture, Frederick University Cyprus,
7, Y. Frederikou Street, 1036 Nicosia, Cyprus
e-mail: art.tp@frederick.ac.cy

P. Touliatos
N.T.U.A., Athens, Greece

H. Cruz et al. (eds.), *Historical Earthquake-Resistant Timber Framing in the Mediterranean Area*, Lecture Notes in Civil Engineering,
DOI 10.1007/978-3-319-39492-3_1

Fig. 1 Just after the strong seismic shock on the 2nd of February 2014, the roof with minimal deformations is still in place, despite the masonry total collapse. It is supported by the timber auxiliary bearing system, although seriously damaged, as well as by the two timber framed internal partition walls

The structure of great historical and constructional interest, withstood successfully three catastrophic earthquakes of 1768, 1867 and 1953 (which destroyed the whole island) until the earthquakes of 26th of January and 2nd of February 2014, during which it partially collapsed (Figs. 1, 2, 3 and 4).

During following months a systematic analysis of the monuments failure mode was carried out proving, finally, that the main reason of the destruction was the wrong repair intervention, after the disastrous 1953 earthquake, which neutralised most of the protective antiseismic techniques realised over 300 years ago (Fig. 2). It seems, also, that, since this building used to be, for a long period of time, the residence and the fortress of Venetian ruler of the Ionian Islands, that the Venetians had observed and understood the benefits of the local constructional system during severe earthquakes, constructing one of the best examples surviving today (Fig. 6).

Apart from the building's indisputable historic value, it actually constitutes a rareexample, for Kefalonia, of double load bearing structure that functions simultaneously and that, in the event that one of them fails, the other bears the full load until the first one is repaired (Fig. 9). This system can be widely found on the island of Lefkas (with some variations as it was recently highlighted), on the island of Mytilene, and features of this system can also be seen in the architecture of Mount Athos etc.

The main concept behind this double anti seismic system of the load bearing structures is the following:

Fig. 2 After a few days additional light column support was added to the remaining in place timber roof, because of the continuing strong and frequent seismic aftershocks. The constructed after the 1953 earthquake, concrete additions of the balcony plate and of the masonry's top zone, were some of the serious failure causes. Internal partition walls

Fig. 3 A saved, 1934, photograph showing all the amount of iron locks, connecting masonry to the floor and roof frame, which, except one, were canceled during the, after the 1953 earthquake, repair

Fig. 4 The structure as it was studied by the author, after the 1953 earthquake repairs and before 2014 earthquakes

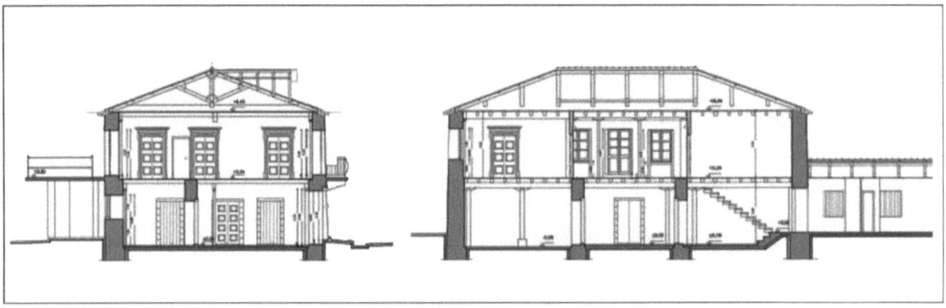

Fig. 5 The cross-section and longitudinal section of the monument, as it was studied by the author before 2014, showing the auxiliary timber frame besides the, load bearing, masonry. The roof wooden frame consists out of the cooperating, but, separate, ceiling and roof load bearing components

Masonry walls made of, usually, rather good quantity stonework, transfer the loads of the wooden structure of the floor and the roof, which were also simultaneously supported by a row of wooden posts placed in parallel position and close to the masonry by means of a corresponding wooden beam which was based on them (Figs. 5, 6, 8 and 9).

When, during the severe and frequent earthquakes in the region of the Ionian Islands, parts of the masonry fail to function properly, then it is this second parallel system of wooden skeleton frame which takes over the load transfer, until the masonry is repaired (Fig. 9).

Moreover, this system has demonstrated very advanced and detailed connections of the wooden load bearing structures, which have pre-scheduled absorption capacity of considerable amount of the dynamic energy released by the earthquake (Figs. 8 and 13). Therefore, this enables the structure to absorb the incoming seismic energy during the rather severe and frequent earthquakes in the region of the Ionian Islands avoiding, however, the building's failure or collapse.

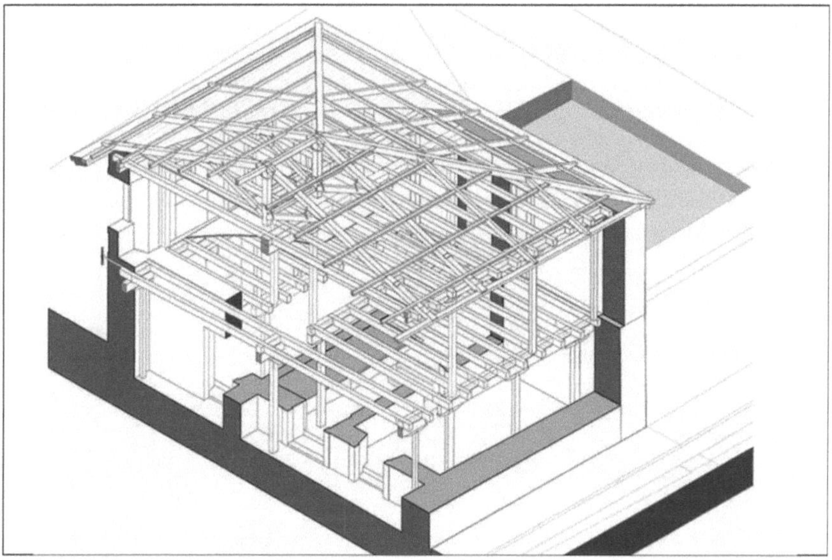

Fig. 6 The axonometric section of the double, timber and masonry, load bearing parallel frames, belonging to the characteristic, hyperstatic, antiseismic traditional constructional system of the, heavily seismically endangered, Ionian Islands of Greece, first studied by the author on the Island of Lefcas, where it still is in common use. The survey and analytical study of this, so called, "Samoli Mansion", which used to be for a long period of time the residence of the Venetian ruler of the Ionian Islands, proved that the Venetians understood the anti-seismic importance of this construction method, adopted it and created an excellent example, already, before 1700

(a) **(b)**

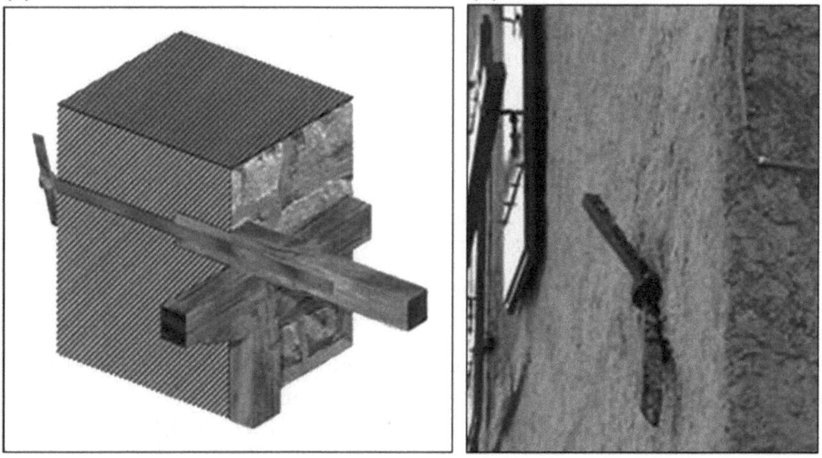

Fig. 7 **a** The axonometric representation of the timber secondary frame bonding to the stone walls. **b** The only one existing and deformed cast iron anchor, after the 1953 earthquake repair, when the rest anchors were eliminated, which, neutralized were found inside the existing roofing and roof load bearing components

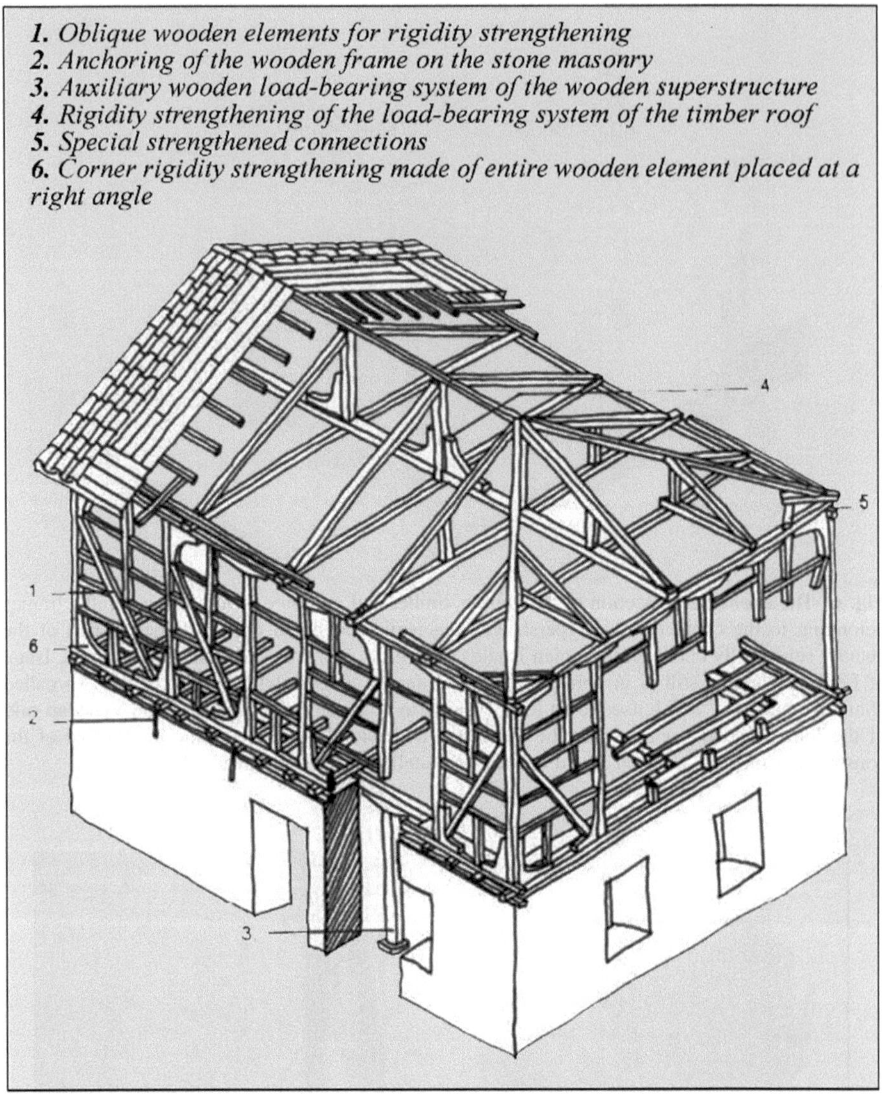

Fig. 8 Typical example of two-storey Lefkada architecture

This system, one of the main European Antiseismic Constructional Historical Methods, as it is nominated by the Council of Europe, PACT 19, uses mainly timber frame in close cooperation with masonry (Figs. 6 and 10)

Fig. 9 Failure mode

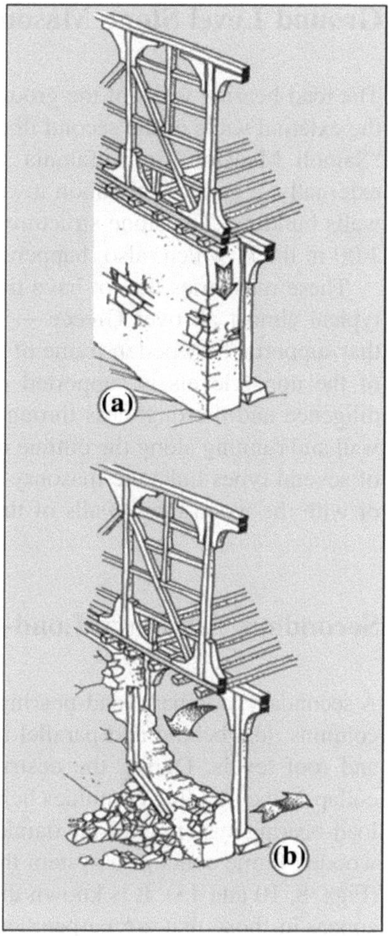

Foundations

The soil on which the city of Lefkada is built is very unstable. For this reason, good foundations were crucial. Unfortunately, it is not easy to locate some examples that would give us a more precise image of the construction. However, from the descriptions of the inhabitants, we know that it is a system of grillage, consisting of tree logs, placed on successive layers at a depth of 0.70–1.0 m or even deeper under the soil surface. The gaps between the logs were filled with sand, non-hammered stones and "portsolana". In the case of the "Samoli Mansion", which is actually built in the shallow water on the eastern shore of the Argostoli Bay, for defensive reasons, the foundation consists out of large stones carefully located in layers on the sand of the sea bottom.

Ground Level Stone Masonries

The load-bearing walls of the ground floor in Lefcas are made of stone. Sometimes the external walls of the second floor are, also, built in stone. That is the case of the "Samoli Mansion" in Kefalonia where, additionally, the ground walls present, externally, a gentle inclination towards inward, relevant to the Venetian manner of walls building. The stone structure is usually up to 3.00 meters high and 60 cm to 1.00 m thick, which, also, happens at "Samoli Mansion" (Figs. 5, 6 and 10).

These masonries do not have timber horizontal reinforcements on their body—typical almost all over Greece—, apart from the heading circumferential binding that supports the wooden frame of the upper levels (Fig. 10). The wooden skeleton of the upper levels is supported on these stone masonries. It is supported with diligence and thoroughness through wooden roof or floor sleepers built inside the wall and running along the outline of the masonries (Figs. 9 and 10). Iron anchoring of several types links the masonry with the wooden frame, of the roof or the floor, or with the wood-fitted walls of the upper level stories (Figs. 3, 6, and 7a, b).

Secondary Auxiliary Load-Bearing Wooden System

A secondary, auxiliary load-bearing system consisting of sparsely arranged wooden columns, just behind and parallel to the exterior stone masonry, supports the floor and roof levels. During the destructive earthquakes, when parts of the masonry collapse, the building continues being supported on the secondary auxiliary wooden load-bearing system until the damage is repaired (Figs. 6, 9 and 10). This auxiliary wooden frame contains a system that increases its rigidity (via proper connections) (Figs. 8, 10 and 13). It is known that the extended failures of the masonries mainly appear in those that are perpendicularly placed to the main direction of the earthquake and the rest, even broken, contribute to the rigidity of the whole construction.

Construction of the Upper-Level Floors

As already mentioned, the wooden load-bearing system of the superstructure's walls is supported on the ground-floor stone masonries. This support is achieved as follows: Along the masonries, parallel to one or both sides of the wall, wooden sleepers are built inside the masonry with cross-sections up to 20 × 20 cm. The beams of the floor, "patomatero", are supported on the sleepers and are approximately 40 cm apart from each other. It is obvious that on the walls that are parallel to the direction of the floor beams, the same extreme beam serves also as the sleeper (Fig. 10). Usually, the floor beams are linked to the wall sleeper via a proper groove. Vertical or slightly oblique handmade iron spikes were completing the

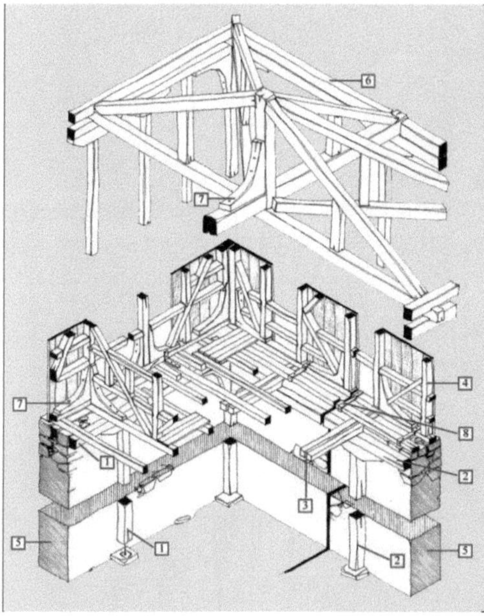

1. Auxiliary wooden load-bearing system next to the ground-floor masonries that support the floor beams.

2. Auxiliary wooden load-bearing system next to the masonries that are parallel to the floor beams.

3. After any failure of part of the masonry, the auxiliary wooden columns bear temporarily the loads of the superstructure through the short projections that are counterbraced by the floor beams.

4. Load-bearing system of the load-bearing wood-fitted floor walls.

5. Load-bearing masonry of the ground floor that supports the wood-fitted walls of the upper lev-els, which at the same time are supported also on the auxiliary system of wooden columns and beams, arranged internally and parallel to the masonries.

6. The load-bearing system of the roof is based on wood-fitted walls.

7. The single-piece wooden element ("elbow", "brazzoli") forming a vertical angle and connecting the horizontal to the vertical wooden elements, thus ensuring the connection and the required rigidity.

8. Anchoring of the sleeper of the wood-fitted wall that is parallel to the beams of the floor. This sleeper that supports the load-bearing wood-fitted walls of the floors is not directly linked to the floor beams, as it happens with the sleeper of the walls that are perpendicular to the direction of the floor beams.

Fig. 10 Support and anchoring of the load-bearing wood-fitted walls of the upper levels in the masonry of the ground floor and the auxiliary, wooden load-bearing system behind it. An example from "Verykios Mansion" at the central square

connection. On these beams, towards the external side of the masonries and parallel to them, are mounted the wooden frames of the upper-level walls. This intention of the builders of this historical structural system to ensure to the maximum possible extent, a longer-lasting support between the masonry and the wooden load-bearing system, is also proved by the oversized iron "anchors" that linked them one to the other (Figs. 6 and 7a). In case the ground floor masonries collapse, they usually fall outwards, and the rest of the building, i.e. the wooden structure, remains supported on the auxiliary woodenload-bearing system (Fig. 9). The "brazzoli" or "corners" are nailed at least on the four corner columns, connecting them to the corresponding beam. These are single pieces of wood made usually of olive trees, which are selected so that they form a natural right angle (Figs. 9, 10 and 13). Their placement ensures the stability of the construction's right angles and its rigidity in the event an earthquake hits. The "column capitals", i.e. equally projecting consoles that are connected to the column in a half-carved way, are mounted on the remaining columns (Fig. 7a). In some, older, structures we even noted some diagonal wooden elements that would render rigid the auxiliary wooden columns of the ground floor.

These details prove the attempts of the traditional builders to ensure the minimum rigidity required on the auxiliary wooden load-bearing system of the ground floor, when part of the masonry fails.

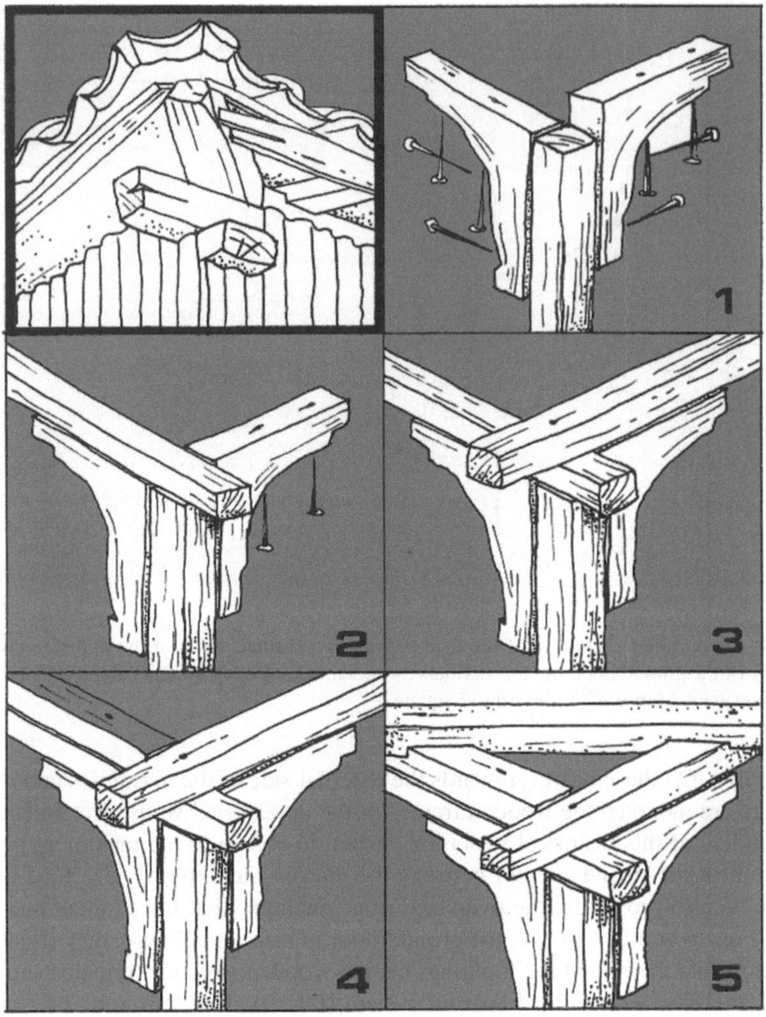

Fig. 11 Typical connection detail of timber roof elements

First-Floor Load-Bearing Walls

The location of the upper floor's external walls was mentioned earlier. These are built on the floor beams and towards the external side of the masonries (Figs. 8 and 10). First a sleeper is mounted along the perimeter. On the small sides, this sleeper is the same, as mentioned, as the extreme "patomatero". On the longer sides, it is placed perpendicular to the "patomatera" and level with the sleeper that lies underneath them. The whole construction ensures the transfer of the loads of the overlying structure to the masonry and the interior columns (Fig. 10).

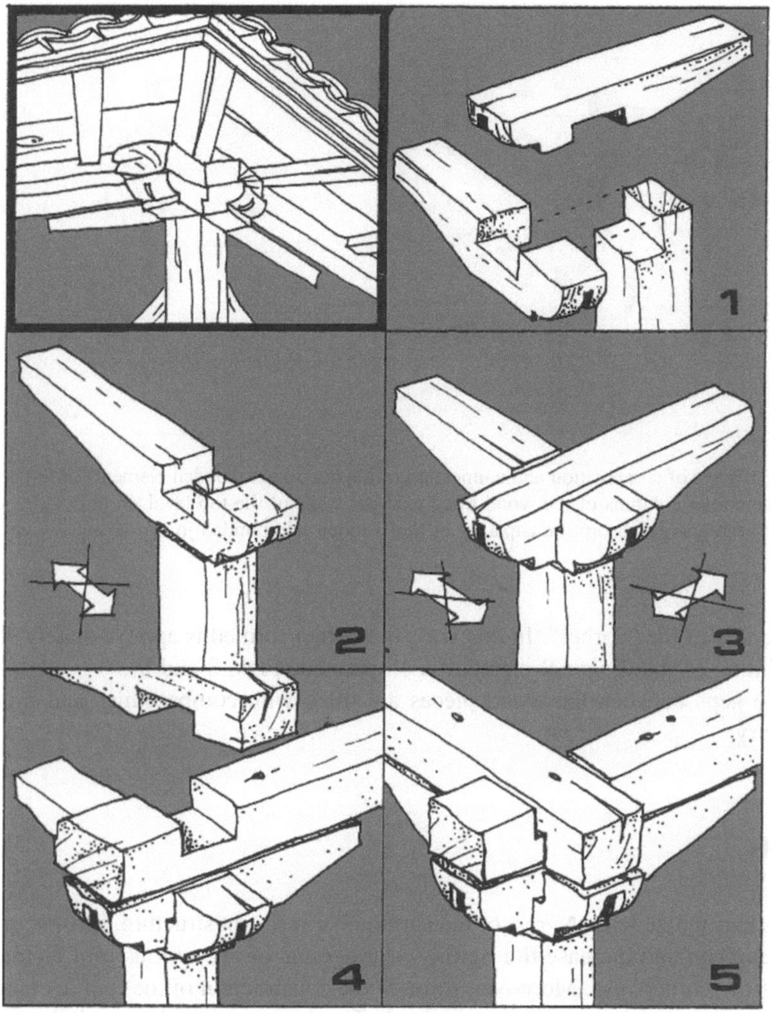

Fig. 12 The hyperstatic "Post and Beam" construction method of the roof

Both, on the upper and lower parts of the corner columns, the "corners" are nailed (Fig. 10). These are fixed using nails and grooves and face alternately inwards and outwards from the building. The upper sleepers of the walls are fixed on them. In the case of a second wooden level, these sleepers support its floor, otherwise they constitute the base for the roof (Fig. 10). The "corners" on the connections of the wall are followed by the "meintania". These are diagonal wooden elements that also contribute to the wall's rigidity. Their proper function is based on their correct placement: the "meintani" usually starts from the base of the upper "corner" of a corner column and ends at the lower part of a middle column, at

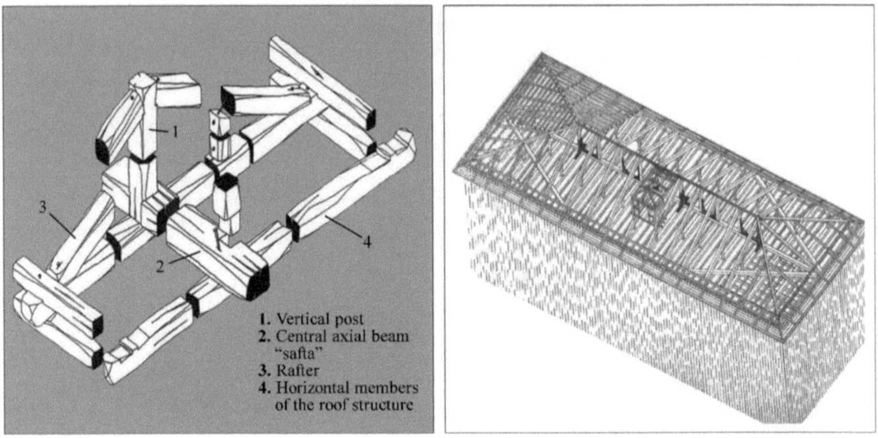

Fig. 13 Ways of composition and connection of the horizontal wooden elements of the perimeter with the corner columns of the wood-fitted external walls of the top level. In both cases effort is given to the energy absorption capacity of the wooden interbonded joints during severe seismic events

the middle of the "corner". In this way, the corner formed is approximately 45° and the sliding of the diagonal element is eliminated (Figs. 8 and 9).

The gaps between the wood pieces are filled with compact tiles and mortar or masonry.

Roofs

Roofs harmonize with the rest of the earthquake-resistant structure. Efforts are made to strengthen and increase the rigidity of the joints of the various roof elements by means of "corner" use. Moreover, through the construction of the roof, a connection between the exterior walls is attempted. The hyperstatic "Post and Beam" construction method is used (Figs. 11, 12 and 13).

Conclusions

This study, once more, proved that the recently well studied traditional Antiseismic Constructional System of Lefcas Island was known and followed, at least 300 years ago during Venetian occupation, not only on that island, but also on the central and the largest of the Ionian Islands, Kefalonia. This constructional method emphasises the significant role of timber in achievement of high structures performance under severe dynamic loading.

It is, also, useful to notice how easily traditional techniques can be forgotten, misunderstood and abandoned during present times, especially under the strong influence of the simplified, highly standardised, isostatic modern concrete structures deontology.

Special effort should be given to detect, analyse and preserve the precious building techniques of the past and to try to repair and strengthen them before it is too late.

General References

- Touliatos P., Report of the Greek Experience Concerning the structural Behavior of Timber Construction in Seismic Zones, Dipartimento di IngeneriaCivile, Universita di Firenze, Commission of the European Communities DG—III
- Touliatos P., Seismic Disaster Prevention in the History of structures in Greece, Seminarion International SociedadPrevencion de Desastres, 23 al 25 de Febrero de 1994, Coordinacion de Humanidades – UNAM
- Touliatos P., Design Problems of Timber Framed Construction in Seismic Zones. 1991 International Timber Engineering Conference London
- Ganted,Mitropoulos M.,Lefkas, Earthquake Building Technology of the 19 Century, 1991 International Timber Engineering Conference London
- Argyrou K., Filippa M., Lefkokiliou S., The Architecture of Lefkada, 1971
- Gante D., Mitropoulos M., Lefkas—Traditional atiseismic structures, Athens 1989
- Papadatou X., The Traditional Architecture of Lefkada, Athens 1978

The Role of a Post-Byzantine Timber Roof Structure in the Seismic Behavior of a Masonry Building—The Case of a Unique Type of Timber-Roofed Basilicas in Cyprus (15th–19th Century)

Marios Pelekanos

Abstract In the Troodos area of Cyprus there is a distinctive type of wooden roofed basilicas, dated from the mid-15th to the late-19th century AD, whose unique constructional feature is the existence of two separate but cooperating parts of the roof, the Inner Roof and the Outer Roof. The Inner Roof forms a wooden rigid triangular prism, with horizontal tie-beams and dense inclined rafters, from which the Outer Roof is literary suspended. The two pairs of tie-beams at the western and eastern end of the church are forming, along with the two composite beams on the northern and southern walls, a full timber circumferential binding. This roof type fully adopts the principle of a uniform distribution of loads on multiple paths, and obtains its overall stiffness through numerous partially semi-rigid joints. To achieve this goal, this particular roof structure was developed through years into a unique and cleverly designed bearing system, retaining a high level of simplicity in all construction details. As far as the overall structure is concerned, the timber-roofed churches of Cyprus present a unique system of co-operation between the rigid roof and the stone masonry, achieving in this way the absolutely necessary stiffness for both.

Keywords Timber · Roof · Masonry · Church · Anti-seismic behavior

Introduction

In the island of Cyprus there is a distinctive type of wooden roofed basilicas, dated from the mid-15th to the late-19th century AD, which, according to many researchers, is unique in the European and Mediterranean area. There are more than 130 churches of this type, all of them located in the Troodos Mountain Area, which

---

M. Pelekanos (✉)
Department of Architecture, Frederick University Cyprus,
7, Y. Frederikou Street, 1036 Nicosia, Cyprus
e-mail: art.pem@frederick.ac.cy

H. Cruz et al. (eds.), *Historical Earthquake-Resistant Timber Framing in the Mediterranean Area*, Lecture Notes in Civil Engineering,
DOI 10.1007/978-3-319-39492-3_2

is the highest of the two mountain regions in Cyprus [1]. As far as typology is concerned, most of the churches are single-aisled (75 %) or three-aisled (20 %) (Figs. 1 and 2). There are also some two-aisled churches (5 %), most probably being the result of an extension of a single-aisled church.

Fig. 1 A single-aisled timber-roofed church, dated in the 18th century (Saints Sergios and Vakhos, Kalopanayiotis village). East façade, showing the opening on the niche of the sanctuary, which is inscribed within the rectangular volume of the church [5]

Fig. 2 A three-aisled timber-roofed church, dated in the 17th century (Ayia Marina, Filousa Kelokedaron village). West façade, showing the wooden shelter over the main entrance [6]

The steep inclination of the roof was associated, mainly morphologically, with the western European roofs and was attributed to the influence by the Franks, who ruled the island from the 13th to the 16th century [2]. As it will be suggested, this association does not take into consideration the findings of a thorough constructional analysis, which strongly links this particular roof with the eastern-type construction systems.

There are many theories about the origin of these churches. Some of them point as the most important factor the need to protect the churches from the heavy rain and snow in the Troodos Mountains [3]. However, this suggestion cannot explain (a) the fact that this church type is found as low as 200 m from the sea level, and (b) the fact that in the Pentadaktylos Mountain Region there is not a single church of this type.

One close look to the geological map of Cyprus reveals that another factor is very important: The margins of the Troodos Volcanic Rock Zone (Fig. 3). Most of the 130 churches are situated within these margins and only 5 of all were constructed, just outside the southern boundary, in the 18th century. The local stone of Troodos is so hard that it cannot be carved and used in arches and domes. It is suggested that this fact, along with the plenty timber from the local pine forests, played the most important role in the evolution of this construction type.

The stone walls are constructed with the volcanic stone of Troodos. The two long walls, which bear the load of the roof, are usually short, about 2.5–3.0 m, while the west wall ends up with a pointed shape up to 6 m in single-aisled and 8 m in three-aisled churches (Fig. 4). To the east we found the niche of the sanctuary, which sometimes is inscribed in the rectangular shape of the church.

The openings are few, usually a door to the west or the south, a small window high up on the west wall and a small opening-slot in the axis of the niche of the sanctuary. Inside, the walls are coated with plaster and in many cases covered with

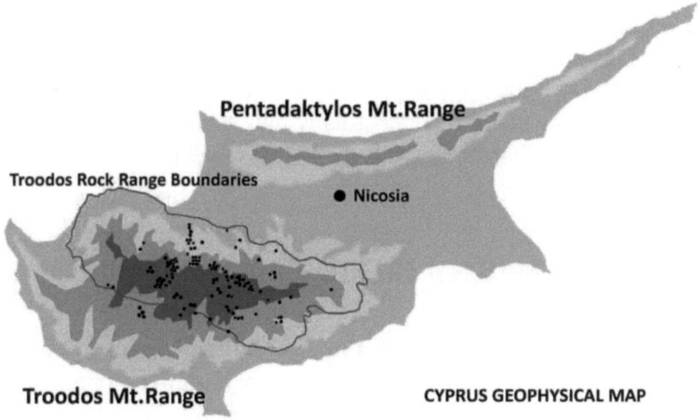

Fig. 3 The location of all timber-roofed basilicas and the relation to the boundary of the Troodos rock range (M. Pelekanos)

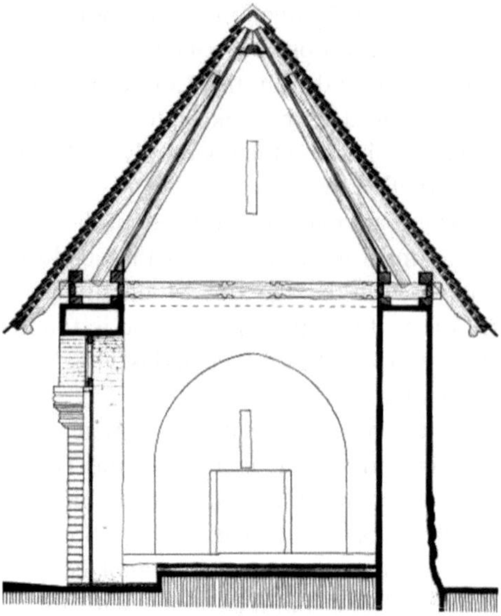

Fig. 4 Saints Sergios and Vakhos, cross section facing east [5]

frescoes. The foundations of the wall are usually very shallow, and in some cases they are founded at different levels, due to an inclined terrain.

The Construction of the Roof

These churches differ strongly from the other churches in Cyprus, not only morphologically, but also due to the uniqueness of the construction system of their wooden roof. The unique constructional feature of the wooden roof churches of Troodos is the existence of two, distinctive, but cooperating parts of the roof. The Inner Roof, a rigid triangular prism, is based on the southern and northern walls. The Outer Roof, which carries the heavy tiles, is actually suspended from the Inner Roof, and it overhangs the two long walls, protecting in this way their upper part from the rainwater. In many cases the roof covers totally the sanctuary niche to the east, while on the west the roof overhangs the wall by 40–50 cm (Figs. 5, 6, 7 and 8).

The Co-operation Between the Roof and the Walls

The load of the roof is totally carried, through the two composite beams, on the southern and northern walls. In relation to the maximum compressive strength of the walls, the static load from the roof is relatively small. The main question is the way, in which the entire structure behaves, when the dynamic load of an earthquake is applied.

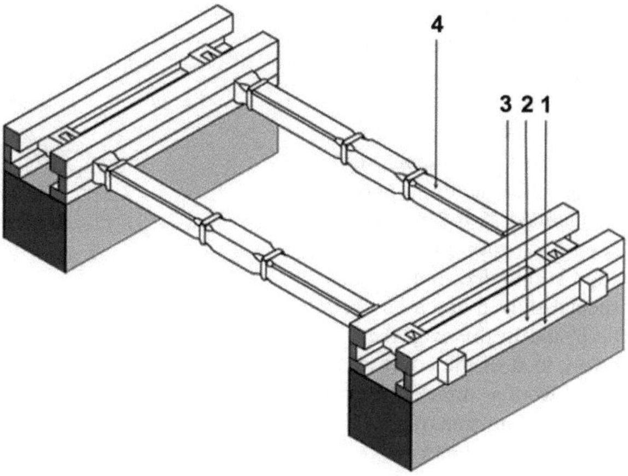

Fig. 5 The construction of the inner roof starts with the placement of two wall-plates (*1*) on which tie-beams (*4*) are connected. Between tie beams, pieces of wood in an upright position (*2*) are placed to keep the right distances. Two square shaped beams (*3*) are then completing and integrating the two composite "box-like" beams, on the north and south walls, to the transverse tie-beams (M. Pelekanos)

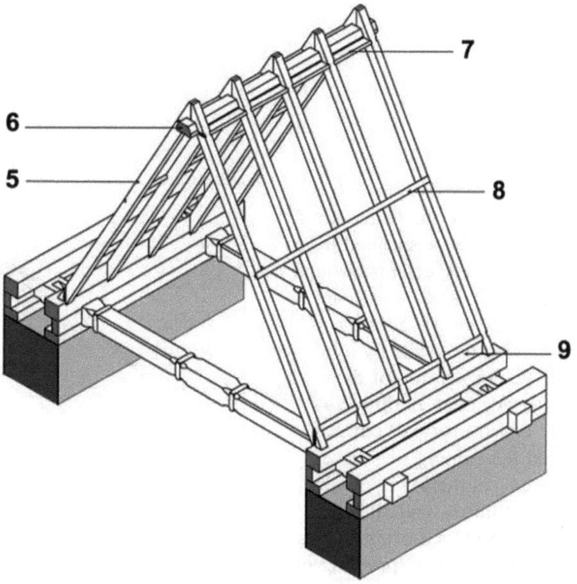

Fig. 6 Rafters (*5*) are placed diagonally every 30–40 cm and hold the connective ridge-purlin ("Carina") (*6*) on the top of the Inner Roof. Several auxiliary wooden sections (*7*, *9*) are placed in order to close firmly the gaps of the interior cover. A special triangular shaped wooden section (*8*) is penetrating the surface of the rafters, in order to connect them firmly to the wooden boards (*10*) (M. Pelekanos)

It is well known that masonry walls cannot resist significant dynamic loads perpendicular to their plane. In terms of vulnerability, the northern and southern walls present the lower (due to their low height and their direct connection to the roof), followed by the wall of the Sanctuary. The lower part of the eastern wall presents significant resistance to out of plane loads, due to the hemi-cylindrical construction of the niche.

Fig. 7 The inner roof is completed with the placement of the wooden boards (*10*) along and over the rafters. A special horizontal beam (*11*) is then nailed temporarily to the rafters and is supported by inclined strut-supports (*12*). Each support's lower edge is founded in a special-shaped notch on every tie-beam. In this way the heavy load of the coming Outer Roof is directed mainly on the strong transverse tie-beams (M. Pelekanos)

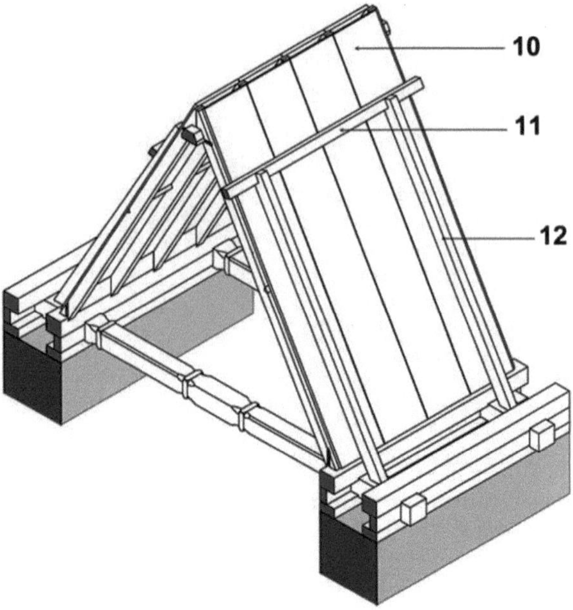

Fig. 8 After the Inner Roof is completed, the Outer Roof is ready to be suspended from it, using the special horizontal beam (*11*). Rafters (*13*) are placed every 35–50 cm and overhang the walls by 50–60 cm. On top of the rafters wooden purlins (*14*) are nailed every 10 cm, to hold the coming tiles. Every tile covers the lower one by its two thirds, and all joints are crossed, achieving in this way the full waterproofing of the Roof (M. Pelekanos)

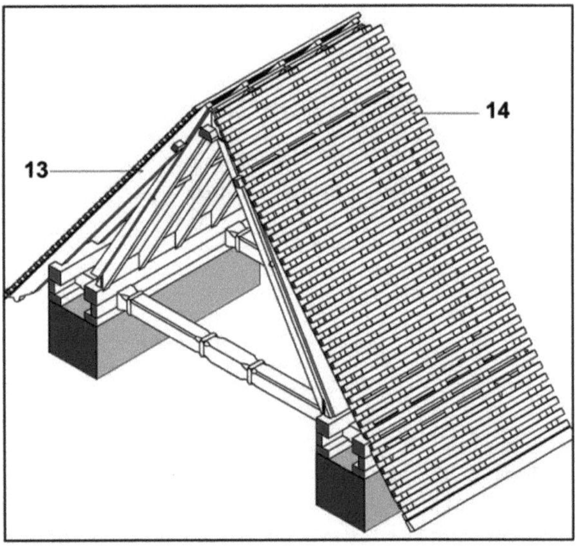

Fig. 9 Church of Panayia (St. Mary) at Kourdali (three-aisled, 16th century). Three dimensional simulation showing the horizontal part of the Inner Roof, with the series of tie-beams in the middle aisle. Two pairs of these tie-beams are embracing the western and the eastern walls, achieving in this way the co-operation of the wooden roof with the stone masonry. Furthermore, these four tie-beams, along with the two composite beams over the series of wooden posts, form a belting system to secure a safe movement of the building during an earthquake [4]

The highest level of vulnerability is presented by the western wall, due to its height, the absence of ribs, the poor connection to the roof and the lack of a significant static load. To confront this weakness, the builders placed two tie-beams in contact with the western wall, one internal and one external (Fig. 9). Despite their awareness about the problems caused by the exposure of the wooden beam in severe weather conditions, the decision was considered critical to achieve the full co-operation of the wooden roof and the stone masonry [4].

Finally, the tie-beams at the western and eastern end of the church are forming, along with the two composite beams on the northern and southern walls, a full timber circumferential binding, well known from the ancient and Byzantine times as "Imantosis" (from the Greek word "imantas" which means "belt"). The important role of these unique tie-beams for the whole structure can be clearly shown in cases, where the failure of the beam leads to a significant crack on the stone masonry's corner (St. Sergios and Vakhos at Kalopanayiotis village) (Fig. 10) [5].

Fig. 10 Church of Saints Sergios and Vakhos at Kalopanayiotis (single-aisled, 18th century). The exposed tie-beam on the western wall was gradually decayed and finally cracked. Subsequently the stone masonry cracked very near the corner connection, due to the lateral thrusts, that were no longer counterbalanced by the tie-beam. It is worth noting that the width at the highest point of the crack on the stone wall has the same width as the crack on the wooden beam. In this case there was no internal tie-beam near the western wall [5]

Case Study—The Church of Ayia Marina in Filousa Kelokedaron

The anti-seismic behaviour of a wooden roof structure is very difficult to be completely simulated and fully analysed by computer models, although this is currently the most accurate scientific method for approaching the truth. For the wooden roof churches of Troodos there is a unique case which confirms, in practice, the important role of the wooden roof in the overall seismic behaviour of the church.

The Church of Ayia Marina is located in the village Filousa Kelokedaron and is dated back to the 17th century (Figs. 11 and 12). It is one of the 26 three-aisled wooden-roofed churches and one of the few that are constructed with the local calcareous stone, as it was built just outside the geological boundaries of the Troodos Rock region. The general dimensions are 12.50 × 17.50 m and it is well known and of great interest because of the inclination of its walls [6].

Fig. 11 Church of Ayia Marina in Filousa Kelokedaron (17th century). South-east façade. The campanile was erected at around 1900 and the four buttresses at around 1997 [6]

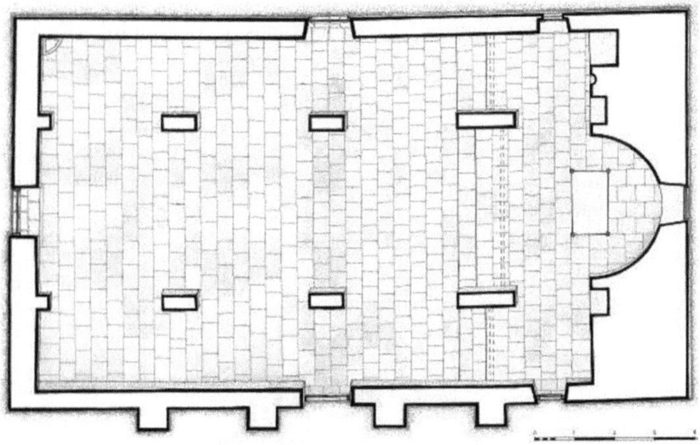

Fig. 12 Plan of Ayia Marina, showing the deformation of the northern and southern walls, due to the inclination towards south-west [6]

At this church a rare phenomenon is recorded, that of a gradual inclination of the entire construction, most probably due to a foundation system differential settlement. The research project carried out by Frederick University in 2012–2013 recorded that some walls are inclined up to 3.8 % (which is more than the inclination of the Pisa Tower in Italy). The construction's inclination seems to be increasing very slowly until today (Figs. 13 and 14).

Some walls and arches of the church present inclination close to the limits to be overturned. Despite that, the church did not overturn until now because of the "box-like" integral behaviour of the whole structure and the corresponding stiffness

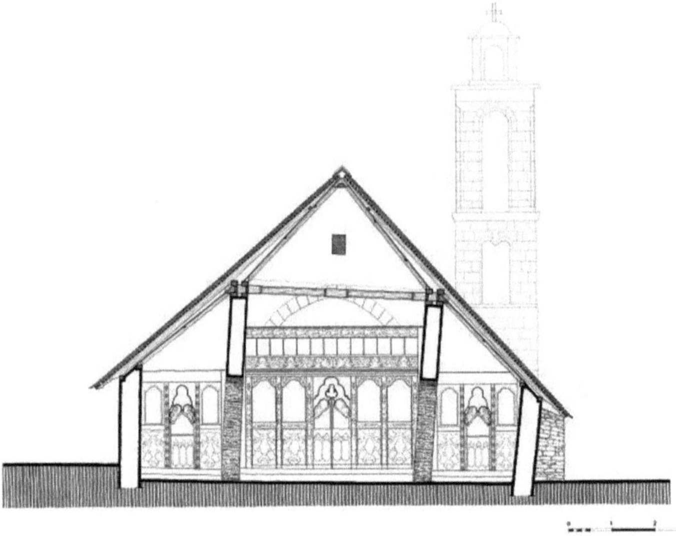

Fig. 13 Cross section of Ayia Marina, facing east. It shows that all walls on the east-west direction are inclined, with the southern being at the limits of overturn [6]

Fig. 14 East façade of Ayia Marina. The lower window was opened at around 1900, when the church was already inclined [6]

Fig. 15 Ayia Marina in Filousa Kelokedaron. Three-dimensional model showing the horizontal structure of the Inner Roof, with the series of tie-beams, and how this structure embraces the western wall, creating at the same time a timber circumferential bracing [6]

Fig. 16 Ayia Marina. The horizontal beam on the Inner Roof, supported by inclined strut-supports. On these special beams the Outer Roof will be totally suspended. Every couple of inclined strut-supports corresponds to a horizontal tie-beam of the Inner Roof [6]

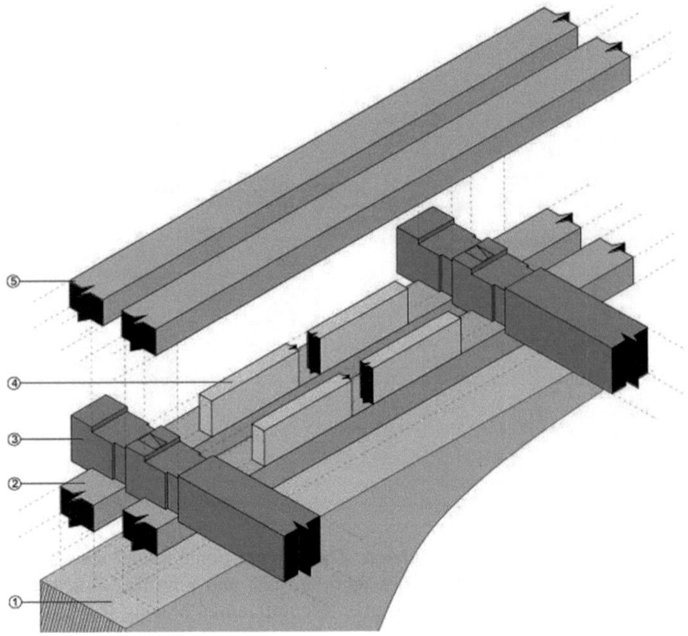

Fig. 17 Ayia Marina. Detail showing the construction of the composite "box-like" beam along the southern and northern walls. The horizontal tie-beams are locked by three couples of beams, without any nailing. The system stays firm due to the special interconnection of the beams and the support by the heavy loads of the Outer Roof [6]

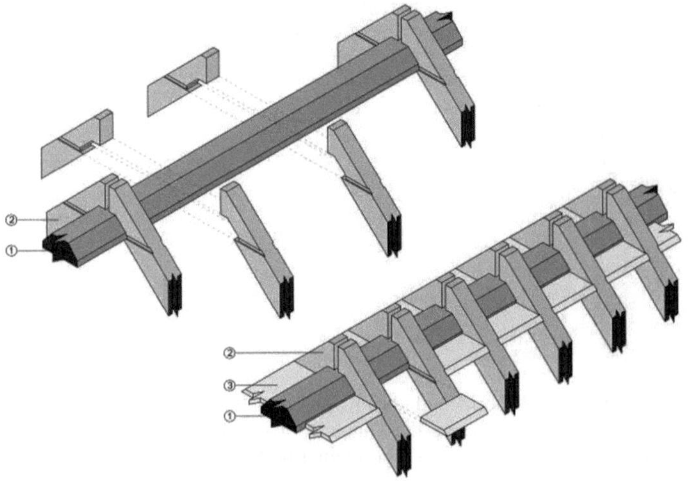

Fig. 18 Ayia Marina. Detail showing the construction of the highest part of the Inner Roof, where rafters hold the connective ridge-purlin ("Carina"). Horizontal auxiliary wooden sections are placed between the rafters in order to close firmly the gaps of the interior cover when the wooden boards are nailed on top [6]

of the wooden roof (Figs. 15, 16, 17 and 18). This case can be recorded as an experimental simulation, in full scale, which clearly demonstrates the important role that this wooden roof system plays in confronting the dynamic loads on the entire structure.

Conclusion

The wooden-roofed churches of Cyprus, which were built during the period from the 15th to the 19th century, present a remarkable uniqueness. Some researchers suggested that the origin of this type derives from the morphologically similar western wooden roofs. This connection, however, cannot be fully adopted, as it is based solely on a superficial observation.

A comprehensive constructional analysis showed that most of the characteristics of this type of construction are similar to the eastern type roofs: The timber circumferential binding ("Imantosis"), the "box-like" integral behaviour of the entire structure, the uniform distribution of the loads throughout the structure, the relative flexibility of the connections and its hyper-static behaviour. Timber roofs in the seismic area of the Eastern Mediterranean are constructed in such a way, mainly to cope successfully with the frequent earthquake phenomenon (Figs. 19, 20 and 21).

The timber-roofed churches of Cyprus, built between the 15th and the 19th century, fully adopt the principle of a uniform distribution of loads on multiple

Fig. 19 Church of Panayia at Kourdali. Three dimensional model showing the structure of the separate inner roofs of the three aisles. The outer roof covers all parts and is suspended from the inner roof of the central aisle [4]

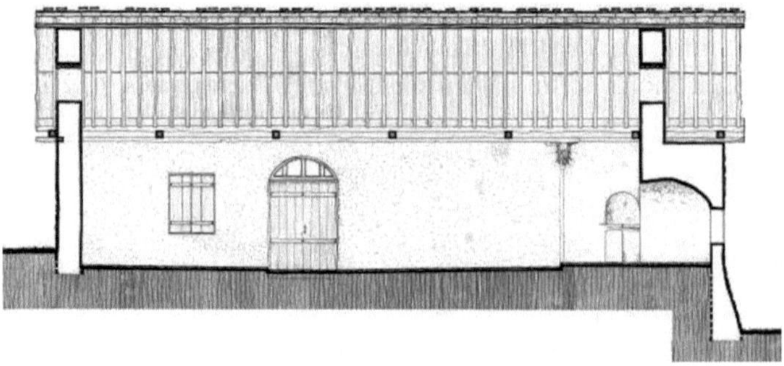

Fig. 20 Church of Saints Sergios and Vakhos at Kalopanayiotis. Longitudinal section facing north. The structure of the Roof is extended over the niche of the Sanctuary which is inscribed within the box-volume of the church. Due to the slope of the ground, from west to the east, special construction techniques were adopted to anchorage the stone walls to the ground. The lower part of the eastern wall was built as a retaining wall, while in several positions the wall was built interlocked with the rocky ground [5]

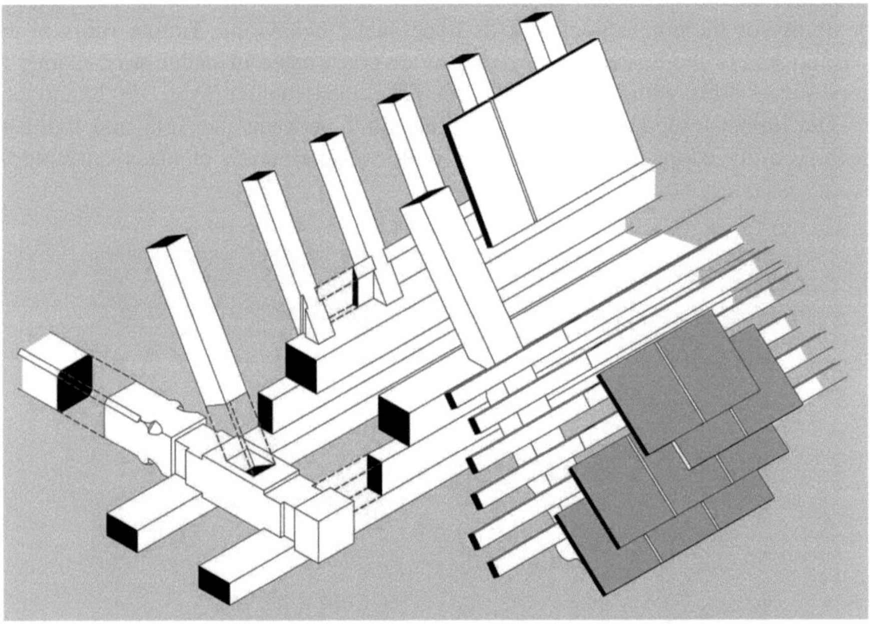

Fig. 21 Church of Saints Sergios and Vakhos at Kalopanayiotis. Construction detail showing the relation between the composite box-like beam, the horizontal tie-beams, the rafters and the wooden boards of the Inner Roof, the strut-support and its connection to the tie-beam, the rafters of the Outer Roof, the wooden purlins and the tiles [5]

Fig. 22 Church of Panayia at Kourdali. General view from north-east [4]

paths, and obtain their overall stiffness through numerous partially semi-rigid joints. To achieve this goal, this particular roof structure was developed through years into a unique and cleverly designed bearing system, retaining a high level of simplicity in all construction details. As far as the overall structure is concerned, the timber-roofed churches of Cyprus present a unique system of co-operation between the rigid roof and the stone masonry, achieving in this way the absolutely necessary stiffness for both (Fig. 22).

References

1. Papageorgiou Ath (1975) The timber-roofed churches of Cyprus. Nicosia, pp 4–22
2. Soteriou G (1931) Early christian and Byzantine monuments of Cyprus. Athens, p 477
3. Jeffery G (1931) A Byzantine monastery in Cyprus. The Builder, no 3349, p 438
4. Pelekanos M (2012) Constructional analysis of the church of Panayia at Kourdali. Frederick University Cyprus (team leader and scientific coordinator)
5. Pelekanos M (2013) Constructional analysis of the church of Saints Sergios and Vakhos at Kalopanayiotis. Frederick University Cyprus (team leader and scientific coordinator)
6. Pelekanos M (2014) Constructional analysis of the church of Ayia Marina at Filousa Kelokadaron. Frederick University Cyprus (team leader and scientific coordinator)

Historic Earthquake-Resistant Constructive Techniques Reinforced by Wooden Logs in Algeria

Amina Abdessemed-Foufa

Abstract The seismicity of northern Algeria is light to moderate and thus a moderate-seismic hazard country. Except the Orléansville 1954, El Asnam 1980 and Boumerdes 2003 earthquakes which happened during the twentieth century, the Algerian country has experienced in the past several moderate seismic events that caused loss of human lives and damage to property in different regions. The historical earthquakes known since ancient time Rusucurru (42) (Central region), Lambaesis (262) (Eastern region), Mostaganem (345) (Western region) and during recent time Algiers (1365, 1716), Cherchell (1732), Dellys (1731), Oran (1780), Blida (1825, 1857), Jijell (1856), Constantine (1858) and Biskra (1869), greatly increases the probabilities of an earthquake occurring and damage in the future. Based on this information, fundamental questions of interest for the architects and engineers are: how did the local population build their houses resisting earthquake loads? What structural techniques did the local population use for building their houses to protect themselves against earthquakes? This paper present the timber-frame earthquake resistant techniques developed such a local seismic culture in northern Algeria cities observed in the Casbah of Algiers and Miliana (Central Algeria), Mostaganem (Western region) and in Batna Highlands region (Eastern Algeria).

Keywords Earthquake-resistant · Timber · Frames · Algiers · Mostaganem · Batna · Algeria

A. Abdessemed-Foufa (✉)
Lab ETAP, Institute of Architecture and Urban Planning,
University "Saad Dahlab" of Blida 1, Blida, Algeria
e-mail: aafoufa@gmail.com

H. Cruz et al. (eds.), *Historical Earthquake-Resistant Timber Framing in the Mediterranean Area*, Lecture Notes in Civil Engineering,
DOI 10.1007/978-3-319-39492-3_3

Introduction

From the religious interpretation, the seismic resistance of the structure strengthened with timber revealed in the book of Sirach (Vulgate or Hebrew texts) translated in Latin by Saint Jerome and later in French by Lemaistre de Sacy in 1759 said "*comme le bois bien lié et attaché ensemble sur le fondement d'un édifice ne se désunit point, ainsi le coeur établi sur un conseil solide demeurera ferme*" [1]. The Bible of Jerusalem translated from the Hebrew in 1973 said: "*Une charpente de bois assemblée dans une construction ne se laisse pas disjoindre par un tremblement de terre; un cœur résolu, après mûre réflexion, ne se laisse pas émouvoir à l'heure du danger.*"Another translation of Hebrew texts by André Chouraki in 1987 said [2] "*Une charpente de bois assemblée, intégrée dans un bâtiment n'est pas ébranlée par un séisme; ainsi d'un cœur appuyé par la pensée d'un conseil, en tout temps, il ne tremble pas.*" All these French translations agree that a timber integrated in a building is not shaken by an earthquake. In the Muslim religion, the earthquake is referred in numerous signs, but no reference to the earthquake resistance of construction is reported.

In Northern Africa and accurately in the Maghreb region, Algiers, Fez and Tunis have developed after the great historical earthquakes respectively in 1716, 1624 and 1758 a seismic culture that resulted the development of earthquake resistant constructive techniques reinforced by timber [3]. The seismicity of northern Algeria is light to moderate and thus a moderate-seismic hazard country [4]. Except the Orléansville 1954, El Asnam 1980 and Boumerdes 2003 earthquakes which happened during the twenty-first century, the Algerian country has experienced in the past several moderate seismic events that caused loss of human lives and damage to property in different regions. The historical earthquakes known since ancient time Rusucurru (42 AD) (Central region), Lambaesis (262 AD) (Eastern Highlands region), Mostaganem (345 AD) (Coastal Western region) [5] and between the fourteenth and nineteenth centuries Algiers (1365, 1716), Cherchell (1732), Dellys (1731), Oran (1780), Blida (1825, 1857), Jijell (1856) in central region and Constantine (1858) and Biskra (1869) in Eastern region and M'sila (1885) in Eastern Highlands [3, 4, 6], greatly increases the probabilities of an earthquake occurring and damage in the future. Based on this information, fundamental questions of interest for the architects and engineers are: how did the local population build their houses resisting earthquake loads? What structural techniques did the local population use for building their houses to protect themselves against earthquakes?

The structural failure of historical buildings during earthquakes has often led to a better understanding of their performance and improvements in their design. This will emphasize on the identification of traditional architecture seismic preventive techniques found in many vernacular buildings in Algeria that made eventually theses buildings resist such several earthquakes. These techniques are the results of a continuous learning circle of trial and error process which gradually improved along time to adapt and resist to the changing requirements of the physical seismic environment. This has allowed local population to develop a local seismic culture in Algeria cities which is observed in the Northern country such as the Casbah of Algiers, Cherchell, Ténès, Koléa, Miliana, Blida and Mostaganem (central Algeria) and in Eastern Highlands such as Batna and M'sila. The earthquake resistant constructive techniques reinforced by wooden logs observed in vernacular architecture are located in Northern and Eastern Highlands Algeria's historical cities.

Earthquake-Resistant Timber Frame in Algeria (Algiers, Miliana and Mostaganem in Central Algeria and Batna and M'sila in Eastern Highlands)

The main local seismic cultures features using wooden logs observed in different region in Algeria are [4, 7–10]:

- seismic isolation technique for foundation,
- reinforced masonry arches,
- reinforced load-bearing walls made of bricks, stone and earth masonry,
- reinforced pillars
- reinforced cupolas.

Some of these constructive techniques were used in Algiers after the 1716 earthquake. It were highlighted first during 2001 until 2004 while in situ investigation during the doctoral thesis research supported by the ministry of higher education and scientific research and later during 2010 when the restoration project of the Citadel began supported by the ministry of Culture. In Miliana, we discovered these techniques in the period 2008–2009 during the restoration project of the great mosque *Sid Ahmed Benyoucef*. During technical visit in 2012 of the ancient city "*Qalaa of Beni Hamad*", Wilaya of M'sila, we observed the reinforced stone and bricks masonry walls with wooden logs of the minaret which is still standing. In Amenthane and Menaa, Wilaya of Batna theses techniques were used first against landslides, but the local population has generalized it because they have noticed that it was effective during earthquake. Despite the moderate seismicity of this mountainous region, the population preferred to use it to avoid damage[1] (2015). We also observed the stone masonry walls reinforced by wooden logs at Mostaganem old nucleus (Coastal Western region) (2016). These techniques highlighted during 2015 and 2016 are the in situ investigation results of the Master students in Architecture and Cultural Heritage.

[1] According to oral tradition 2015.

Fig. 1 Seismic isolation foundations (Palace of the Dey, Citadel of Algiers) © Abdessemed-Foufa 2010

Seismic Isolation Technique

The seismic isolation technique was observed only at Algiers at the Palace of the Dey (Citadel of Algiers). It was used by putting two or three layers of cypress (Cupressaceae) wood logs called "*thuya*" under the foundation of some buildings, especially for monumental buildings such as the "Citadel", mosques and palaces. This system of superposition of two materials, one rigid and the other flexible guarantees, due to the slip movements or rolling, a good resistance to shear during earthquakes (Fig. 1).

Reinforced Masonry Arches

In Algiers and Miliana, arch presents a rather particular constructive technique. In fact the gallery arches are reinforced by logs of wood. It is found at the level of the articulation of the arc with the column, on the top of the capital, on the level of the abacus and at the departure of the arch. There is a superposition of three logs of *thuya* to a base of two brick layers, (Fig. 2) and perpendicular superposition of two lines of three logs of *thuya* to brick layers (Fig. 3).

This system of superposition of two materials, one rigid and the other flexible guarantees due to the slip movements or rolling which is a good resistance to shear during earthquakes. The system of arcade is thus preserved and this preventive

Fig. 2 Reinforced arches with one layer of cedar (Palace of the Dey, Citadel of Algiers) © Abdessemed-Foufa 2005

Fig. 3 Reinforced arches with 2 layers of cedar logs (Citadel of Algiers) © Abdessemed-Foufa 2005

construction technique contributes to the resistance of the earthquake loads. Beside we can observed another system used between two arches inside the bricks masonry. In fact, at the joint between the departure of two arches, two layers of three logs of *thuya* are superimposed and inserted to avoid the cracks during the earthquake (Fig. 4).

Fig. 4 Reinforced masonry between 2 arches with 2 layers of cedar logs (Citadel of Algiers) © Abdessemed-Foufa 2005

Reinforced Load-Bearing Walls

In Algiers, the masonry is made of brick walls of variable sizes; 3 × 10 × 20 cm, 3 × 112 × 22 cm, 4 × 12 × 22 cm; and thickness; 60–80 cm; bound by a sand and lime mortar. The wall presents a stratification with different layers between which is inserted, in regular intervals, logs of *thuya*, not squared, of 10 cm in diameter. These timbers are embedded along the length of the walls and regularly distributed at every 80–120 cm. Timber does not exert any tension force (Fig. 5). The same techniques were highlighted in Mostaganem historical nucleus [11, 12] where the masonry is of two facings of stones (Figs. 6 and 7).

In the Eastern Highlands region in Amenthane and Menaa, Wilaya of Batna, in traditional houses [13], they were logs of cypress (Cupressaceae) wood called juniper which is inserted in the depth of stone or earth masonry walls (Figs. 8 and 9). Such sub-divisions into the walls of small panels make the walls resist better the shear force and thus prevent large cracks and collapse. We also find in this eastern region a wooden system whose articulation helps measuring soil movements due whether to earthquakes or landslides. This traditional warning system is located inside the mosque, it announces the population soil movements.

This combination of rigid and flexible materials as seen on Figs. 5, 6, 7, 8 and 9 allows absorption of the shear force during the earthquakes. In addition, the walls show very few cracks and do not collapse. In fact, according to the dynamics of the structures, the elements in masonry play a significant role in the earthquake response of the building. The lateral earthquake loads tend to deform the panel of masonry in parallelogram shape, causing the formation of a diagonal rod of

Fig. 5 Reinforced bricks masonry walls by logs of *thuya* Algiers (Palace of the Dey, Citadel of Algiers) © Abdessemed-Foufa 2005

Fig. 6 Reinforced stones masonry walls by logs of wood (Mostaganem) © Abdessemed-Foufa 2016

Fig. 7 3D restitution of the Reinforced stones masonry walls (Mostaganem) © Maraf 2012

Fig. 8 Reinforced earthen masonry wall by logs of Juniper (Menaa-Aures) © Abdessemed-Foufa and Adjali 2015

Fig. 9 Reinforced earthen masonry wall by logs of Juniper (Menaa-Aures) © Abdessemed-Foufa and Adjali 2015

compression which acts at the level of the angles [14]. However, the walls are split in several parts, three or four parts according to the height of the wall; this appears in the walls of the palace of the Dey as well as in the houses in Algiers and in vernacular houses of Menaa and Amenthane and in ottoman historical houses of Mostaganem. Thus, the lateral loads are distributed with each layer where there is a difference of materials. This will prevent the walls undergoing significant deformation. These walls respond favorably to the earthquake excitation, it is thus certain that these walls were designed so as to be able to resist the earthquake loads. This disposition of materials avoids shearing and the effect of rod in the wall because the latter is subdivided in several parts.

Links of the Walls

The masonry walls are also consolidated in the angle. Effectively they are achieved by alternating crossing by wood logs of *thuya*. In this matter, Carette [15] described this system observed at Algiers by: "*…j'ai remarqué dans les anciennes maisons mauresques en démolition une précaution excellente prise par les constructeurs pour consolider les angles. Elles consistent à placer horizontalement, de cinquante en cinquante centimètres de hauteur, des pièces de bois d'environ deux mètres de longueur. Ces pièces noyées dans la maçonnerie, se prolongeaient alternativement suivant chacun des deux murs et venaient se croiser dans l'angle. J'ai vu des maisons sapées à la base et à moitié démolies, se soutenir encore grâce à cet artifice de construction*"[2] [15].

This horizontal reinforcement quoins by logs of wood in stones masonry was highlighted in Amenthane and Mostaganem during respectively 2015 and 2016 following an in situ investigation by our student of the master degree of Architecture and Cultural Heritage in Constructive Culture (Figs. 10 and 11).

This system of linkage at the angles, in the absence of any vertical element, constitutes a traditional technique of reinforcement of the angles to prevent the vertical walls from tearing apart. They are willing to places where the greater fragility of the walls is due to combinations of bending effects, shear and torsion. To make a traditional masonry a seismic construction, it is necessary to perform a chaining system which forms a three-dimensional network mechanically continuous at corners where no free masonry board should appear [14, 16].

Pillars

The large masonry pillars are reinforced with cedar logs. The timber is inserted into the entire width of the pillar. The different rows of masonry and wood alternate regularly. The timber beds are arranged in both directions alternately (Fig. 12).

---

[2] *I noticed in the old Moorish houses in demolition, an excellent precaution taken by the builders to consolidate the angles. It consists in placing horizontally, every fifty centimeters in height, pieces of wood of approximately two meter long. These parts incorporated in the masonry, were prolonged alternatively according to each of the two walls and came to cross in the angle. I saw houses sapped at the base and half demolished, but still standing due to this artifice of construction.*

Fig. 10 Consolidating angles by juniper wood (Amenthane-Aures) © Kais D. 2011

Fig. 11 Consolidating angles by logs two perpendiculars layers (Mostaganem) © Abdessemed-Foufa 2016

Fig. 12 Timber insert in the depth of the pillar (Citadel of Algiers) © Abdessemed-Foufa 2005

Dome and Cupola

Algiers' dome and cupola are constructed of bricks masonry. Each masonry coverage based on cedar logs. These logs are playing the role of shock absorbers (Fig. 13).

Discussion

This research work undergone initially for the doctorate thesis discussed in 2007 [3] has allowed us on behalf of the ministry of culture to elaborate the manual of architectural, constructive and architectonic typologies of the Casbah Permanent Safegarding Plan (2008–2010) [17]. This manual is today a guide for architects working on the restoration of the built heritage of the Casbah of Algiers. Besides that, we highlighted during the investigation of 2010 the foundations and the copula inside the citadel of Algiers. These techniques were added to the manual of Algiers. These techniques are considered from 2008 as a transferable indigenous knowledge (TIK) in the Asian Disaster Reduction Hyperbase (Asian DRH) [8].

As the part of the university research project CNEPRU 000420140037 untitled: Methods and Instruments of Landscape Heritage Conservation, the case of Wadi Abdi Valley (Aures) and South of Piedmond (Ziban), these highlighted constructive typologies in Wilaya of Batna will constitute the rehabilitation manual of the built cultural heritage in the Aurès and Sahara border regions.

Since the enactment of the 98-04 law on cultural heritage [18], Algeria is building knowledge on tangible heritage. This research project is a collaboration for the identification of the empirical traditional constructive typologies to avoid earthquake loads. Hence since 2014 to advance the constructive culture of vernacular architecture in Algeria through research and investigation, the graduate students of the master of Cultural Heritage of Blida 1 University make in situ investigation throughout all the territory and in this way the Wilaya of Batna and Mostaganem were examined during 2015 and 2016.

Fig. 13 Logs of thuya inserted in the cupola bricks masonry (Mosque of the Dey, Citadel of Algiers) © Abdessemed-Foufa 2010

Conclusion

The masonry structures strengthened with wooden in Algeria come from an ancestral knowledge. Indeed by empiricism, population repaired, reinforced and rebuilt what was damaged or destroyed by recurring earthquakes. So the timber was used as a stiffener and damper in walls, foundations and domes. Thus in the central region as well as in the eastern region of the highlands, wood was the material associated with masonry bricks, stones or earth to reinforce it to withstand earthquakes.

The earthquake resistant buildings techniques highlighted are the ancestral intelligent knowledge that has been used as a response to natural hazards. Elaboration of the rehabilitation manual highlighting the know-how thanks to the local seismic culture will contribute to the preservation of the built heritage in areas where earthquake are frequent.

References

1. Lemaistre de Sacy LI (1990) La bible de port royal, Editions Robert Laffont, coll. «Bouquins»
2. Chouraki A (1987) La Bible, Ed De Brouwser D, Paris
3. Abdessemed-Foufa AA (2007) Contribution pour la redécouverte des techniques constructives sismo-résistantes dans les grandes villes du Maghreb Alger, Fès et Tunis durant le 18ème siècle. PhD thesis, unpublished, Ecole Polytechnique d'Architecture et d'Urbanisme (EPAU) Algiers, Algeria, 420 p

4. Benouar D (1994, July) Materials for the investigation of the seismicity of Algeria and adjacent regions during the twentieth century, Special issue of Anali Di Geofisica (vol XXXVII, no 4)
5. Ferdi S, Harbi A (2013) Roman literary and epigraphic spurces for a study of historical seismicity in Algeria Circa 42-420 AD. J Seismol doi:10.1007/s10950-013-9400-5. Published online 26 Sept 2013, 11 p
6. CRAAG (1994) Les séismes en Algérie de 1365 à 1992
7. Abdessemed-Foufa A (2005) Contribution for a catalogue of earthquake-resistant traditional techniques in northern Africa: the case of the Casbah of Algiers (Algeria). Eur Earthq Eng J 02 (05):23–39
8. Abdessemed-Foufa A (2008) Rediscovery and revival of traditional earthquake-resistante techniques in Algeria. The case of the Casbah of Algiers, disaster reduction hyperbase. DRH asian application. drh.edm.bosai.go.jp/database
9. Abdessemed-Foufa A, Benouar D (2010) Investigation of the 1716 Algiers (Algeria) earthquake from historical sources: effect, damages and vulnerability. Int J Architectural Heritage Conserv Anal Restor 4(3):270–293
10. Abdessemed-Foufa A, Benouar D, Terki Y (2015) Local seismic culture around the word. Case study: local seismic culture in vernacular architecture in Algeria. Seismic retrofitting. In: Correia M, Lourenço PB, Varum H (eds) Learning from vernacular architecture. CRC Press, pp 101–102
11. Maraf Z (2012) Evaluation de la vulnérabilité sismique des deux Casbahs de "Tijditt et Derb Tobbana". Master of Civil Engineers, Major Risque Managment. Department of Civil Engineers, Faculty of Sciences and Technology, University Abdelhamid Ibn Badis, Mostaganem, unpublished, 200 p
12. Sadallah S (2016) Les techniques de constructions parasismiques vernaculaires renforcées par le bois dans les deux Casbahs de "Tijditt et Tobbana à Mostaganem. Master of Architecture and Cultural Heritage, Culture Constructive, Institute of Architecture and Urban Planning, University Saad Dahleb of Blida 1, Algeria (unpublished)
13. Adjali S (1986) Habitat traditionel dans les Aurès. Le Cas de la Vallée de l'Oued Abdi. Annuaire de l'Afrique du Nord. Tome XXV. Ed CNRS. pp 271–280
14. Betbeder-Matibet J (1991) Construction Parasismiques. Techniques de l'Ingénieur, Traité de construction, C3290C, Paris, 22 p
15. Carette AEH (1850) Algérie, l'Univers ou Histoire et Description de tous les peuples de leurs religions, moeurs, coutumes, Paris Ed Firmin Dido Frères
16. Zacek M, Construire Parasismique, Risque Sismique, Conception parasismique des Bâtiments, Règlementation, Ed Parenthèse, Marseille
17. Abdessemed-Foufa A (2010) Le manuel de Réhabilitation comme outil de conservation dans le cadre du plan de sauvegarde de la Casbah d'Alger. World Congress for Middle Eastern Studies. WOCMES, Barcelona
18. Law 98-04 of June 15, 1998 on protection of cultural heritage, the official journal of Algerian Republic of June 17, 1998
19. Minsitère de la Culture (2011) De Terre et d'Argile. Catalogue de l'exposition dédiée aux architectures de Terre dans le monde, Photographies de Kais Djilali, Algérie

A Comparative Evaluation of the Results of Two Earthquakes: Istanbul and Lisbon Earthquake in 18th Century

Şevket Erşan

Abstract Earthquakes affect both urban texture and social structure of cities in history. It can be seen that in two scales such as urban scale and building scale. In this context, two earthquake examples such as the 1755 Lisbon Earthquake and the 1766 Istanbul Earthquake will be analyzed in different scales. This paper is giving introduction to the main urban texture in 1700s consisting of timber housing structures and damages of these earthquakes. Construction system of Istanbul and Lisbon will be analyzed. In the methodology of this study a comparative evaluation of the 1755 Lisbon Earthquake and the 1766 İstanbul Earthquake will be presented. By looking at the archive records, the development of urban structure after earthquakes has different tendencies in Lisbon and Istanbul. The first inclination is to remove the old city and re-build completely the new city as in Lisbon. By doing this renovation the new regulations are settled down. The second inclination is reproduction of the old housing construction system in Istanbul. It could be seen from both the local archive documents and European travelers' records. The aim would be a comparative approach in order to develop and preserve the historic urban areas and historical buildings that has a potential to make an evaluation by the reflection of past to the present.

Keywords Lisbon earthquake · Istanbul earthquake · Timber housing

The Cities of Lisbon and Istanbul in the 18th Century

At the middle of the eighteenth century, Lisbon, capital of Portugal, was one of the finest, wealthiest, and most beautiful and luxurious towns in Europe. It was Europe's fourth largest city, after London, Paris, and Naples. Situated on the edge of the Atlantic Ocean, at the latitude that—by courtesy of sea breezes—provided an

Ş. Erşan (✉)
Faculty of Architecture, Restoration Department Istanbul,
Yildiz Technical University (YTU), Istanbul, Turkey
e-mail: sevketersan@gmail.com

H. Cruz et al. (eds.), *Historical Earthquake-Resistant Timber Framing in the Mediterranean Area*, Lecture Notes in Civil Engineering,
DOI 10.1007/978-3-319-39492-3_4

agreeable climate, it was Europe's major maritime gateway to the world, through which a substantial part of the European expansion and colonization had passed.

In the eighteenth century, Istanbul, imperial capital of the Ottoman Empire, was one of the most intriguing cities. In depicting the 18th century Ottoman world, one of the key words must "change". For the Ottoman Empire; 18th century was a period when western conquest stopped and started to follow the western development. The sultan's grand vizier and son-in-law Nevşehirli Damat Ibrahim Pasha sought to broaden the empire's contacts with western Europe by sending ambassadors to Paris, Vienna, Moscow and Warsaw. The first Turkish ambassador to Paris was Yirmisekiz Çelebi Mehmet Efendi, who in 1720 was instructed by Ibrahim Pasha 'to visit the fortresses, factories and works of French civilization generally and report on those that might be applicable' to the Ottoman Empire. [1]

1755 Great Lisbon Earthquake and 1766 Istanbul Earthquake

The 1755 Lisbon earthquake, also known as the Great Lisbon Earthquake, struck at 09:40 a.m. on Sunday, November 1, 1755, the Catholic holiday of All Saints Day. That was the time, when most of the godly citizens of Lisbon were at prayer. The earthquake lasted between 3 and 6 min. Contemporary reports stated that the earthquake ruined a great deal of the city edifices and caused gigantic fissures 5 m wide to appear in the city center. Approximately, 40 min later, a huge tsunami wave entered the mouth of the Tagus River and hit the coastal quarters of Lisbon and other settlements located along the river estuary banks; the first wave was followed by two more waves. Overthrown hot stoves in many kitchens initiated numerous local fires that resulted in vast and all fusing flaming inferno which completed a near-total destruction of Lisbon. The time of the earthquake and its position in an area of fairly shallow water were the main factors that contributed to such devastating effects. The narrow, winding streets did not enable the inhabitants to escape quickly enough. Panic overtook many of the survivors, including the King, who sought refuge away from the city. Eighty-five percent of the Lisbon houses were ruined or heavily damaged. The event was one of the most destructive and deadly earthquakes in European history. Of Lisbon's population of ~275,000, 40,000–60,000 people were killed. [2]

The 1766 Istanbul earthquake, struck at May 22, 1766, the another holiday Festival of sacrifice. But the difference to the Lisbon earthquake was that mosques were empty since prayers and most of people in Istanbul were outside. Therefore, the disaster results were lighter from Lisbon earthquakes. Of Istanbul's population of ~600,000, 4000–5000 people were killed.

Building Systems in the Ottoman Empire and Portugal in the Middle of the 18th Century

By the time of Lisbon's 1755 earthquake and Istanbul's 1766 earthquake, timber framed buildings existed already for many centuries in Europe and Ottoman Empire (for instance, "fachwerk" houses in northern Europe), as a common construction system, not related to earthquake resistance. In other areas where earthquakes are frequent, as in the Balkans, timber framed walls (for instance, "bondruk") had become widespread because, after each earthquake, people observed that they had resisted better and copied them.(Fig. 1)

In Istanbul is the main material used in the construction of wooden housing until the 20th century. According to who came to Istanbul at the end of the 18th century zoologist Olivier "In Istanbul, will help to make solid construction, despite the ease of finding any material such as stone, brick and lime, from whether rich or poor community, somehow always almost all the houses were made of wood" (Fig. 2).

Fig. 1 "Bondruk" timber frame wall

Fig. 2 The oldest Bosphorus mansion "Amcazade Hüseyin Pasha Yalisi" 1699

Reactions After the Earthquakes

After the 1755 earthquake the construction methods in Lisbon (and elsewhere in seismic areas of the country) changed substantially. The Marquis of Pombal, a Government Minister, remained in the city and organized emergency measures. He invited a military engineer, Manuel da Maia, to oversee the reconstruction of the city. Along with his team of architects and engineers, da Maia in many ways personified the Age of Enlightenment; he was, above all, rational and practical. He presented reconstruction proposals to the Senate in three stages, and at each stage, from a number of approaches, the one chosen was the boldest and the most rational. [3] As a result, in May 1758, a license was granted to begin reconstruction of the city, including the Pombaline quarter. The reconstruction of this area was based on a completely new rectilinear plan (Figs. 3 and 4) and on a strict set of rules governing the design and construction of the new buildings. A sturdy tri-dimensional interconnected timber bracing system was erected by the carpenters before the masons started their work. The imposing cage-like structure gave rise to the designation "gaiola" which thereafter describes this type of construction.

The three-dimensional timber structure of the 'gaiola' in the interior walls above the first floor can be identified in Fig. 5, as well as other interior walls (partition walls), which are wooden panels that do not have structural functions. Façades are made of masonry without the 'gaiola'. The openings (doors and windows) in these walls define the geometry of the masonry structural elements. [4] Roofs are made with timber trusses and ceramic tiles and may include window openings. Floors are timber slabs and should be considered as flexible diaphragms. Ground floor interior walls are masonry walls supporting a system of vaults made of blocks of ceramic

Fig. 3 Buildings where grouped in very regular blocks

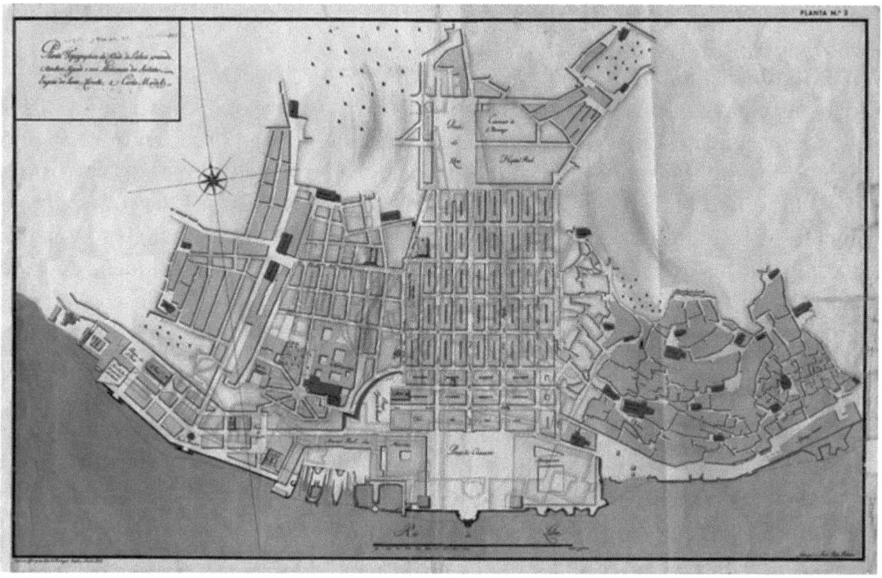

Fig. 4 After the earthquake of 1755: the adopted plan of the reconstruction

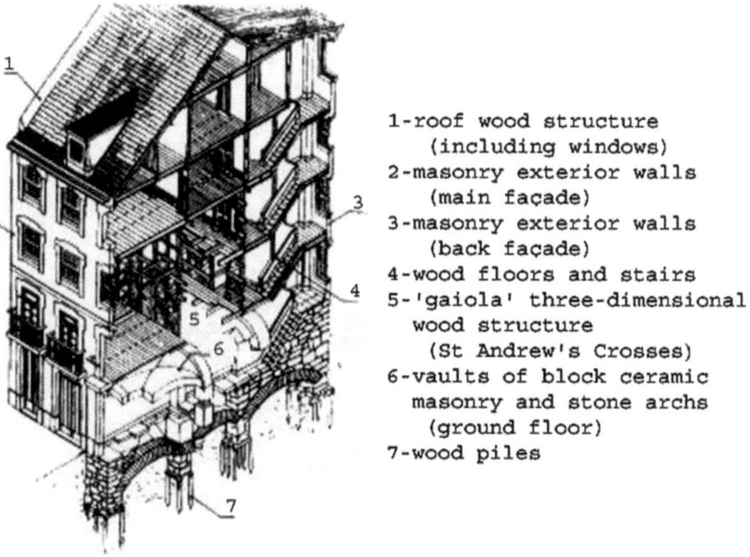

Fig. 5 Structural details of a 'Pombalino' building (adapted from Mascarenhas [6])

masonry and stone arches. Foundations include short and small diameter woodpiles connected by a timber grid. [5]

The timber structure of the 'gaiola' is like a birdcage made of vertical and horizontal elements braced with diagonals. The masonry of the exterior walls is made of irregular blocks of calcareous stone and lime mortar with very poor strength capacity. Masonry infill of the 'gaiola' can be stone (rubble) or clay bricks. It is usual to find both type of masonry at interior walls. Most of the times, timber elements of the 'gaiola' are notched together or connected by nails or iron ties, according to historical information regarding construction techniques. A more detailed description of 'Pombalino' Buildings can be found in Cardoso et al. [7].

In Fig. 6, where the masonry of the main façade wall was made invisible, several connectors are shown, both in iron (A, B) and timber (C). Masonry confinement between interior timber grid and exterior ashlars is also noted. [8] Basic tri-dimensional bracing unit: diaphragm floor and bracing walls in two directions (Fig. 7).

After the 1766 Istanbul Earthquake, It is seen that only the repair of monuments was considered by the Ottoman Empire. The scientific method did not appear to be followed regarding demolished building. [10] Scale of repairs of the houses and mosques remained only building scale. People have accepted their fate and rebuilt

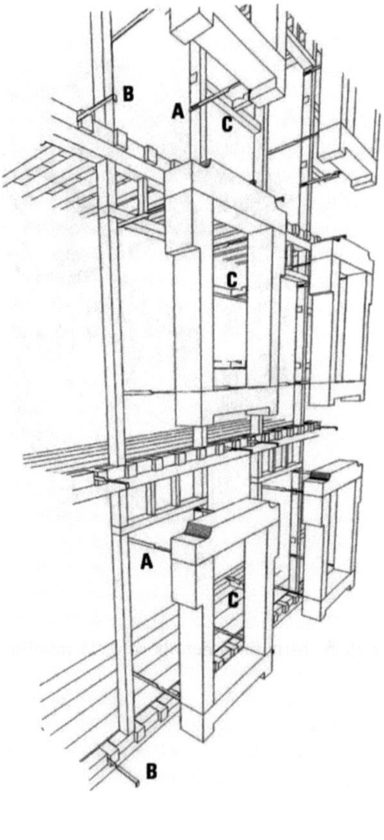

Fig. 6 Details of Pombalino buildings [9]

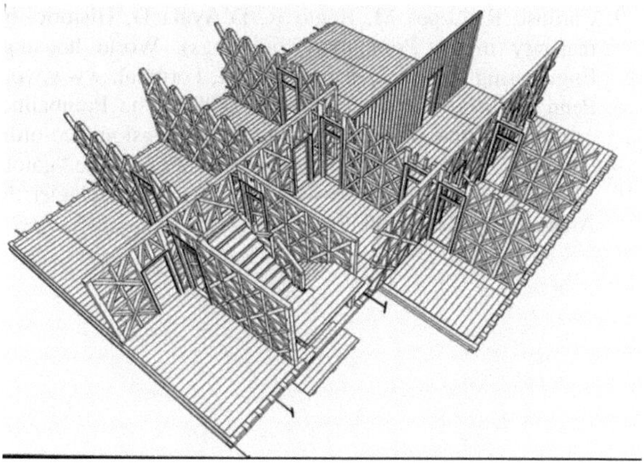

Fig. 7 Construction details of Pombalino buildings [9]

their houses fastly with same details. This behavior perhaps shows us the habits of a community. This behavior is judged by foreign travelers of the period. The foreign eyes saw that fatalism was an important place in Turkish community.

Conclusions

The 1755 Lisbon Earthquake has opened the way for many innovations in architecture, it has caused new approaches in the field of construction technology and showed us how important is the preparation of laws and regulations on urban planning and to follow them. Compared to the 1766 Istanbul earthquake, seen from this perspective, it is not perceived in a similar way in the world of Ottoman architecture. After the 1766 Istanbul earthquake when considering the scale of repairs of buildings, it reveals change in the building and urban scale not been considered.

References

1. Freely J (2011) History of Ottoman architecture. WIT Press
2. Kozak J, Cermak V (2010) The illustrated history of natural disasters. Springer Publishing, London
3. Augusto Franqa J (1983) Lisboa Pornbalina e o Illurninisrno, 3rd edn. Lisbon, pp 59–76
4. Cardoso R, Lopes M, Bento R (2004) Earthquake resistant structures of Portuguese old pombalino buildings. In: 13th world conference on earthquake engineering Vancouver, B.C., Canada
5. Cardoso R, Lopes M, Bento R (2004) Seismic assessment of Pombalino buildings. In: 13th world conference on earthquake engineering Vancouver, B.C., Canada
6. Mascarenhas J (1994) “Pombaline downtown, some technical innovations”, 2nd ENCORE. LNEC, Lisbon (in Portuguese)

7. Cardoso R, Lopes M, Bento R, D'Ayala D, Historic, braced frame timber buildings with masonry infill ('Pombalino' buildings). World housing encyclopedia report, Earthquake Engineering Research Institute EERI, Portugal. www.world-housing.net
8. Penn R, Wild S, Mascarenhas J (1996) The Pombaline quarter of Lisbon: an eighteenth century example of prefabrication and dimensional co-ordination. Construction history, vol 11
9. Coisas V, Characterization and preservation of the "gaiola" construction. An overview
10. Mazlum D (2011) 1766 İstanbul Depremi Belgeler Işığında Yapı Onarımları. Istanbul Araştırmaları Enstitüsü, İstanbul

Some Examples of Turkish Houses with Wooden Frame in the Seismic Zone Anatolia

Ayten Erdem and Uzay Yergün

Abstract Wooden frame Turkish houses in the area encompassed by the Ottoman Empire have their own special characteristics from the point of view of the order of the rooms, the design plan and the construction system. The earliest examples of these houses have been dated to the 17th century Today, there are few remaining examples of the first Turkish houses with an open hall "hayat", wooden frame construction system, adobe fill, double rows of windows and mostly directed towards their own courtyard. In Turkish houses that passed through various developmental stages, the open hall was brought into a more protected arrangement and beside the types of open hall plan, the plan with inner hall "sofa", corner hall and central hall were created; however, the use of the wooden frame construction system continued. The wooden frame construction system was preferred because it was the continuation of a tradition and because it suited the geography of Anatolia, located in an earthquake zone and rich in forested areas. In the wooden frame method; the assembly details were basic and the house could be put together like a tent, generally the frame was filled stones, adobes and bricks and sometimes wood and sometimes there would be no fill at all. The importance of the wooden frame houses did not receive much attention and were preferred to concrete apartment buildings until the 1970s, but their importance was noticed in our day even as old, important examples were torn down. So, using existing examples, the purpose of this study is to examine, document and make wooden frame Turkish houses known according to historical development.

Keywords Turkish houses · Wooden frame construction · Anatolia · Earthquake zone

---

A. Erdem (✉) · U. Yergün
Architectural Faculty, Yildiz Technical University,
Barbaros Boulevard, 34349 Istanbul, Turkey
e-mail: aerdem@yildiz.edu.tr

U. Yergün
e-mail: yergun@yildiz.edu.tr

H. Cruz et al. (eds.), *Historical Earthquake-Resistant Timber Framing in the Mediterranean Area*, Lecture Notes in Civil Engineering,
DOI 10.1007/978-3-319-39492-3_5

Introduction

When the Turks first entered Anatolia in the 11th century, they used the kind of houses found there, but it is seen that when, in the 16th century, they began to expand even further and entered the Balkans, they began to construct courtyard homes with exterior/open air *hayat*s and *sofa*s/halls [1] (Figs. 1, 2 and 3). It is believed that while these exterior/open-air sofa home types had been used as early as the Hittite and Hellenistic periods, over time the Turks adapted these styles of houses to meet their own needs [2]. Influenced by a combination of their nomadic life style, their adopted religion of Islam, and the local cultures of Anatolia, this unique Turkish style of courtyard house remained popular up to the mid-nineteenth

Fig. 1 The view of *hayat*

Fig. 2 The view of *hayat* and iwan/*eyvan*

Fig. 3 Koşut House from courtyard, Kütahya

Fig. 4 *Hayat* from the outside and innerside

century [3]. This housing space includes a *hayat*, a semi-open space extending in front of the rooms of the house, providing both shelter and a natural environment (Figs. 4 and 5). The upper level rooms are also accessed through this *hayat*. The essential units of this Turkish house—the rooms—are actually equivalents of the original Turkish home, the tent, and each is arranged to be multi-functional and to serve as a family home in its own right [4] (Fig. 6). These semi-open hall or *hayat* houses were traditionally varied according to choices of *eyvan/*iwan and kiosks (Table 1; Fig. 7), but beginning from the 18th century, the spaces between these rooms began to be enclosed, thus protecting them from the exterior elements, resulting in plans that included "inner" and "middle" sofa spaces (Table 2). Arranged to accord with the patriarchal Turkish-Islamic family structure, up until the 19th century the Turkish house was accessed through its courtyard (Fig. 8). This courtyard, which was isolated from the street by high stone walls, included a wooden staircase that provided access to the rooms of the upper level. The courtyard played

Fig. 5 *Hayat* from the outside and innerside

Fig. 6 Murat house, the guest room

an important role in the daily lives of the women of the house and included spaces and elements used for such activities as preparing and cooking food, baking bread, washing and drying the laundry, and caring for animals, etc. The lower level of the house was entirely isolated from the street. Used for the home's service and storage needs, the lower level was organized according to the arrangement of the courtyard, while the upper level constituted the actual residence and consisted of the spaces where family members led their daily lives. While there are variations according to plan types, Turkish houses always were constructed with a wooden framework. Multi-floored Turkish houses of Anatolia with sloped roofs and wide eaves were mostly constructed in forested areas on the seashore, in inner regions near the shore and in regions where the central government was in control (Fig. 9).

Table 1 Plan types with an outer, open hall/*hayat* [10]

Bursa, Murat House 17th Century	Yenişehir, Şemaki House 18th Century	Birgi, Çakır Ağa House 18th Century
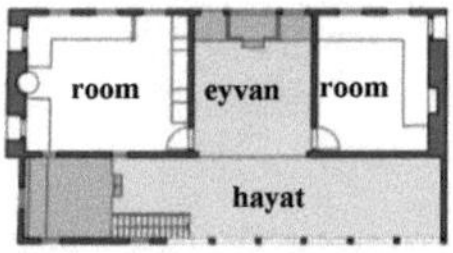	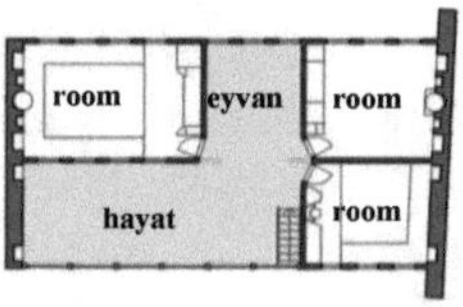	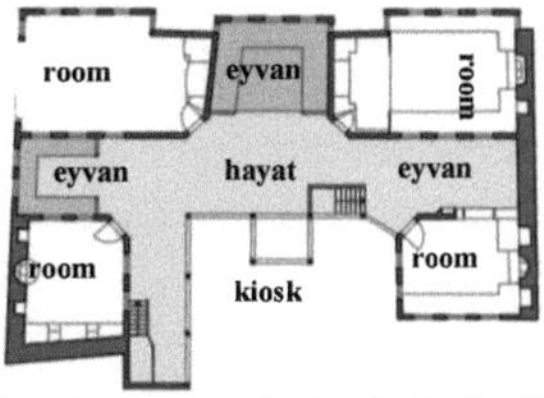
Main Floor Plan 0 2 4 6 8 10 m	Main Floor Plan 0 2 4 6 8 10 m	Main Floor Plan 0 2 4 6 8 10 m
The schema with two rooms and one eyvan.	The schema with corner hayat, three rooms and one eyvan.	U plan schema, three eyvan, one kiosk.

Fig. 7 The view of *hayat* with *eyvan*

Table 2 Plan types with a closed hall/*sofa* [11, 12]

Göynük, Müderrisoğlu House 18th Century	Cihangir, Kadiri Lodge House 18th Century	Boyabat, Abidler House 19th Century
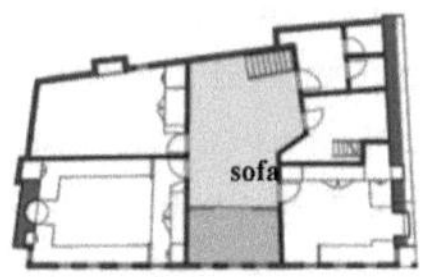	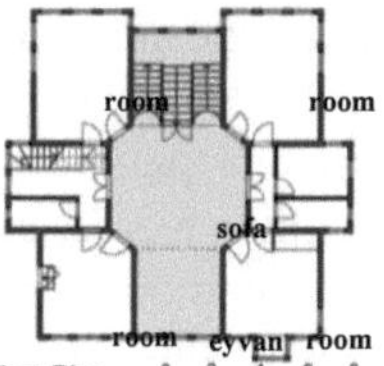	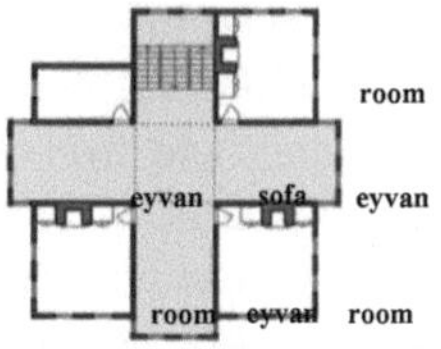
Main Floor Plan 0 2 4 6 8 10 m	Main Floor Plan 0 2 4 6 8 10 m	Main Floor Plan 0 2 4 6 8 10 m
The plan with an inner hall.	The plan with a central hall and two eyvans.	The plan with a central hall and four eyvans.

Fig. 8 The entrance from the street

Fig. 9 Kula houses

Turkish Houses in Processes of Development and Change

There is available only very little information and statistics relative to the Anatolian Turkish houses that existed between the 13th and 17th centuries. Some general ideas and limited information can be gleaned, however, from documents surviving from foundation and court records, laws and regulations passed by the Imperial *Divan*/Council, and from the miniature paintings representative of the painting art form of the Islamic world [5]. In other words, extant historical records tell us that the Turkish House with *hayat* developed during the 13th and 16th centuries, but we have no firm clues as to the stages of this development. The earliest examples of Turkish Houses with *hayat* that survived to this day belong to the 17th century.

The Houses with *Hayat*

However, based on the documented examples of 17th century Turkish homes that remained standing until the mid-20th century,[1] we understand that the outer walls of the 16th and 17th century houses were built of stone or mud-brick, that the interior walls of the upper level rooms of these two story structures were wood framed, with open-air halls/*hayat*s, and primarily consisted of two or three rooms and sometimes *eyvan*. These houses had a single facade that overlooked the courtyard. We can describe these houses as being isolated from the street (Fig. 10). As the techniques involved in wood-frame construction advanced, the exterior facades of the upper story began to be built with wooden frames filled-in with mud-brick. The formerly blind walls facing the streets also began to be fitted with windows [6] (Fig. 11). The windows facing both the street and those overlooking the courtyard were built in double rows. The windows in the lower row had wooden shutters, but were devoid of glass. Those in the upper level were, however, covered with small pieces of colored glass. The outer facades were not plastered; the mud-bricks, however, were laid-out in zig-zag or herringbone patterns and these walls were decorated with sun or cypress depictions (Fig. 12). Unfortunately, only a small number of these houses have survived to the present. One of those is the Murat House located in Bursa and known today as the Ottoman House Museum. In reference to the plan and the decorations of this two-room house with *eyvan*, it is assumed that the house dates to the late 17th century (Table 1; Figs. 11, 12, 13, 14 and 15).

The 18th century is regarded as the classic period of the *hayat*-style Turkish home [7]. In this period, the load bearing walls and the walls of the ground floors of

Fig. 10 Maliki house [8]

[1]Among the countless photographs and drawings of Turkish homes compiled and published by one of Turkey's most important representatives of 20th century architecture, Sedat Hakkı Eldem (1908–1988) and his students.

Fig. 11 Murat house, street facade

Fig. 12 Detail of facade, Murat house

Fig. 13 Murat house, street facade

Fig. 14 Murat house from courtyard

Fig. 15 The cupboard of Murat house

these large homes were constructed of stone or mud brick, while the multi-roomed three floors of the upper levels had wood framed walls, a *hayat*, *eyvans*, and kiosks, with facades overlooking the courtyard and the street. One of the other Bursa examples of homes dating from that period, the Yenişehir Şemaki House, also today serves as a museum. The upper floor of this two-storey mansion consists of three rooms and a *hayat* (Table 1; Figs. 16, 17 and 18). The third room, which was built along the short side of the *hayat,*[2] reflects some of the unique features belonging to that century. On the other hand, the rooms built on both of the short sides of the *hayat* of the Birgi Çakır Ağa Mansion in Birgi are more protected (Table 1; Fig. 19). In this century, the tradition of having two rows of windows on the facades continues, however, the facades have begun by this date to be plastered and

[2]Because during the restoration on the Murat and the Şemaki houses; *hayat*s were enclosed with glass, the half-open space feeling can't be felt.

Fig. 16 Şemaki house from courtyard

Fig. 17 Şemaki house, *hayat*

Fig. 18 Şemaki house, guest room

Fig. 19 Çakırağa mansion with kiosk

Fig. 20 Inner hall, Müderrisoğlu house

while the overhanging protrusion and windows of the upper wood framed levels open the building to the outside, the stone ground floor continue to isolate the structure from outside.

Inner-Hall/*Sofa* Houses

By the middle of the 18th century we see both a continuation of the *hayat*-plan house and the introduction of houses in which rooms have been built at both sides of the *hayat*. This development both isolates the rooms from the outside and introduces an inner-hall/*sofa* plan (Table 2; Fig. 20). This inner *sofa* plan, considered to represent the second stage in the development of the Turkish house, quickly became the most commonly used plan; its popularity most likely stemming from the fact that this plan was both more economic and more protected [9]. This plan type varies according to the windows put into one or two of the *sofa* facades. Most prevalent was the plan that included windows on both facades, was commonly referred to as the *karnıyarık*, or "cut-in-the-middle." In this development, a mid-sofa plan house was achieved by beveling the corners of the rooms and extending the *sofa* with are *eyvans*.

Central-Hall/*Sofa* Houses

The changes that began in the 19th century in urban centers also affected changes in the form of the *hayat* house and led to the eventual disappearance of the former more, semi-rural type home. Homes accessed directly from the street began to be built on smaller land parcels, *hayat* became a central location surrounded by the rooms. The *eyvans* were either enclosed, making them into rooms, or were discarded altogether. To provide illumination in the *sofa*, *eyvan*s were positioned at the corners of the rooms, or between the rooms. Those homes with four eyvans became the most developed type of the central-*sofa* type house, becoming a cross-patterned *sofa* plan (Table 2; Figs. 21 and 22). The development was completed by beveling the corners of the square centralized sofa or by providing it with a round or elliptical shape. Also, by the 19th century the facades of the houses now employed only a single row of windows, apertures that also changed in shape and in size. The outer facades were now either plastered over or sided with wood cladding, developments that began to reflect more western styles (Figs. 22 and 23).

The Construction System of the Turkish House

Wood always represented the primary building material of the Turkish house, with a preference for wood framing (Fig. 24). This system was preferred for reasons such as its being light, providing the opportunity of multi-stories being built,

Fig. 21 Central hall and the iwan

Fig. 22 Abidler house, Boyabat

Fig. 23 Armutçular mansion, Mudurnu

Fig. 24 Wooden frame system, Ünye

creating variety in architectural solutions, providing climate control, necessitating fewer wooden materials in comparison with the massive wooden system, rapid construction and suitable for Anatolia which is in an earthquake zone. Alongside wooden material, mud-brick and stone were used in the construction of the houses. In Anatolia, generally the space in the wooden frame system was filled with mud-brick, stones or bricks and these buildings were called *hımış* (Fig. 25). Sometimes the houses were built in which wood and tree branches were used for fill (Fig. 26) and sometimes the inside of the wooden frame was left empty. In wood frame wall that was not filled, on the inner and outer facades over the frame, the construction was nailed down thin half-batten/*bağdadi* horizontally and at intervals and plastered (Fig. 27) or the interior facade was made covered with *bağdadi* and plaster and the outer facade was covered with wooden boards (Fig. 28).

The foundations of the houses that don't go very deep, the ground floor walls, the one or two main walls that rise from the ground to the roof and courtyard walls were made with unhewn stone. Mud mortar and lime mortar were used as binder.

Fig. 25 *Hımış* building with mud-brick

Fig. 26 Wood frame was filled wood

Fig. 27 *Bağdadi* and lime plaster

Fig. 28 Houses were covered with buttresses in the stone wall

Wooden horizontal beams at regular intervals, sometimes wooden posts and buttresses were set at in the stone walls without plaster (Fig. 29). Mud-brick was also used in place of stone on the ground floor and in the main walls. However, the external facades of mud-brick walls were always plastered. The ground floor walls, generally made without windows, both provided intimacy and increased the wall's sturdiness. In homes with *hayat*, the wooden posts with capitals, set at the ground level on stone bases at regular intervals, and supporting the *hayat*, were present also at the upper level, and supported the roof of the *hayat* (Fig. 4). The flooring materials between the floors were made up of round beams, set at intervals of 40–60 cm, and of boards with a width of 2–3 cm. set on the beams, but in the opposite direction (Fig. 30).

The wooden frame was made up of posts set at regular intervals on bases (Fig. 31), of buttresses supporting the posts. In order to make the wooden walls of the upper floors, most of the time the posts 10–14 cm thick were preferred (fir, pine). In the houses with old dates the posts rested on capitals and the post

Fig. 29 The wooden posts and wooden boards

Fig. 30 The posts under the *hayat*

Fig. 31 Relation with the post and base

interval was kept wider (45–70 cm). The thickness of the wooden frame walls—including mortar—was built between 15 and 20 cm. and the corner posts were supported with buttresses. Between the posts in houses of recent date, when the interval posts that were half as thick as the thickness of the main post were set up, the distance between posts was reduced. While elements like the post that creates wooden frame walls, the diagonal brace and the bases, etc. relied on pressure and tension, diagonal braces were used in the corners against side thrust pressure and horizontal forces (Figs. 24 and 25). The methods used at the floor level are called the single base method, double base or single base on one side and double base one on the other. The projections that have been created underneath by the beams being extended outside are supported with diagonal brackets/*payanda*, consoles and curved struts/*eli böğründe* (Figs. 25, 28 and 32). The wooden frame walls were first inside and outside, plastered with mud mortar to which straw was added, and then with lime mortar or the exterior facades were covered with wooden board

Fig. 32 Curved struts/*eli böğründe*

Fig. 33 Hipped roof, alaturca tiles

(Figs. 22 and 23). Wooden material was used in all the elements like the staircase, cupboard, door, window, flooring and ceiling which made up the Turkish houses.

The houses were completed generally with wide eaves and hipped roofs. In the past the roof covering was created out thin pieces of wood known as *pedavra* and set with stones, saying, 'Don't let it fly away in the wind'; because of fires later Turkish style tiles were used (Fig. 33).

Conclusion

Anatolia is in the second degree Alps-Himalayan earthquake zone which is one of the active earthquake zones [13]. It is understood that there were the wooden frame buildings on account of the examples that have been able to survive until today of the type of building most suitable to the region that has experienced innumerable earthquakes throughout history. What a pity it is that the 16th century examples of partition frame Turkish houses with *hayat* have not survived until today because their importance still was understood in our country at the beginning of the 20th century; however, the studies that Sedat Hakkı Eldem and his students did in relation to these houses has provided documentation for them.

In the wooden frame construction system, the buttress that supported the post and base crosswise was especially set in the corners, the frame elements like the base, post, beam and buttress were attached to one another with iron nails both dove-tailed and pounded, with the help of the bases the connection established between the wooden frame system and the stone walls increased the sturdiness of this system against earthquake. In addition wood as a light and elastic material reduced the loss of life during an earthquake from the damage created because of material being old, rotten and neglect. In conclusion wooden frame buildings made of wood that is a natural material that can be renewed and turned for centuries both resisted against earthquake and provided for the continuation of traditional building techniques.

References

1. Kuban D (1995) *Türk Hayatlı Evi*, İstanbul: Ziraat Bankası Yayinlari, the Turkish hayat house. Eren Yayincilik, İstanbul, p 48
2. Eldem SH (1968) Türk Evi Plan Tipleri. İTÜ Mimarlık Fakültesi Yayınları, İstanbul, 2. Baski, p 25
3. Kuban D (1995) p 66
4. Günay R (1999) Türk Ev Geleneği ve Safranbolu Evleri, İstanbul: YEM Yayinlari, tradition of the Turkish house and Safranbolu Houses. YEM Yayinlari, İstanbul, p 46
5. Kuban D (1995) p 48
6. Eldem SH (1984) Türk Evi-Osmanlı Dönemi I/Turkish Houses-Ottoman Period I. TAÇ Vakfı Yayinlari, İstanbul, p 43

7. Kuban D (1995) p 57
8. Eldem SH (1984) p 65
9. Eldem SH (1968) p 25
10. Kuban D (1995) pp 57, 59, 63
11. Erdem A (1999) Göynük. Müderrisoğlu Evi, YTÜ Dergisi, Sayı 2(1999):73–85
12. Seçkin N, Özakın R, Erdem A, Yergün U (2004) Boyabat (Sinop) Kentsel Kültür Varlıkları Envanteri 2003, TÜBA Kültür Envanteri Dergisi. J Cult Inventory, Sayı 2004/3, pp 141–169
13. https://tr.m.wikipedia.org/wiki/Deprem_kuşaklari

Traditional Timber Housing Structure in Zeyrek

Aylin Akcabozan

Abstract Zeyrek, as a World Heritage Site and one of the important residential areas of the old city of Istanbul, has a significant number of historical timber houses belonging to late 19th–early 20th century. Due to its topographical, climatic, social, economic and cultural features as well as its nearby wooded environment, local materials and traditional technology, Zeyrek has reflected a precious architectural pattern consisting of many of timber-framed buildings for a long time. This paper examines traditional architecture, presents and promotes the experience with timber housing restoration in Zeyrek, Istanbul, aiming to analyze the existing wooden housing pattern in the area and to investigate the restitution and restoration project of a traditional timber house in detail. The conservation and restoration process in the traditional structure of wood extracted positive lessons of history. In this context, the classification of physical characteristics commonly found in these timber housing settlement has been done considering the method of construction, the façade proportions of the elements, plan schemes, articulation of different forms, the conservation and restoration projects in the site. The present study follows methodological principles of historical research about Zeyrek. A case study about traditional timber house structure consists of field surveys and documentation. According to observation and analysis of the traditional housing settlement, the main classifications were obtained, along with photographical survey showing the current situation of the timber housing characteristics of the site. The paper focuses on this transformation by putting forward integrated conservation approach.

Keywords Traditional housing · Timber structures · Integrated conservation · Zeyrek

A. Akcabozan (✉)
Yildiz Technical University, Istanbul, Turkey
e-mail: aylin.akcabozan@gmail.com

H. Cruz et al. (eds.), *Historical Earthquake-Resistant Timber Framing in the Mediterranean Area*, Lecture Notes in Civil Engineering,
DOI 10.1007/978-3-319-39492-3_6

Introduction

Zeyrek, as a World Heritage Site and one of the important residential areas of the old city of Istanbul, has a significant number of historical timber houses belonging to late 19th–early 20th century. Due to its topographical, climatic, social, economic and cultural features as well as its nearby wooded environment, local materials and traditional technology, Zeyrek has reflected as a precious architectural pattern consisting of many of timber-framed buildings for a long time.

This paper examines traditional architecture, presents and promotes the experience with timber housing restoration in Zeyrek, Istanbul, aiming to analyze the existing wooden housing pattern in the area and to investigate the restitution and restoration project of a traditional timber house in detail. The conservation and restoration process in the traditional structure of wood extracted positive lessons of history. In this context, the classification of physical characteristics commonly found in these timber housing settlement has been done considering the method of construction, the façade proportions of the elements, plan schemes, articulation of different forms, the conservation and restoration projects in the site.

The present study follows methodological principles of historical research about Zeyrek. A case study about traditional timber house structure consists of field surveys and documentation. According to observation and analysis of the traditional housing settlement, the main classifications were obtained, along with photographical survey showing the current situation of the timber housing characteristics of the site. The paper focuses on this transformation by putting forward integrated conservation approach.

This study consists of 5 main chapters including introduction and conclusion. The first part (introduction) includes general information about Zeyrek World Heritage Area. In the second part of the study, the aim of the study and a brief history of conservation of the site will be explained. The third part includes swot analysis in Zeyrek World Heritage Area and integrated conservation works related to the area. The fourth part includes the case study on a survey of the traditional timber structure in Zeyrek.

History of Conservation in Zeyrek

One of the most famous and symbolic settlements of the Turkish era is the Fatih district, originally developed around the first large mosque and imaret of Istanbul. In the southwest it is connected to Aksaray with the Etmeydanı and Horhor quarters. To the east it is circumscribed by Saraçhanebaşı, Şehzadebaşı, and toward the Golden Horn Zeyrek, Çarşamba and Sultan Selim, in the Edirnekapı direction. The area where the quarter is located has retained its status as a religious symbol since the foundation of the city.

The quarter was devastated by the fire in 1908, while thousands of buildings in the region were consumed by the Cibali fire of 1918. In the period prior to World War I

Fig. 1 General view of Zeyrek Conservation Area [6]

small-scale, two or three-storey apartment houses gradually began to replace the wooden ones based on a grid-iron plan. Zeyrek is located at a key point in the histories of both the Byzantine and the Ottoman Empires. Zeyrek is situated on the fourth hill of Historic Peninsula. One of the most significant example of Byzantion Architecture is Pantokrator Monastery, founded by Empress Eirene in the 12th century and timber houses of Ottoman Architecture gives the unique identity to Zeyrek [1].

Zeyrek's historical legacy, unique architecture and well-preserved neighbourhood helped it gain an Urban Conservation Area status in 1975; in 1985, it entered the UNESCO World Heritage Project list. Needless to say, the district survived many catastrophes until then. Fires that started in the workshops at the Golden Horn shores spread through, fuelled by the North winds, causing immeasurable damage in the wooden buildings, creating the need for reconstruction. These fires impoverished the residents, changing Zeyrek's texture and architecture (Fig. 1).

As quoted from the Istanbul Project: Istanbul Historic Peninsula Conservation Study, the opportunities of the area and contribution to the conservation of immovable cultural beings. Zeyrek Conservation Area boundaries were first determined in 1974 and ratified by General Statute in 1987. A total of 469 buildings are surveyed in the study, that 51 % of them are listed. 86 % of the listed buildings, are defined as civil architecture. Molla Zeyrek Mosque is one of the most important monumental structures in the region, and the most symbolic of it [1].

Although many public and private institutions have made great efforts to emphasize the importance of the issue, the conservation of cultural beings is an ongoing problem in Turkey. Zeyrek World Heritage Area is analysed in detail in terms of conservation problems threats and potentials.

Effective conservation and integrated conservation approaches are necessary to enhance the environments and to create better income opportunities for the people living in the area by re-functioning regulations along with residential use. In Zeyrek, attempts are presently directed towards upgrading the surroundings of the Pantokrator with the intention of raising awareness for the preservation of the area and attracting the attention of investors or sponsors [1].

Location of the House

The house is located in the Haydar Avenue, end of the Haydar Bostanı Street, blind alley. There are lots of historical timber houses on Haydar District. City block: 2188, parcel no: 33 (in the Pervititch Map, parcel no: 153), no: 7. The adjacent

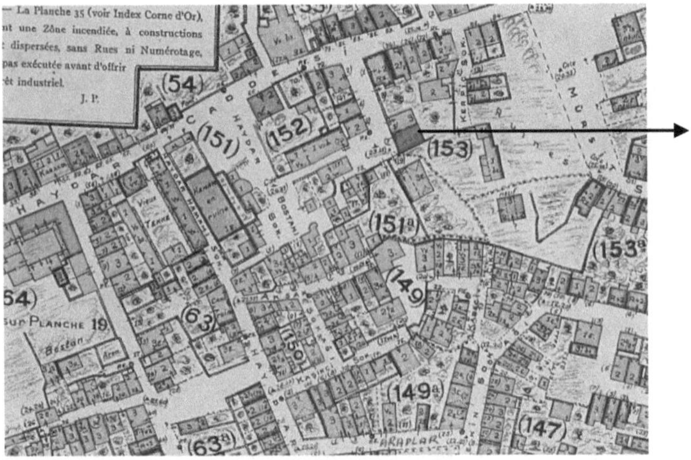

Fig. 2 Pervititch maps, Haydar Bostanı Street, no: 7 [7]

building is a similar 19. Century style house which has a jetty. One façade of the house is on the street and the back façade looks towards a garden. There is a fire wall on the west side of the house. At the end of the street there is a big timber house with garden (Figs. 2, 3, 4 and 5).

Fig. 3 Façade of the house [6]

Fig. 4 Historical buildings in Zeyrek, Haydar District

Fig. 5 Details of the timber house in Zeyrek [7]

Survey

Plan: There are 4 floors in the building. The basement floor is an unused floor, the other three floors are used by different tenants. The basement floor plan: As the pavement has been elevated from the initial level, the basement floor can be reached by descending three steps. The floor is painted with green paint, and there are lots of garbage near the door. There are two rooms opening to the entrance. The rest of the basement floor cannot be used because of the bad smell, moisture, bug infestation.

The entrance floor plan: The entrance provides passage to the first flat and a staircase by which we can reach the upper flats. The first flat is separated in two zones. The first zone is the original house limits. The second zone is the part that is constructed later. The original part has two rooms and a bathroom, and the new part has two rooms in addition. The ex-projection has become a light well.

The first floor plan: The first floor can be reached by a staircase. This staircase construction is original but the steps are covered with an unconvenient linoleum material. Some steps are damaged and there is some structural failure and bendings. The second flat is separated by a wall which must have been constructed later, and the entrance of the flat leaves passage to three volumes:

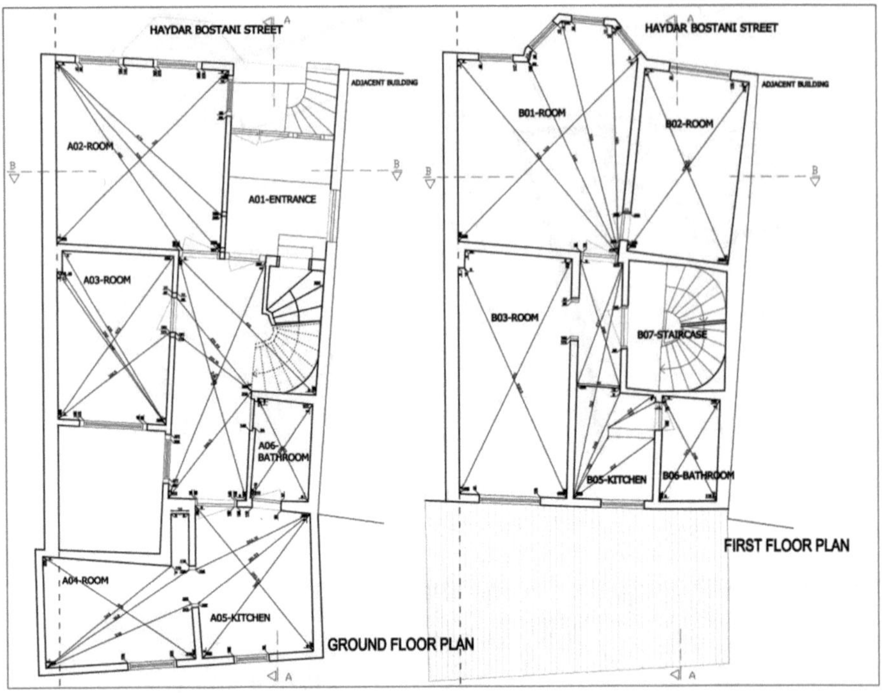

Fig. 6 Survey: ground floor plan, first floor plan [5]

The first volume is the living room, larger than the other rooms and with a jetty. The ceiling is decorated with a more elaborate style. The floor is bending, the jetty also has some structural failure. There is another room linked to this living room which is used at present as children's room. There is no door between these spaces, just an opening leads passage.
The second volume is the parents' room, there is also some modern windows with no historical value. The third volume is the kitchen and the bathroom which are separated from the rest of the house with a 1 m high wall (Fig. 6).

The second floor plan: The third flat's door is situated on the first step of the stairs rising to the second floor. These stairs are covered with linoleum, as well. There are three volumes opening to the staircase, a kitchen, a bathroom, and a living room. Passing through the kitchen, we can also reach the bedroom, which is opening to the 'cihannümâ' (the space over the jetty, which is used as a balcony, and covered by a roof) (Figs. 7 and 8).

Façade: The firewall is very damaged and has also some structural failure, because of the roof loads. The roof is not original and is constructed on two parts, one of these part covers the main building, the other part covers the "cihannüma". The ridge of the main roof is higher than the ridge of the other part, and they are not one the same line. The doors and windows on the façade are original but the

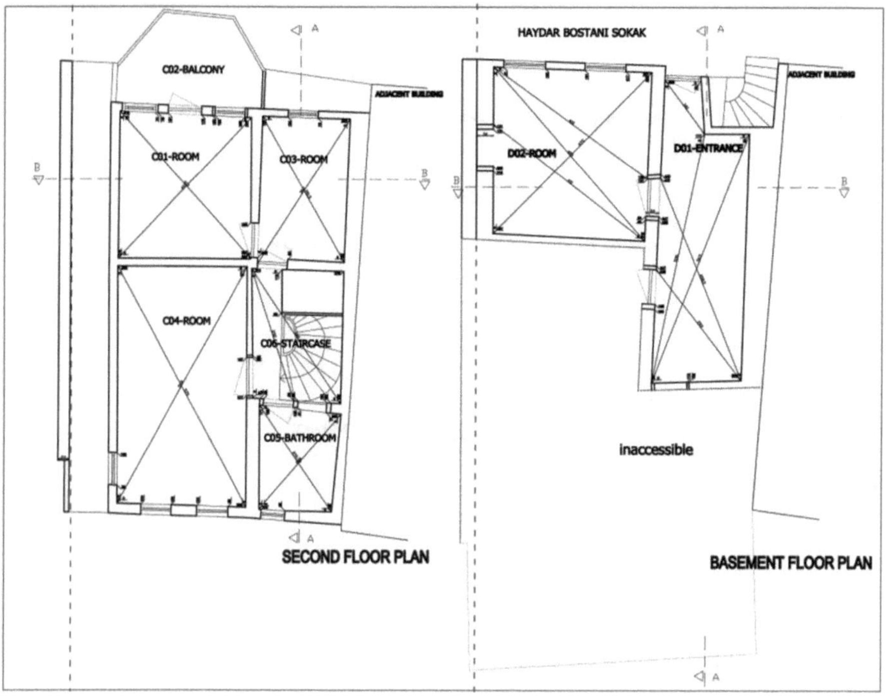

Fig. 7 Survey: second floor plan, basement floor plan [5]

wooden cladding is damaged with some moisture. There are also some material loss on the steps, plaster cracks on the basement floor, plaster loss around window openings, and some ornaments losses on window-guards (Fig. 9).

Restitution

Plan: The Basement floor plan hasn't been changed. Restitution study shows that it has encouraged its original state.

Entrance floor: Ground floor plan hasn't been the original state. The rectangular mass at the back side has been added probably by the present users. The entrance floor was probably used as a dining and living area. There also could have been a kitchen located on the basement floor.

First floor: Timber house which was designed for a family are using three different families so at the end of the stair case a wall was added. However, this staircase construction is original. Because of this additional wall two door which opened to sofa had been closed. The first floor must have been used as private rooms such as the masters bedroom and several other bedrooms and bath.

Fig. 8 Survey: A–A Section [5]

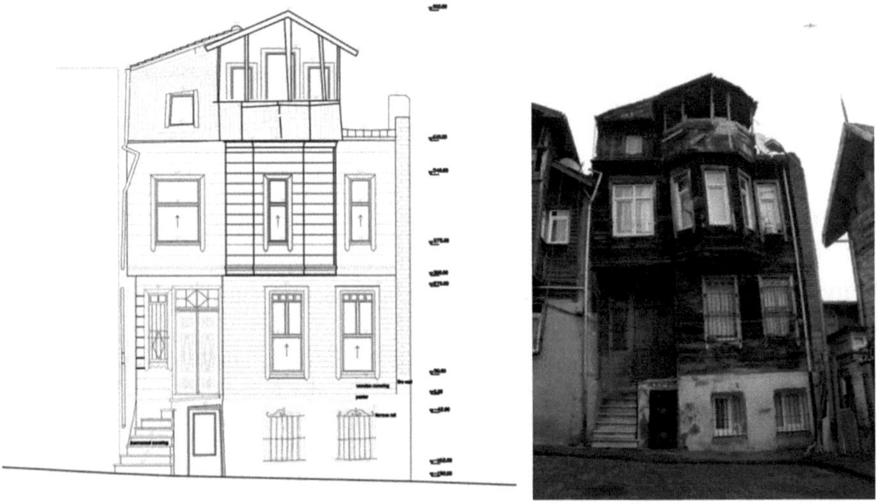

Fig. 9 Restitution façade drawing [5] and façade photo [6]

Second floor: There was an additional door opened sofa. There might be two volumes using façade collusion but after repairing roof this volumes disappeared. Presumably there also could have been a storeroom using as a bath.

Façade: The mahya also had been leveled on the same line but today it has been changed. The roof could have covered all façade.

Restoration Proposals

The following restoration proposals are represented on the plans. The function of the building is house, in which designed for one family. There are three flat and a basement floor in the house. There are some changes on the floors, where restoration or maintenance work is needed.

Basement floor: There is a study room in the basement floor and in other parts could be a storage or cold room.

A. Windows and doors: Repairation of window and door openings, frames, sash, doors, lintels, sills and shutters whenever necessary.
B. Exterior finishes: Basement floor structure is load bearing system. Repairation of walls and maintenance of exterior finishes. The cracks and the deformations on the stone are repaired by the grouting.

Entrance floor: There is a kitchen, a dining room and wc on the entrance floor. A door near the wc is opening to the garden of the house. Windows and doors: Repairation of window and door openings, frames, sash, doors, lintels, sills and shutters whenever necessary. Exterior finishes: Repairation of wooden claddings.

First floor:

A wall which is separating living room and children room is canceled by protection of structural elements and designed a larger living room. There is a room and a bathroom on the first floor.

A. Windows and doors: Repairation of window and door openings, frames, sash, doors, lintels, sills and shutters whenever necessary.
B. Exterior finishes: The exterior finishes of the jetty is repaired.
C. Interior features: Retaining and discovering original paint colors and other decorative colors, when necessary, replacing them with colours. Repairation of decorative wooden ceiling elements.

Second floor:

There is a living room, a bedroom opening to the balcony, an attick room and a bathroom. An appropriate porch roofing is designed for the balcony.

A. Windows and doors: Repairation of window and door openings, frames, sash, doors, lintels, sills and shutters whenever necessary.

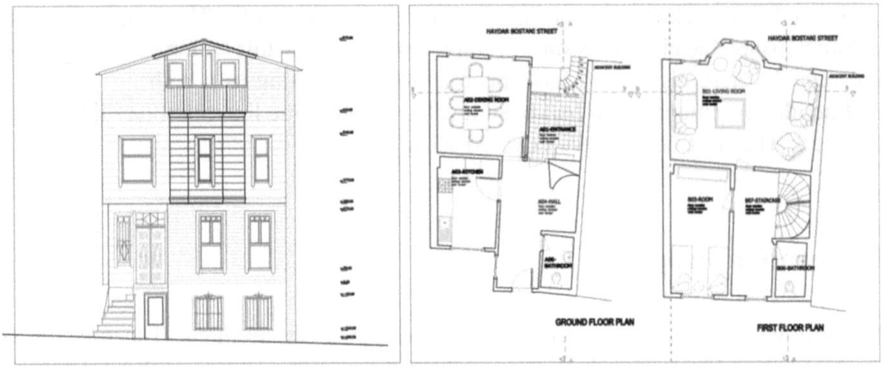

Fig. 10 Restoration façade drawing and plan proposals [5]

Façade:

- An appropriate roof designed and the mahya's are leveled on the same line
- The balcony above the jetty (cihannüma) is repaired. An appropriate balustrade is designed for the balcony.
- The timber covering on the second floor's façade is repaired.
- The structure of the jetty is repaired.
- The bending on the first floor is repaired.
- The marble blocks of the stairs are changed because of the erosion.
- The cracks and the deformations on the stone are repaired by the grouting.
- An appropriate porch roofing is designed for the balcony.
- Decorative elements on windows are repaired.
- Structural failure of the fire wall is repaired.
- The concrete repair on the entrance is changed with an appropriate material.
- The pipes and cables are cancelled on the façade.
- The gap on the roof floor is closed (Fig. 10).

Integrated Conservation

Integrated conservation can be defined with the Declaration of Amsterdam. The main objective of the integrated conservation of architectural heritage are mentioned in the Resolution 76 as the conservation of monuments group of buildings and sites, and integration of these into the physical environment of present day society through revitalization and rehabilitation programs. With this resolution, it is also emphasized that social groups and property owners should be integrated into the conservation process [2]. The Declaration of Amsterdam (1975) emphasized the extent of the conservation and the roles and responsibilities of the public authorities and the society [3]. It is also mentioned in the document that "to help meet the cost of restoration, adaptation and maintenance of buildings and areas of architectural or historic interest, financial support and fiscal relief should be made available to private owners" [4].

Following an integrated approach to urban conservation and historic revitalization combining a number of actions that address environmental, social and economic concerns facing world heritage sites of universal concern.

Conclusion

Vernacular timber housing in Zeyrek quarter was recognized as vulnerable at the time of incription. Despite the threat of pressure for change, many efforts have been carried out in order to conserve and strengthen the timber structures within the site.

Therefore, changes of the social structure in the area have also affected the use of those structures. Loss of historical monuments, architectural integrity of the city, traditional materials and techniques. Additionally, socio-economic reasons and high costs of restoration can decelerate the conservation process. By the integrated conservation process; social groups and property owners may collaborate and all timber houses in the Zeyrek World Heritage Area could be saved. The revival of the Zeyrek quarter is a long term Project which demands a detailed process of cleaning, conservation and restoration.

The aim of this study is to analyze the existing wooden housing pattern in the area and to investigate the restitution and restoration project of a traditional timber house in detail. The conservation and restoration process in the traditional structure of wood extracted positive lessons of history.

References

1. Gülersoy NG (2008) Istanbul Project: Istanbul Historic Peninsula Conservation Study, Istanbul Technical University Faculty of Architecture and ITU Urban and Environmental Planning and Research Center
2. Zengin US (2010) Urban conservation as an ownership problematic: a thesis submitted to The Graduate School of Natural and Applied Sciences of Middle East Technical University
3. Enlil Merey Z (1989) Keeping a historic neighborhood lively: integral preservation approaches in Paris and Bologna. Yerleşim Yazıları, Şehir ve Bölge Planlama Bölümü Yayını, Yıldız Üniversitesi Matbaası, İstanbul
4. European Charter of Architectural Heritage (1975)
5. Akcabozan A, Tunalive S, Hacihasanoglu I (2008) Architectural Survey and Restoration Studio: Timber House at Zeyrek, Istanbul Technical University
6. Akcabozan A (2015) Zeyrek Photograps, Istanbul
7. Pervititch J (2000) Istanbul in the insurance maps of Jacques Pervititch, The History Foundation of Turkey, Istanbul

Possible Precursors of Pombalino Cage

Alice Tavares, Aníbal Costa and Dina D'Ayala

Abstract Pre-Pombalino buildings in Lisbon nowadays represent a rare alive witness to a major historic European earthquake event: the 1755 Lisbon earthquake with tsunami followed by a great fire. More substantial interest is however shown by researchers in the so-called Pombalino cage [1, 2], a structural typology created after the 1755 earthquake to rebuild the city, and considered by many authors, as a first engineered solution designed specifically to be resilient to earthquakes. The Pombalino cage is a composite timber-framed masonry construction with a wide use of St Andrews crosses and specific connections. In comparison very modest attention has been paid to clearly identify in history possible precursors of this technological solution, despite the many examples of vernacular masonry infilled timber framed buildings, which are common to many Mediterranean regions exposed to seismic hazard. An additional difficulty in tracing with certainty the historic development of the Pombalino cage is the lack of original drawings for this construction typology. This paper attempts to determine the conditions at the time, in terms of existing technical know-how and current construction methods that would have made the development of the Pombalino cage, possible. The work carried out in this research encompasses in situ surveys of 18th century buildings, documentation from several archival libraries, analysis of ancient treatises, and previous research by the authors. This manuscript presents first a review of: (i) traditional timber-framed construction systems belonging to earthquake-prone regions; (ii) then connects these to influential contemporary knowledge on the 1755 Lisbon earthquake; (iii) evidence from Portuguese legislation, regulations, and notices published after the earthquake; (iv) and finally the academic education and

A. Tavares (✉) · A. Costa
RISCO, Civil Engineering Department, University of Aveiro,
Campus Santiago, Aveiro, Portugal
e-mail: tavares.c.alice@gmail.com

A. Costa
e-mail: agc@ua.pt

D. D'Ayala
University College London, Gower Street, London WC1E 6BT, UK
e-mail: d.dayala@ucl.ac.uk

H. Cruz et al. (eds.), *Historical Earthquake-Resistant Timber Framing in the Mediterranean Area*, Lecture Notes in Civil Engineering,
DOI 10.1007/978-3-319-39492-3_7

training of Portuguese military civil engineers and architects (before and after the earthquake) who developed the Pombalino Cage; (v) the paper concludes with a discussion of whether these elements can be identified as the necessary precursors of the Pombalino cage.

Keywords Pre Pombalino · Pombalino cage · Military architecture and engineering · Timber-framed structure · Treatises · Historical earthquakes

Traditional Timber-Framed Construction Systems Belonging to Earthquake-Prone Regions

The analysis of historical earthquake catalogues show a significant number of destructive earthquakes throughout the Mediterranean region in the 500 years prior to the 18th century. After the Sicilian earthquake in the twelfth century, several other strong earthquakes occurred in the south of Spain, Italy, Greece, and Portugal up to the seventeenth century [3]. Among those the 1356 and 1531 Lisbon earthquakes (with tsunami), and the 1693 Sicilian earthquake, can be considered as the most significant. In the same period several catastrophic earthquakes also took place in Latin America, and such events were also reported in the technical literature in 18th century Portugal. Communities along the Mediterranean coast have developed traditional composite construction systems with timber and masonry. Previous studies by the authors [4] have shown some links between these systems and some others present, for example, in highly seismic prone regions such as Turkey and Kashmir. The use of composite timber structures in buildings was mentioned as early as by Vitruvius in *Ten Books on Architecture* (1 BC), was referred to by Alberti (15th century) [5] when discussing his childhood, and it is present in stamps of Diderot's and D'Alembert's *Encyclopédie* (18th century) [6], giving a sense of a technique without boundaries, even to these days. In Europe, this type of construction is present in earthquake-prone regions and has similarities to the same type of constructions in other non-seismic regions. They present a common feature: timber elements as the main part of the load-bearing structure. However, the key issue for evaluating and distinguishing them is their capacity to withstand lateral actions in addition to vertical loads [4]. Some of the timber-framed systems which could have been known to the post-1755 Portuguese engineers among others are: the Spanish post-and-beam (central region), the Italian (Herculaneum), the Portuguese pre-Pombalino (Lisbon), the Turkish *himis* (northern region), and the Greek systems. The Spanish post-and-beam construction system (Fig. 1) is characterised by horizontal timber beams and vertical timber uprights confining adobe or brick masonry walls. Despite its similarity with the antique Herculaneum (Italian) construction system, the Spanish system has bigger spacing between the timber elements and sometimes an additional horizontal short piece in the beam support area. Although the characteristic of regularity is present,

Fig. 1 Spanish timber-framed (*credits* A. Tavares)

Fig. 2 Prepombalino building (*credits* A. Tavares)

Fig. 3 Greek timber-framed system (*credits* A. Tavares)

these are not resilient earthquake systems due to missing diagonal elements, a lack of proper dimensioning and connections between construction components. However, it is an easy and inexpensive way to build. The pre-Pombalino (Fig. 2), the Greek (Fig. 3), and the *himis* system (Fig. 4) present a cantilevered upper floor

Fig. 4 Turkish timber-framed system (*credits* A. Tavares)

Fig. 5 Pombalino cage wall, after recovery (*credits* A. Costa)

(timber-framed walls) supported by timber beams above stone masonry load-bearing walls at the ground floor. In each system diagonal timber elements are present near corners and openings in the ground floor masonry walls. In spite of the fast-disappearing and deep alterations of pre-Pombalino buildings [7], parts of the old timber frame layout are still visible, as the case studies of Alfama buildings (Lisbon) have shown. The alterations during the nineteenth century mainly introduced balconies to the upper floor, increased the size of the windows, and demolished the middle longitudinal stone masonry wall. Nevertheless, the interior shows parts of the old composite timber structure with brick, which differs from the Pombalino cage (Fig. 5), but has similarities to the *himis* system [8], even in the dimensioning and spacing of the uprights (0.25–0.35 m). This timber structure is composed of regularly spaced timber uprights with at least three levels of horizontal timber elements and diagonal elements near the walls cross direction. Although, this system could be more resilient to an earthquake compared with other traditional techniques as rubble masonry, it probably revealed some problems. From the survey assessment the following problems can be noted: (i) the lack of connections between interior walls and gable walls and between floor timber beams with those walls; (ii) the cantilevered upper level breaks the vertical continuity in terms of stiffness and mass which compromise the whole structure; (iii) the collapse of the front façade. These were vulnerabilities that needed to be mitigated to provide a resilient structural system, and which, we can assume, were probably observed by the military engineers responsible for the reconstruction of Lisbon when they conceived the Pombalino cage. Although the use of St. Andrews crosses was known, they were not used in construction with the same systematic approach or regularity in quadrangular shapes involving a wide range of connections as that present in the three-dimensional (3D) structural concept of the Pombalino cage. Nevertheless, pre-Pombalino buildings are a valuable multicultural heritage that played an important role in the decision to adopt a composite system with timber.

Influential Knowledge Contemporary to the 1755 Lisbon Earthquake

The topic of seismic activity has been debated by learned people since antiquity, involving mainly the search for the reasons (through religious or scientific approaches) behind the events. The 1755 Lisbon earthquake that also affected Spain and Morocco triggered a large number of publications, manuscripts, and debates in the scientific community (i.e. J.J. Rousseau, F.M. Voltaire and I. Kant) [9] and society in general. From these, only information that could have influenced the Portuguese decision-makers is presented in this section. Among the technical literature widely divulgated at the time, two relevant treatises known by the Portuguese engineers were Vitruvius' *De Architectura* and Alberti *De re Aedificatoria* (1443–1452) [5]. In those treatises, the concept of the building as a body in which each part has its own function, which contributes to the global behaviour, is a key point. Moreover, both suggest the need for a deep knowledge of materials, the improvement of robustness of connections between construction components, and necessary measures to ensure the durability of the construction [4]. Rules for good construction were expressed in terms of geometric and proportional relations between architectural and structural elements. Alberti advises readers to learn from the ancients, valuing their experience and solutions, in addition to learning from the visual observation of old buildings [5]. This capacity for taking lessons from existing buildings is a key element of learning, well developed in military experience and training and well suited to the pragmatic approach of the Portuguese military architects and engineers. After the 1755 earthquake, several Spanish manuscripts [3], authored by influential academics of the time, such as Rector of the University of Seville J. Cevallos (1726–1776), Mathematics Professor at Salamanca University O. Gallardo de Villarroel, Oviedo University Professor B. Feijoo y Montenegro (1676–1768), tried to explain the event investigating the natural causes triggering the occurrence of seismic shaking [3]. Later in 1760, John Michel attempted to establish a link between seismic phenomena and the propagation of electric waves, trying to find a reason for the almost simultaneous earthquake effects in regions faraway from an earthquake's epicentre. However, while the scientific debate on the origin of seismic shaking continued in Europe, in Portugal, as a direct consequence of the devastation created by the earthquake, the debate focused on new rules for reconstruction. Indeed, there is evidence of Portuguese military engineers dictating overall changes to the construction system, creating an engineered solution, far from the traditional system in use at that time.

The devastation of the 1693 Sicilian earthquake (known by the Portuguese) had triggered the development of new rules in urban planning, as shown in the rebuilding of Noto [10]. A similar approach, for a new engineered construction system, as it will be seen in Portugal, also took place in the southern Italy Borbon State, following the devastating 1783, Calabria earthquakes, which gave rise to the use of the *Baraccata* system. The extent to which these examples, or the debate on reconstruction of cities in the Spanish Viceroyalty of Peru' [11] were known to the Portuguese, needs further investigation. It is however clear that the need for the creation of the Pombalino cage was not argued using as reference historic or contemporary architectural treatises, such as those by Vitruvius, Alberti, Serlio, Bibiena and others, despite them being known to the Portuguese. Nevertheless, more regular urban planning, occurring in Europe at the time, was also a basis of influence for the choices adopted by the Marquis of Pombal when rebuilding Lisbon.

Another important element, influencing urban planning in the 18th century was the new interest in the reduction of diseases and health of the population in line with the Enlightenment principles. The *Tratado da Conservação da Saúde dos Povos*by the Portuguese physician Antonio Ribeiro Sanches (1699–1783), published in 1756 in Paris and marketed in Lisbon at the same time [12], is part of this new field of scientific literature of the time. He presents a deep reflection on the urban requirements for the promotion of good public health, citing the treatises of Vitruvius and Alberti, and quoting the Frenchman Buffon regarding historical earthquakes. Sanches left Portugal due to the Inquisition, and knew the Marquis of Pombal during the latter's time in Paris as the Portuguese Minister of Foreign Affairs [9]. It is very interesting to observe that his concerns and some of his advice are also included in the proposals of the renowned military engineer, Manuel da Maia, leading the technical committee for the reconstruction effort in Lisbon. Sanches addresses the topics of improvements that should be made in urban construction to allow for better ventilation and lighting, but most of all the establishment of public sanitation and the widening of the streets. In line with this, he presented several recommendations to guide the reconstruction of Lisbon that were specifically related to buildings. Several of these concerns can be found in the later Portuguese legislation of 12th June 1758, where urban and construction rules are cast [13]. The relevance of this work is proven by the inclusion of an annex entitled "Considerations about earthquakes", where Sanches briefly describes the historical effects of earthquakes and the Lisbon earthquake with a specific chapter *Notícia do Terramoto*. Contained in this annex are some interesting considerations, including the appropriate ground soils for the foundation of cities, the lowest construction height (mentioning the case of Peru and Jamaica), and the need for good connections in corners and foundations [12]. The relevance of this treatise is that it raises the pragmatic background of the Pombalino cage, including concerns about urban control, regularity, and the improvement of safety and living conditions above concerns about aesthetic values or urban speculation.

Legislation and Regulations Published After the Earthquake

The analysis of the legislation during the reconstruction of Lisbon [13, 14] stems from an interest in understanding what resources or procedures were needed to mobilise and possibly influence the characteristics of the Pombalino cage design and its realisation. Immediately after the earthquake, security measures were implemented to defend the coast near Lisbon and the interior of the city from thieves. This was implemented using the army, exploiting the military's experience in imposing public order measures, experience developed during the war with Spain and the colonisation ventures in Africa, America, and Asia. The first step of the reconstruction was published on 29th November 1755 with clauses regulating the future Master Plan, which promoted the survey of existing buildings, for which the military architects and engineers were mobilised. Further rules were published (December 1755; May 1758; June 1758) to maintain the control of the urban reconstruction process. Any new construction in the pre-established perimeter of the centre of Lisbon was forbidden until the survey was completed; if this was not observed, the construction had to be demolished at the owner's expense (February 1756; June 1758; April 1759). However, owners were allowed to repair buildings (if they were still in useable conditions). For this reason, it is possible to see buildings of this period in some of the central rebuilt areas. The announcement of the Master Plan's implementation was published in 10th February 1756, at that time announcing the concepts of the uniformity to be used in each city block, the increase in the width and regularity of streets, and the limitation on the height of buildings. Lisbon's large-scale seismic destruction led to the imposition of pragmatic measures for its reconstruction, also in terms of construction methods used and availability or trade of materials. At the same time, there were two huge challenges that needed to be overcome: necessary improvements to the traditional construction system and to production of construction materials. It is in response to these challenges that the uniformity and regularity of construction was given such importance, above any decorative or aesthetic motivation according to purely architectural treatises. Other subsequent guidelines and laws promoting the national trade of timber (May 1756), its manufacture, and its circulation (September 1756; January 1758), offering tax exemptions show that the decision to promote a composite timber structure had been made at an early stage in the reconstruction process. The many edicts promulgated in this period show government's commitment to purchase surplus materials (29th July 1757) tospur the production of other construction materials as lime and bricks.

Throughout the years of reconstruction, there were two relevant moments in the publication of provisions and government charters. The first was the 12th May 1758 edict, when together with the announcement of specific construction rules inside the urban Master Plan, citizens were given the opportunity to enter into a contract committing them to the purchase (or build) a new building within a maximum period of five years. The most impressive point of this advertisement is the particular reference to the potential advantage for owners to 'benefit from fewer risks of future seismic and fire damage' due to the new type of construction system in rebuilt areas. If there is any doubt about the point at which the application of the Pombalino cage began, this precise charter establishes the confirmation of its use at least from this moment as well as the specific purpose of the resilience of the new construction system. The second relevant document, published 12th June 1758, is the government's provision of some construction details, which however still fell short of a complete description of the entire construction system. This document emphasises the need to ensure a dovetail connection between the joists and the timber beams placed along the perimeter at the floor and attic levels. Further guidelines on the proportional relation between the height of the storeys and the dimensions of the openings, prescribed for each street or square, were presented. Almost one year later in 15th June 1759, a further charter was published with other guidelines, about the prevention of fire risk through the imposition of a whole stone masonry gable wall and an increased height of eight palms above the slope of the roof. This fire risk prevention preoccupation also justifies the complete plastering of the timber structure in this construction system compared with other European cases, while the specific mention of the materials of the gable walls points out that the traditional construction was mixed with timber post and struts, as evidenced by some pre-pombalino houses in Alfama. The detailed prescriptions of this charter are in accordance with improvements for seismic resilience, which were probably based on the observation of the 1755 earthquake's effects on buildings; for example, the charter included the imposition of a uniform height for buildings, the regularity of façades, and the prohibition of the use of balconies or others salient elements on façades [13].

Finally, in order to understand the background of the Pombalino cage, the large number of timber constructions built during the reconstruction process should be considered. They were everywhere at the time, and were built due to the huge demand for housing. Due to humanitarian reasons, building in public squares, navy land, open urban plots, and other open spaces was allowed. The growing number of timber constructions jeopardised the normal course of urban redevelopment, which led to the publication of a series of edicts imposing their demolition (1760; 1769) [14].

Portuguese Military Architects and Engineers—Academic Education and Military Experience

The Pombalino cage is an engineered solution that was conceived by Portuguese soldiers skilled in architecture and engineering. The association between military knowledge and architecture or engineering, through military academies, is evident from the earliest point in the history of the independent Kingdom of Portugal and it was not only confined to the design or the coordination of the construction of fortifications. These military technicians had a wide range of responsibilities at the time, depending on royal decisions including in situ surveys of land and buildings, defining new urban settlements, and the planning and control of all the stages of civil, military, or religious buildings construction [13, 14]. In addition to these duties, the military master architects or master civil engineers were responsible for the choice of materials (supported by master masons and carpenters) for a complete estimated budget proposal (16th January 1689). Usually, the most experienced military technicians taught new appointees in small groups (four per year, according to 1689 legislation), which accompanied the military master during surveys, building inspections, measurements, the laying of foundations, and the evaluation of the condition state or quality of existing constructions [14]. At the end of their academic studies and training, the best tutees could be put in charge of royal or important constructions, as part of technical teams, until they had achieved appropriate experience [13, 14]. This procedure was practiced until the period of the 1755 earthquake, besides the regular calls to foreign architects or engineers for specific commissions. From the end of the seventeenth century, the training of the Portuguese military technical staff was preferred over the invitation of foreign military expert. Although the exchange of experience and knowledge involving foreign treatises was rather common, it should be highlighted that the Portuguese military engineers also elaborated on and used their own treatises, some of which were published in print and others as manuscripts. The citation and commentary on foreign authors by Portuguese military technicians is present in some of the treatises, sometimes criticising the nonsense of practical applications, sometimes valuing the long experience of the military revealed by his proposals. Close to 1755 earthquake, examples of these publications included *Architectura militar e fortificação* (1680) and *Methodo Lusitano* by L.S. Pimentel (1613–1679); *Arquitectura Militar* by Dionísio de Castro (1729–1749) and *O Capitão de Infantaria Portugueza* by G. Lourenço Debrie(…–1755) show pragmatic concerns about the application of military construction theories. The interest in deeper knowledge of construction materials was expressed in these treatises as in academic training course structure. In the seventeenth century, Portuguese academic training was developed at the Military Academy of Santiago da Espada and S. Bento de Avis (1615) [15]. The Fortification and Military Architecture Academy replaced it on 1647 during the reign of D. João IV was later established in other places, until the reign of D. Pedro II. These show the preparation of military forces to defend Portugal's frontiers as a result of the conflict with and independence from Spain

(1640–1668). These military technicians had a theoretical knowledge based on fortification treatises of the time, but also on other topics such as settlements of urban areas due to the expansion of the colonies [14]. This was the contemporary Academy of the 1755 earthquake event. The technical preparation for the reconstruction of Lisbon should also be addressed here.

The College of Nobles was established in 1761 and the Army Artillery Regiment in 1763. In the College of Nobles, mathematics, military architecture, civil architecture, physics, and design, as well as French, Italian, and English were studied. This last issue shows that their learning involved foreign military treatises/sources from these three countries, but also teaching by foreign military technicians. The academic degree established three years of study of mathematics, divided into broad topics, including arithmetic, geometry, trigonometry, statics, calculus, hydraulics, geography, topography, astronomy, optics, design, and architecture. The physics discipline involved the demonstration of theories through experimental tests, a very practical point of view: 'teaching only what is demonstrable by geometry, calculus, or qualified experimental tests' [13]. The military architecture included general rules of fortification, the various regular and irregular methods of creating square fortifications, the methods of construction, the defence of a site, and creation of camps and army fortifications. The structure of the course had several common topics with foreign treatises, such as the French treatises by S. Vauban (1643–1715) *De L'attaqueet de la defense des places* (1704) and F. Bélidor–*La Science des ingénieurs dans la conduite des travaux de fortification et d'architecture civile* (1729) and J. Muller's book—*Treatise of artillery* (1757). These treatises show a wide range of military knowledge about constructions that could also be used to help understand the behaviour of structures during seismic events. An example is trying to identify a building's centre of mass through the correspondence of the charge capacity of structures with their geometry. The civil architecture discipline was about the rules, principles, and reasons behind the measures and proportions proposed in the treatises. In this course, 3D models of construction elements made of timber were used to explain the role of each component [13]. This was the preparation given to military engineers and certainly used in the reconstruction of Lisbon.

The main team nominated for the reconstruction of Lisbon were mostly military technicians, both architects and civil engineers: Manuel da Maia (the coordinator of the team), Carlos Mardel, Eugénio dos Santos, and Elias Sebastian Poppe. In addition to their links to the Military Academies as professors or students, they had also worked together in relevant constructions before the Lisbon earthquake. They followed the established professional training of the time, which involved surveys, design plans, measurements, and overseeing construction. From these experiences, this research considers their participation in the construction of Mafra convent (40,000 m^2) as specifically relevant. For the construction of this great monument, which was designed to house 300 monks, it was necessary to employ 52,000 persons [15]. This fact imposed the need for temporary housing, a hospital, and a

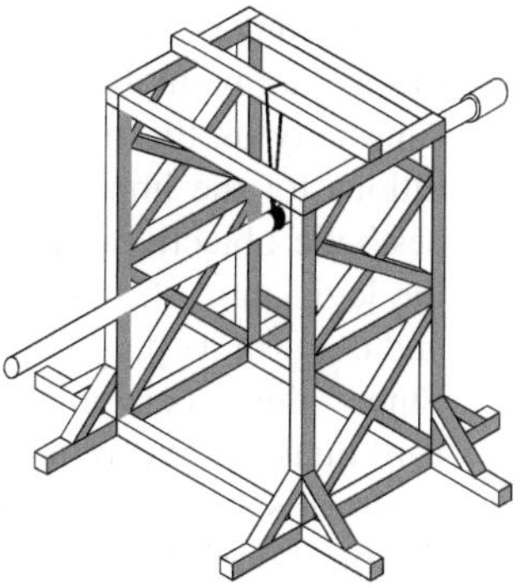

Fig. 6 Military battering ram based on drawing of L'Encyclopaedie (Art Militaire. Planche XII), *credits* A. Tavares

church, all of which were built with timber. The complementary structures used in the construction of the Mafra convent (1717–1730), the 'timber island' as it was known, was a very important experience for acquiring a solid knowledge of the behaviour of timber structures. This particular fact shows common links with the Lisbon reconstruction period: the use of accessible material, with fast building methods, which were progressively standardised. Nevertheless, the timber military siege battering ram, some of which were depicted in the Diderot Encyclopaedia's voice "Art Militaire. Fortification" [6] (Fig. 6), should also be considered as a relevant precedent in the Portuguese engineer's knowledge, due to their design and mechanical characteristics. As a simple timber structure, it needs to combine the dynamics of the movement of the central balanced timber beam in both directions to induce the impact against the door of fortification walls, while ensuring the stability of the timber support cage with diagonals. Some other secondary military structures made of timber, such as siege towers for visual control over the enemy or to attack walls, again combined a main timber structure that needed to support the movements of the soldiers and eventual horizontal loads due to defence actions or movement in the field. Some examples of bridges made of timber, with St Andrews crosses in their lateral parts, belonging to military camp constructions and the previously mentioned structures are just a few examples of military structures that could be precursors of the Pombalino cage. The design solution of the Pombalino cage—the regularity and wide use of timber St. Andrews crosses, the improvement of the connections, and the 3D behaviour expected—reveal this knowledge.

Conclusions

The eighteenth century engineered construction system of the Pombalino cage is one of the most relevant inheritances of seismic engineering. However, a gap still exists regarding its precursors. From the results of the analysis of documents and surveys of buildings, this research concludes:

- the precursors of the Pombalino cage did not rely on architectural treatises; instead, a favourable ideological and reasonable sense of defining regularity and simplicity as the main characteristics of urban architecture provided the conditions for the implementation of a new construction system, which subsumed aesthetic objectives and avoided projected singularities in the urban block;
- the legislation shows conditions for the implementation of the Pombalino cage and confirm its purpose to guarantee resilience to seismic events and fire hazard. Since the decision was made to maintain the capital in the same earthquake-prone region, it was necessary to give the population a renewed sense of safety and confidence;
- the pre-Pombalino buildings made from timber composite structures, in comparison with other constructions made with rubble masonry, showed better behaviour, even with some limitations, in the 1755 Lisbon earthquake; the observation of this and the identification of how to overcome these buildings' vulnerabilities could be precursor lessons;
- the massive use of timber buildings during the construction of the Mafra convent (40,000 m^2), which employed and housed more than 50,000 persons, and the increasing use of timber buildings in Lisbon immediately after the earthquake provided the test bed and training necessary to deepen the understanding of timber structures behaviour for the military engineers involved in these two major construction events of the time. This is a very likely precursor, but further investigation is necessary;
- the military experience, including knowledge about ancient siege timber structures used in war, for example in siege strategies, could be a relevant precursor;
- the association of the military architects and civil engineers with the Academy, where a pragmatic but also an experimental approach with a deep knowledge about materials was pursued, created the necessary background and knowledge base to develop the type of timber structure chosen (the Pombalino cage), which had the capacity for an approximate 3D behaviour, concerning the dynamics loads involved during their use.

Acknowledgments Alice Tavares is gratefully and acknowledge to FCT—Fundação para a Ciência e Tecnologia due to postdoc research support (SFRH/BPD/113053/2015).

References

1. Mascarenhas J (1994) Pombaline downtown, 2nd ENCORE. LNEC, Lisbon (in Portuguese)
2. Monteiro M, Lopes M, Bento R (2005) Dynamic behaviour of Pombalino quarter. In: Proceedings of 250th anniversary of the 1755 Lisbon Earthquake, 1–4 Nov 2005, LNEC, Portugal
3. Vallina U (2010) El terremoto de Lisboa enEspaña, Spain. ISBN 978-84-8010-185-1
4. Tavares A, D'Ayala D, Costa A, Varum H (2014) Construction systems (chapter 2). In: Structural rehabilitation of old buildings, vol 2. Series "Buildings pathology and rehabilitation". Springer, Berlin. doi:10.1007/978-3-642-39686-1

5. Rykwert J, Leach N, Tavernor R (1996) Leon Battista Alberti on the art of building in Ten Books. The Mit Press, Cambridge, pp 70–85. ISBN 0-262-01099-2
6. D'Alembert JLR, Diderot D (1751–1772) Architecture et parties que en dépendent, and Art Militaire. Fortification L'Encyclopedie, Paris, France (in French)
7. D'Ayala D, Spence R, Oliveira C, Silva P (1996) Vulnerability of buildings in historic town centres: a limit-state approach. X1 world conference on earthquake engineering, Acapulco
8. Cobancaoglu T (2001) "Himis" construction system in traditional Turkish wooden houses. In: Lourenço P, Roca P (eds) Proceedings of historical constructions, Guimarães, Portugal
9. França JA (1989) A reconstrução de Lisboa e a arquitetura pombalina, Bib. Breve, Lisbon
10. Tobriner S (1989) La genesi di Noto: una città siciliana del Settecento. Ed. Dedalo
11. Quinn N, D'Ayala D, Descamps T (2015) Structural characterisation and numerical modelling of historic Quincha Walls. Int J Arch Heritage. doi:10.1080/15583058.2015.1113337
12. Sanches AR (1756) Tratado da Conservação de saúde dos povos. In: Gendron P (ed) Paris e comercializado em Lisboa by Bonardel e DuBeux, Mercadores de Livros
13. Silva AD (1828-9) Legislação de 1775 a 1790 e Legislação 1763 a 1774, in Collecção de Legislação Portugueza. Lisboa: Typografia Maigrense. Portugal (in Portuguese)
14. Silva AD (1842) Supplemento à collecção de Legislação Portugueza Ano 1750 a 1762. Typografia de Luiz Correa da Cunha, Lisboa, Portugal
15. Pereira J (1985) Arquitetura Barroca em Portugal, Bib. Breve, Portugal. ISBN 972-566-171-0

Local Seismic Cultures: The Use of Timber Frame Structures in the South of Portugal

Javier Ortega, Graça Vasconcelos, Hugo Rodrigues and Mariana Correia

Abstract The well-known timber frame *Pombalino* construction system was devised after the 1755 Lisbon earthquake by the Portuguese government and its use was mandatory for the complex reconstruction process carried out in the city. It can be considered as the first technical regulation regarding seismic resistance. However, its use was not limited to Lisbon and the system was also embraced by local communities, which adopted it as a model of earthquake resistant construction. Its use thus spread around the country and eventually took root in the vernacular way of building of the country, becoming part of the Portuguese Local Seismic Culture. Nowadays, *frontal* walls can be identified in many vernacular constructions scattered across the country. This paper studies the use of *frontal* walls in vernacular buildings in the South of Portugal, which is the area with highest seismic activity within the country, focusing on the city of Vila Real de Santo António, which was chosen as the main case study. The paper also discusses the possible negative effects of the abandonment of this distinctive technique resulting from the Portuguese Seismic Culture on the overall vulnerability of vernacular constructions.

J. Ortega (✉) · G. Vasconcelos
ISISE, Departamento de Engenharia Civil, Universidade do Minho,
Campus de Azurem, 4800-058 Guimarães, Portugal
e-mail: javier.ortega@civil.uminho.pt

G. Vasconcelos
e-mail: graca@civil.uminho.pt

H. Rodrigues
RISCO, Escola Superior de Tecnologia e Gestão, Instituto Politécnico de Leiria,
Campus 2, 2411-901 Leiria, Portugal
e-mail: hugo.f.rodrigues@ipleiria.pt

M. Correia
CI-ESG Research Centre, Escola Superior Gallaecia, Largo das Oliveiras,
4920-275 Vila Nova de Cerveira, Portugal
e-mail: marianacorreia@esg.pt

H. Cruz et al. (eds.), *Historical Earthquake-Resistant Timber Framing in the Mediterranean Area*, Lecture Notes in Civil Engineering,
DOI 10.1007/978-3-319-39492-3_8

Keywords Pombalino construction system · Vila Real de Santo António · Local seismic culture · Seismic vulnerability · Structural alterations

Introduction

Local Seismic Cultures and traditional earthquake resistant constructions and techniques arise in seismic prone regions as a reaction to earthquakes. People continuously exposed to seismic hazard eventually had to learn how to protect themselves and developed preventive measures for earthquake mitigation. These construction efforts made by local populations gave rise to the development of what was defined by Ferrigni [1] as Local Seismic Culture.

In Portugal, there is a particular well-known and widely studied type of timber frame construction that can be highlighted as the most representative example of Portuguese Seismic Culture: the *Pombalino* construction system [2, 3]. This system was devised by the government and its use for the complex reconstruction process carried out in Lisbon after the 1755 earthquake was mandatory. It can be considered as the first technical regulation regarding seismic resistance. Furthermore, local communities adopted this constructive system as a model of earthquake resistant construction and its use widely spread around the country, eventually becoming endemic of the vernacular way of building of the country.

Given the fact that using timber frame structures has proven to improve the seismic behavior of vernacular architecture, as has been reported in past earthquakes in many countries [4–7], its preservation as a traditional earthquake resistant practice is important. This paper firstly intends to evaluate whether the use of timber frames as a traditional seismic resistant technique for vernacular architecture in the South of Portugal, traditionally a seismic region, is still active. Secondly, the city of Vila Real de Santo António was selected as a case study because it also followed a *Pombaline* development contemporary to the reconstruction of Lisbon. The plan included the provision of timber frame partition walls for some of the buildings and, thus, an overview of the type of constructions originally conceived is provided. Finally, the alterations done in the original constructions and the current state of the city center are described and the effect of these changes on the seismic vulnerability of the city is discussed.

Structural Timber Frames and Local Seismic Cultures

Structural timber frames are a very common vernacular practice that can be observed in many seismic prone areas. Traditionally, the use of timber reinforcement elements embedded within the rubble, mud bricks or ashlar masonry walls consists, in many cases, of just a few vertical or horizontal timber elements inserted within the walls or a rough timber grid of horizontal timber trunks or tree branches lying longitudinally and transversally at different levels of the wall. However, in some other cases, these timber reinforcements were sophistically arranged constituting the main load-bearing element of the building and a proper structural timber frame.

The use of timber frames filled with masonry is clearly a strengthening method applied because of the excellent tensile properties of the timber and also because the timber elements constitute successful slip planes and both vertical and horizontal shock absorbers, helping to dissipate much more energy. This vertical, horizontal and, in many cases, diagonal reinforcement, also improves the connections between the different structural elements among themselves, tying the building and enhancing its 'box-behavior'. Furthermore, it can restrain the out-of-plane bending and in-plane shear effects because, by confining portion of the walls, it enhances the masonry compressive strength, its shear strength and its deformability properties [8].

Different types of structural timber frames can be observed as seismic reinforcement for vernacular constructions in many countries, such as Greece [8], Turkey, where is known as *hatil* [9], Northern Pakistan, where is known as *cator and cribbage* [5], or in Kashmir, where there are two main types known as *taq* and *dhajji-dewari* [6]. Also in Central and South America, there are timber frame constructions used since Pre-Hispanic times in rural houses, known as *bahareque* in Colombia, Ecuador and El Salvador, *quincha* in Peru and *taquezal* in Nicaragua.

The Portuguese Case

There are several particular cases throughout history in which devastating earthquakes induced the development of building regulations for the reconstruction of the city through a post-earthquake concerted response that involved the government of that time. Generally, these regulations intended to prevent future losses of life and property by providing rules concerning urban planning and architectural practices, establishing height restrictions, fireproofing systems and sometimes even creating new seismic resistant structural systems. This happened in Portugal, after the most destructive earthquake in the history of the country in 1755, which destroyed large areas of Lisbon, including the downtown.

A complex reconstruction process was organized by the Marquis of Pombal and introduced new urban, architectural and structural concepts. It was strictly enforced and the constructions built according to them are known as *Pombalino* buildings. The most relevant seismic resistant provision was the inclusion of a three-dimensional braced timber frame structure named *gaiola pombalina* as the internal structure of the building. The *gaiola* is a resistant and flexible cage, whose walls are composed by horizontal, vertical and diagonal timber elements with different geometries, usually filled with rubble or brick masonry and plastered. These walls are known as *frontal* walls. The timber elements are connected using traditional solutions, such as dovetail or mortise and tenon, and are sometimes reinforced with nails or metal elements. The external walls are made of stone masonry and the *frontal* walls, act as the shear walls of the building, enabling it to resist horizontal loads by dissipating substantial amounts of energy and by providing a bracing function, avoiding the premature out-of-plane collapse of the exterior masonry walls. A minimal timber skeleton can be also present in the inner face of the masonry walls, facilitating their connection with the floors and the inner timber frame shear walls.

Timber Frame Structures in the South of Portugal

As previously commented, the *Pombalino* structure developed in Lisbon was also exported to surrounding villages and cities in Portugal where it became part of the vernacular way of building, as a clear evidence of the local builders' seismic concern, who acknowledged that the system actually improved the seismic behavior of their constructions. As a consequence, nowadays, *frontal* walls can be identified in many vernacular constructions scattered around the country, including the South of Portugal (Fig. 1). A relevant example is the case of Benavente (Fig. 1a), a city in the South of Lisbon, where a simplified version of this seismic resistant construction system was chosen for the reconstruction works of the city after the 1909 devastating earthquake, based on the Lisbon experience, and all the rebuilt districts of the town present buildings with a timber skeleton and stone masonry or adobe infill [10].

Nevertheless, the use of this technique has been gradually abandoned. Since there is a close correlation between the earthquake frequency and the development of seismic resistant building practices, when earthquakes are sparse and there are large periods of time without seismic events, larger than an average generation life time, the function of different techniques implemented after an earthquake are forgotten and ultimately abandoned. This adoption of changes in behavioral

Fig. 1 Examples of *frontal* walls in Portuguese vernacular architecture: **a** Benavente [10]; **b** Alcácer do Sal [11]; and **c** SamoraCorreia [11]

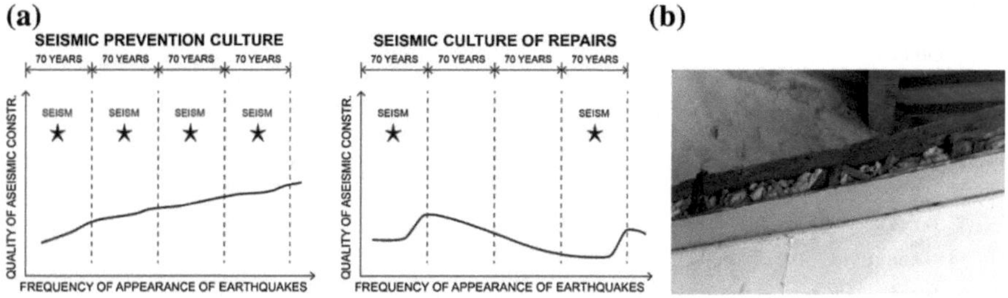

Fig. 2 **a** Evolution of the quality of aseismic construction in relation with the earthquake frequency (adapted from [12]); **b** remaining hidden *frontal* wall in Lagos

patterns due to a loss of collective memory of past events eventually leads to the abandonment and erosion of Local Seismic Cultures (Fig. 2a).

The loss of seismic awareness inevitably means an increase in the seismic vulnerability of communities. An illustrative example of this phenomenon took place in Lisbon, where the seismic resistant practices used in the *Pombalino* buildings were progressively abandoned and a new typology appeared: the *gaioleiro* buildings. This new typology consisted of four or five stories high stone masonry buildings that respected the *Pombalino* buildings formal characteristics but had a worse construction quality and neglected the initially devised seismic resistant solutions, such as the structural timber frame, worsening their seismic behavior and increasing their vulnerability [13]. Nonetheless, similar cases occurred in other cities of the country, such as Lagos, in the Algarve region. This city was also very much affected by the 1755 earthquake and reconstruction works introduced structural solutions aiming at improving the seismic behavior of the buildings, such as *frontal* walls [14]. However, the use of those constructive solutions was eventually abandoned, as people lost awareness of the seismic risk and nowadays only few remains of this seismic culture can be observed in the city (Fig. 2b).

Vila Real de Santo António

The reconstruction of Lisbon was not an isolated event. Vila Real de Santo António is located in the region of Algarve, the southernmost area of Portugal, which has a significant seismicity. It was much affected by the 1755 earthquake and was practically abandoned at the time. However, the Marquis of Pombal considered that this region had a considerable economic potential and enacted an official recovery program during the 1760s and 1770s that included the setting-up of a completely new planned town. The city was built at the end of the 18th century, contemporary to Lisbon, and followed similar urban, architectural and construction solutions, including the use of seismic resistant measures, such as the use of the characteristic timber frame *frontal* walls. The main reason for planning an entire new city was the

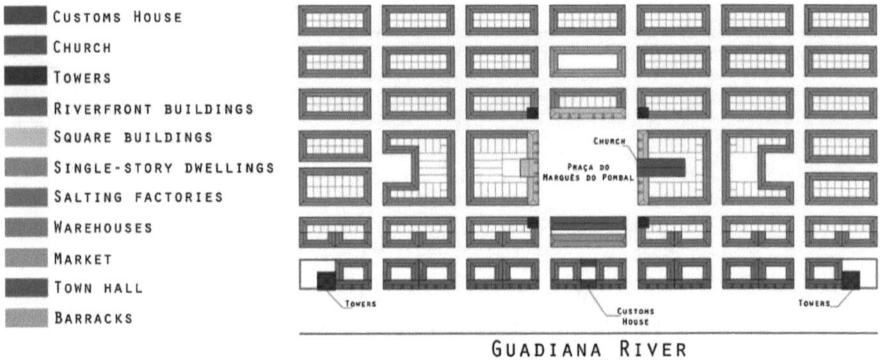

Fig. 3 Original plan of Vila Real de Santo António and main building types (adapted from [16])

attempt to boost the Algarve local economy through industrial development [2] but also was created for the control of port transactions and as a display of political power, because of its strategic position, at the extreme South of the Algarve facing the Spanish border [15].

The new *Pombaline* city plan was formed by a rectangular area with one of the long sides placed along the Guadiana River, facing east, and consisted of a grid of seven by six urban blocks organized around a big central square (Fig. 3). The whole plan is characterized by its strong regularity, where all the blocks are approximately 53 m long and 22 m wide, with the exception of the central one, which is slightly longer, approximately 55 m. All the streets are 9 m wide. There are essentially four distinct architectural types that defined a clear hierarchy at an urban level: (a) buildings in the riverfront, which have two main stories and a third attic floor; (b) buildings in the main square, which also have two main stories; (c) single story dwellings, characterized by their small scale and simplicity; and (d) single story factories and warehouses, which basically consists of a system of masonry arcades perpendicular to the façade walls and organized around a patio. Besides these strongly defined types, there are also some specific unique buildings: the Customs House, the church, and the 'towers'.

In terms of construction and materials, a similar prefabrication process already applied for the Lisbon reconstruction plan, was also applied in the construction of Vila Real. Ashlars used for the quoins or opening frames arrived already cut and worked, ready to be placed, as well as the wooden components, such as doors, windows, beams and floorboards [15]. Stone masonry was used for the load bearing exterior and party walls, which are the main structural resisting elements of all buildings. Timber was used for the roof rafters and trusses, floor beams and timber frame partition walls in upper floors. The roof was simply covered by wooden boards on which ceramic tiles were laid. Some ground floor rooms had vaulted ceilings supporting the first floor as a fire prevention measure, as in Lisbon [2]. Ground floor partition walls were usually built in solid brick masonry. The buildings were plastered with lime and sand and whitewashed.

Use of *Frontal* Walls

With respect to the seismic resistant constructive solutions introduced in the buildings at the time of their construction, the most significant elements were the timber frame structures used as partition walls in the upper floor of the two-story buildings (Fig. 4). They are similar to those used in the *Pombaline* quarter in Lisbon but, given the low height of the buildings, the use of the whole *gaiola* seismic resistant construction system developed for the reconstruction of Lisbon, was not necessary. Nevertheless, the timber roof structure and the timber floor structure were connected through these *frontal* partition walls [15]. These walls consisted of vertical, horizontal and diagonal timber members filled with stone masonry coming from neighboring quarries in the mountains.

Some of these timber frame structures were still observed in some of the buildings at the present time. For instance, one of the most representative buildings of the city: the *Alfândega* or Customs House still present several examples of timber frame partition walls in the upper floors, showing different geometries (Fig. 5a, b).

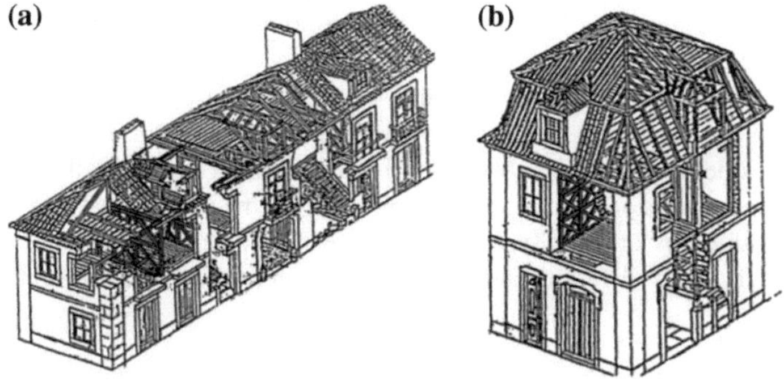

Fig. 4 Use of *frontal* walls in two-story buildings in Vila Real de Santo António (adapted from [2]): **a** riverfront buildings; **b** one of the 'towers' surrounding the central square

Fig. 5 Different type of *frontal* walls observed in Vila Real de Santo António: **a** *Alfândega* building; **b** *Alfândega* building [17]; **c** *Alfândega* building; **d** South riverfront 'tower' [17]

Some of them lack the diagonal timber members (Fig. 5c) but all of them use nails to strengthen the connection between elements. The two 'towers' located in the riverfront also presented *frontal* walls in the upper floors at the time of the construction, with a stone masonry infill similar to the masonry used for the main load bearing walls (Fig. 5d).

The Current State of the *Pombaline* Core of Vila Real de Santo António

There has been a transformation process in the city and the works carried out in the *Pombaline* core have led to an important depravation of the historical built-up fabric. Only very few buildings still possess *Pombaline* characteristics, such as the original masonry walls or the opening distribution (Fig. 6a). Generally, the transformation process has been characterized by a massive occupation of the blocks, leading to an extreme densification of the urban fabric, since patios were continuously occupied by additional constructions. Single story buildings have been systematically exposed to demolitions, substitutions and large modifications, but even the best preserved buildings still show important alterations in relation to their original conception.

The main alterations that can be found in the original constructions involve the addition of new floors, the opening of new windows and doors, the enlargement of the original openings or the substitution of the original roof and floors (Fig. 6b). Most of these alterations are a normal consequence resulting from the changes of use of the buildings and the new needs of the users. However, they are not only detrimental in terms of loss of authenticity of an important architectural and urban

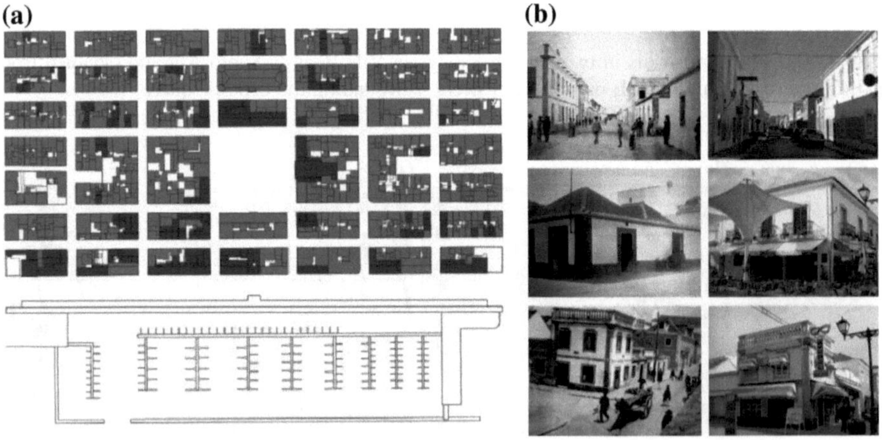

Fig. 6 **a** Current plan of VRSA (*highlighted in orange* those buildings that still preserve some original characteristics); **b** common alterations undergone by the original constructions [18]

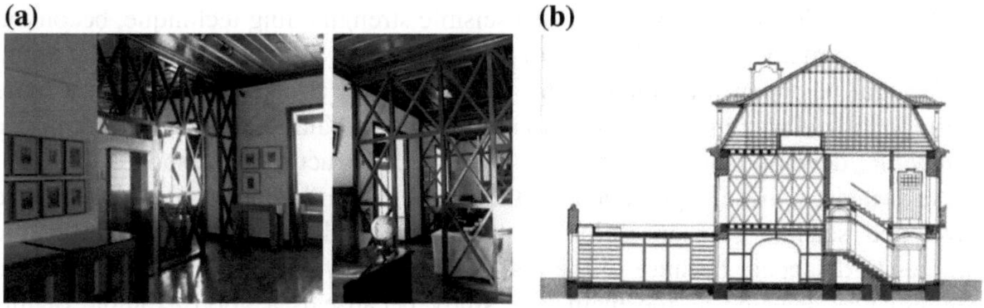

Fig. 7 **a** Aesthetic reconstruction of *frontal* walls in the south riverfront 'tower'; **b** cross section of the rehabilitation project of the South riverfront 'tower' [17]

heritage. They also jeopardize the safety of the buildings by directly affecting several parameters that are crucial in the seismic response of the building and introduce new sources of vulnerability.

The Substitution and Removal of *Frontal* Walls

Another systematic alteration that the buildings of Vila Real de Santo António have undergone is the elimination of the original *frontal* walls, revealing again a loss of seismic awareness by ignoring a distinctly seismic resistant feature of the original buildings. When buildings are reconstructed or rehabilitated, these *frontal* walls are obliterated or just kept as a vestige, ignoring the actual structural function as a seismic resistant element, which may have contributed to increase the seismic vulnerability of the buildings. This is well exemplified by the rehabilitation of the South riverfront 'tower', where the *frontal* walls were reconstructed but just as a formal element (Fig. 7).

Conclusions

Portugal has a moderate seismicity characterized by small events of order of magnitudes smaller than five, but several devastating earthquakes have sporadically struck the country throughout its history. The seismic history is thus distinguished by large periods of time without seismic events and has led to the development of a reactive response behavior in which local communities just respond to earthquakes in the intermediate aftermath of the event. In this way, important measures and seismic resistant construction techniques were devised and implemented after large earthquakes. They were sometimes mandatorily imposed by the government, such as the use of the *gaiola* construction system and *frontal* walls. This efficient technique was developed in Lisbon but soon widely spread across the country, as inhabitants understood its potential as a seismic strengthening technique, becoming traditional in many regions. The use of *frontal* walls in several regions in the South of Portugal has been presented in this paper, as well as its application in a more institutionalized way in the construction of the city of Vila Real de Santo António, a *Pombaline* development contemporary to the reconstruction of Lisbon.

Nevertheless, these large periods of times without earthquakes have caused those developed seismic resistant solutions to be eventually abandoned and seismic cultures to disappear because of the loss of seismic awareness. Vila Real de Santo António was selected as the main case study because is a good example of this progressive abandonment of traditional strengthening techniques enforced after important seismic events. The seismic concern that arose in the whole country after the 1755 earthquake materialized in several *Pombaline* developments such as the creation *ex novo* of Vila Real de Santo António and resulted in an effective urban plan and carefully constructed architecture. However, the deep mischaracterization of the historical city center due to the multiple alterations suffered, led to the oblivion of the developed reinforcement measures, including the use of *frontal* walls and, as a result, the seismic safety of the buildings may have been compromised. The adequacy and efficiency of these traditional strengthening solutions has been successfully tested over time and, subsequently, they are worthy to be studied. The re-adoption of some of these techniques can help in preserving and retrofitting surviving examples of vernacular architecture without prejudice for its identity.

Acknowledgments The authors wish to express their gratitude to the Portuguese Science and Technology Foundation (FCT) for the scholarship granted in the scope of the research project 'SEISMIC-V—Vernacular Seismic Culture in Portugal' (PTDC/ATP-AQI/3934/2012).

References

1. Ferrigni F (1990) À la recherché des anomalies qui protégent, Actes des Ateliers Européens de Ravello, 19–27 Novembre 1987. PACT Volcanologie et Archeologie & Conseil de L'Europe, Ravello
2. Mascarenhas J (2004) Sistemas de Construção—V. Livros Horizonte, Lisbon, Portugal
3. Cóias V (2007) Reabilitação estrutural de edifícios antigos. Argumentum, GECoRPA, Lisboa, Portugal
4. Vintzileou E (2011) Timber-reinforced structures in Greece: 2500 BC-1900 AD. In: Proceedings of the Institution of Civil Engineers (ICE), structures and buildings, vol 164 (SB3), pp 167–180
5. Hughes R (2000) Cator and Cribbage Construction of Northern Pakistan. In: Proceedings of the international conference on the seismic performance of traditional buildings, Istanbul, Turkey
6. Langenbach R (2009) Don' tear it down! Preserving the earthquake resistant vernacular architecture of Kashmir. UNESCO, New Delhi, India
7. Gülhan D, Güney IÖ (2000) The behaviour of traditional building systems against earthquake and its comparison to reinforced concrete frame systems; experiences of Marmora earthquake damage assessment studies in Kocaeli and Sakarya. In: Proceedings of the international conference on the seismic performance of traditional buildings, Istanbul, Turkey

8. Vintzileou E (2008) Effect of timber ties on the behavior of historic masonry. J Struct Eng 134(6):961–972
9. Langenbach R (2007) From “Opus Craticium” to the “Chicago Frame”: earthquake-resistant traditional construction. Int J Arch Heritage 1(1):29–59
10. Vieira RS (2009) Do Terramoto de 23 de Abril de 1909 à reconstrução da Vila de Benavente. Um processo de reformulação e expansão urbana. Câmara Municipal de Benavente, Benavente, Portugal
11. Correia M, Merten J (2001) Preliminary report of the local seismic culture in Portugal, in Taversism project—Atlas of local seismic culture. EUCCH, Ravello, Italy
12. Ferrigni F, Helly B, Mauro A, Mendes Victor L, Pierotti P, Rideaud A, Teves Costa P (2005) Ancient buildings and earthquakes. The local seismic culture approach: principles, methods, potentialities, Centro Universitario Europeo per i Beni Culturali, Edipuglia srl, Ravello, Italy
13. Mendes N, Lourenço PB (2010) Seismic assessment of masonry “Gaioleiro” buildings in Lisbon, Portugal. J Earthq Eng 14:80–101
14. Paula RM (1992) Lagos, Evolução Urbana e Património, Câmara Municipal de Lagos, Lagos, Portugal
15. Correia J (1997) Vila Real de Santo António. Urbanismo e Poder na Política Pombalina. PhD Thesis, Faculdade de Aquitectura da Universidade do Porto, Portugal
16. Rossa W (2009) Cidades da razão: Vila Real de Santo António e arredores. Monumentos 30:16–31
17. Figueiras R (1999) Vila Pombalina: Vila Real de Santo António. Câmara Municipal de Vila Real de Santo António, Vila Real de Santo António, Portugal
18. Gonçalves A (2009) Vila Real de Santo António. Planeamento de pormenor e salvaguarda em desenvolvimento. Monumentos 30:40–53

8. Vintzileou E (2008) Effect of timber ties on the behavior of historic masonry. J Struct Eng 134(6):961–972
9. Langenbach R (2007) From "Opus Craticium" to the "Chicago frame": earthquake-resistant traditional construction. Int J Arch Heritage 1(1):29–59
10. Vieira RS (2009) Do terramoto de 23 de Abril de 1909 à reconstrução da Vila de Benavente. [illegible]
11. [illegible]
12. [illegible]
13. [illegible]
14. [illegible]
15. [illegible]
16. [illegible]

The Use of Wood with an Anti-seismic Function in the Architecture of Palermo During the 18th Century

Tiziana Campisi and Federica Scibilia

Abstract The study analyzes the interventions in the historical buildings after the earthquakes that struck Palermo in 1726 and 1751, with a particular attention to the anti-seismic reinforcements. The research, based primarily on archival documents, focuses on a series of interventions in the architectural heritage, in which the use of wood for structural elements, was able to guarantee for an "antiseismic" reconstruction. Although wood was not commonly used as other materials (iron, for instance) in the post-earthquake reconstruction process of Palermo it was certified its use as an anti-seismic system in the reinforcements of the damaged buildings. It is documented, for instance, the insertion of wooden beams, used similarly to iron chains, both in monumental buildings, and also in the less valuable ones, useful to provide a better cohesion for the masonries, or the introduction of wooden and lightweight partition walls; or, even, the substitution of stone vaults with vaulted wooden structures or ceilings and the realization of staircases using wood for the entire main structure.

Keywords Sicily · 18th century · Earthquake-resistant frames · Wooden structures

Both Prof. Tiziana Campisi and Arch. Federica Scibilia authored the introduction, whereas the second paragraph is to be attributed to Arch. Scibilia, and the third paragraph to Prof. Campisi.

T. Campisi (✉) · F. Scibilia
Dipartimento di Architettura (d'ARCH), Università degli Studi di Palermo,
viale delle Scienze, edificio 8, 90128 Palermo, Sicily, Italy
e-mail: tiziana.campisi@unipa.it

F. Scibilia
e-mail: federica.scibilia@unipa.it

H. Cruz et al. (eds.), *Historical Earthquake-Resistant Timber Framing in the Mediterranean Area*, Lecture Notes in Civil Engineering,
DOI 10.1007/978-3-319-39492-3_9

Introduction

18th century reports illustrate the amount of the damages the city of Palermo suffered as a result of earthquakes (particularly, those occurred in 1726 and 1751), violently perceived by citizens because of the geological characteristics of the subsoil and the intrinsic features of historical buildings. Most of the historical building areas, in fact, rose on buried riverbeds (the *Kemonia* and *Papireto* streams), or on the reclaimed area of the ancient port, thus revealing the inadequacy of many foundations and amplifying the seismic effects. Moreover, the considerable building density of the old town constituted an aggravating factor that increased the risks associated with potential collapses.

The disastrous earthquake which had devastated Eastern Sicily in 1693 was still a fresh memory, and urged engineers in Western Sicily to highlight the most delicate technical features. This inaugurated a debate about the building static conditions after an earthquake—especially stability and seismic safety.

The widespread building system, like the monumental one, was affected by additional raisings, necessary because of the inability of the old town to grow outside its walls. The continuous reconfigurations transformed small buildings into large ones; palaces, behind their façades embellished with plaster and stuccoes, often concealed poor housings with different structural characters, which were not always able to ensure a sufficient level of safety.

Earthquakes and their related problems, from the safety of houses and monuments to the reconstruction phases of what had been compromised or had collapsed, constituted an opportunity for the definition of seismic design criteria. This laid a fertile cultural ground for the experimentation of new construction methods in an environment resistant to any innovation or modification of well-known constructive practices. After all, those practices were sufficiently guaranteed by centuries of satisfactory results, in which the adoption of wooden construction systems prevailed and took hold.

Reconstruction was primarily based on the guidelines of reparation and structural consolidation known so far and handed down by tradition. This meant preventing any danger of future collapse by eliminating all endangered constructive elements. Therefore, overhang elements (cornices and balconies, for instance), together with dangerous or unstable structures, were replaced, regardless of history and consistency in the appearance of architectures. Indeed, if earthquakes sometimes resulted in traumatic demolitions and reconstructions for the historic urban districts, leaving unchanged its typological and original matrices, for the monumental buildings, instead, earthquakes became a pretext for radical volumetric and stylistic changes.

Builders drew upon that technical knowledge which had been repeatedly tested for the construction of temporary structures, adopting wooden frames during festivities or for temporary theatres, roofs, and also light wooden partition walls. Special teams coordinated by carpenters, architects, and engineers were created to quicken safety works, primarily in order to demolish the wooden elements of floors and roofs: trusses, beams, and planks, by selecting the material according to their quality and size. Then, they immediately proceeded to create supports, secure precarious parts, and provide for temporary constructions.

Wood was an essential resource: it could be easily disassembled and reassembled, it was light, reusable, and versatile in order to make the necessary reinforcements and to chain up systems.

Despite technicians and designers could count on a solid theoretical and practical expertise, they still were not able to deal with structural prevention and repairs, before and after earthquakes. In most cases, the contingency of those traumatic seismic events led to the design and the application of new and original techniques of demolition, for the safeguard and the consolidation of damaged buildings. Those techniques depended on the constitutive materials, the technical and constructive features of the building itself, and the type of damage it had endured.

"False" Wooden Vaults, Domes and Chains

The earthquakes that struck Palermo in 1726 and 1751 forced technicians to reflect upon the seismic validity of traditional constructive solutions, thus developing some interesting attempts of "seismic improvement" [1].

Although wood was not as commonly used as other materials (iron, for instance) in the post-earthquake reconstruction process in the city of Palermo, archival documents had certified its use as an anti-seismic system in the reinforcements of the damaged buildings. Its lightness and elasticity, indeed, made it a versatile material, suitable for different applications. An interesting example is represented by the construction of "false" vaults, characterized by the use of the so-called *incannucciata* technique, formed by a wooden framework (instead of stone), generally made with poplar ribs which may or may not be independent from the upper roof covering (Fig. 1).

A similar intervention was documented in Palermo after the earthquake of the 1st of September 1726, during the dome reconstruction of San Carlo Borromeo church. The so-called *capitoli di fabbrica* (a sort of technical documents) define the ways to build this "false" dome [2]. Reed mats (*incannucciato*), made with large reeds (*cannoni*) and fixed with iron nails, covered the intrados of the wooden structure, and were then covered with layers of mortar and plastered. These layers had to be applied according to the prescriptions contained in the building yard document, by respecting predetermined time intervals.

For architects, the realization of light-material vaults and domes in post-earthquake interventions became a valid alternative to the consolidation of stone domes using iron tie rods and chains. Nonetheless, the adoption of metallic elements was privileged, since they could guarantee the maintenance of the original structure.

The execution of lightweight wooden structures completed with reeds and plaster is documented in local areas, especially for the realization of "false" vaults

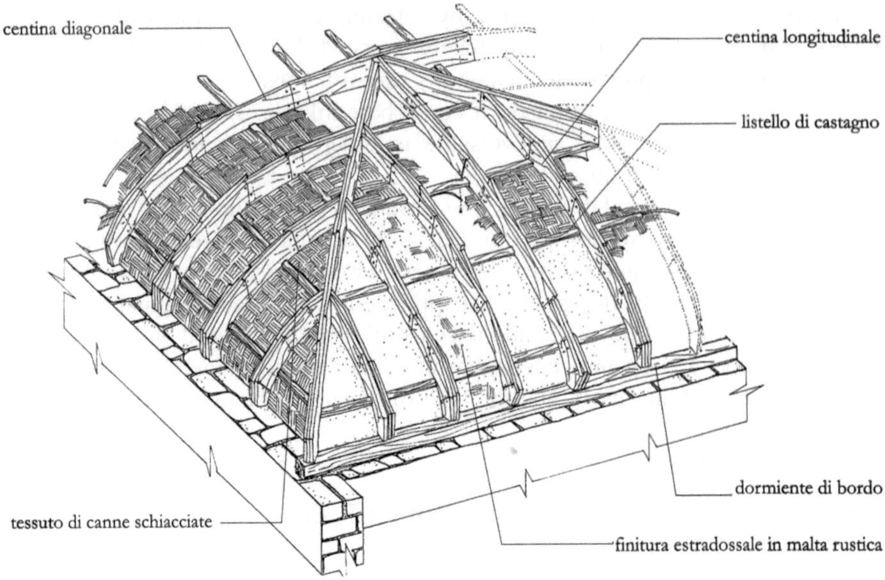

Fig. 1 Detail of a wooden ceiling: axonometric view of the constructive system [18]

in palaces. Less frequently, however, as in this case, they could be applied in the context of religious architecture. In churches, the technique had been tested since the 16th century, and during the twenties of the 18th century it became more widespread in the south-east of Sicily (where it was used, in particular, for the roofing of centric plan buildings) [3]. The use of this constructive technique with anti-seismic purposes is documented in the second half of the 18th century in some projects of architect Rosario Gagliardi from Noto [4], who was the key player of the reconstruction in the Val di Noto area after the disastrous earthquake of 1693. He used a "reeds and plaster" ceiling in the internal oval dome of Santa Chiara church in Noto (1750 approx.); the vault, unlike San Carlo church in Palermo, had a self-supporting structure independent from the upper wooden framework, as it is showed in the longitudinal section in the project, which also included the plan of the roofing structure (Fig. 2).

The same architect also designed a lightweight structure for the elongated oval vault of San Michele church in Scicli (1750). Architect Gagliardi supported the necessity that the *dammuso* (the vault) should have a timber structure with large, plastered reeds above and below, since he was well aware that this kind of system would have been more effective in the occurrence of an earthquake [5].

The adoption of lightweight structures with anti-seismic functions in Sicily is also to be found later, as documented, for example, by the reconstruction of the "reeds and mortar" vaults of the western dormitory in the Benedictine monastery in Catania, damaged by the earthquake that struck the city and its surrounding areas in 1818 [6].

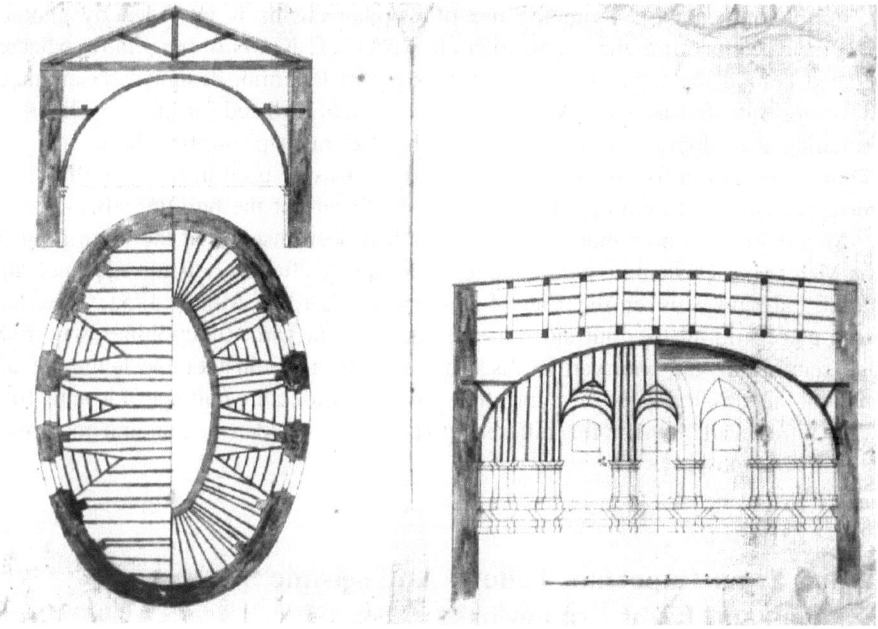

Fig. 2 R. Gagliardi, constructive drawing related to the wooden vault of Santa Chiara church in Noto [4]

The proven tensile strength of wood, associated with the use of hard wood species having a high mechanical resistance (which could endure rot degradation, especially if the beams were included into masonries), induced the architects to use wooden chains in post-earthquake reconstructions in order to improve structural safety. The adoption of wooden chains as reinforcing elements had indeed a long, local tradition, as it is demonstrated by a series of cases documented during the Middle Ages in countries of Roman influence [7], such as Sicily itself [8]. Wooden chains, generally derived from *ruvolo* (oak), although considerably less common than iron chains, found many applications in the reconstruction process following the earthquakes of 1726 and 1751.

Documents return slight but significant traces allowing to highlight this aspect. The presence of these elements is demonstrated, for instance, by the consolidation of the Cathedral of Palermo, designed by architect Giovanni Amico from Trapani. One of the most successful architects in Sicily at the time, Amico worked extensively in Palermo after the disruptions caused by the earthquake of 1726. The *misura e stima* relation (a final building yard document) dated 15th of October 1729, relative to the construction of the top of the new bell tower (replaced with a neo-Gothic structure), indicated that wooden (oak) and iron chains had been both adopted, but unfortunately the document does not specify either the size of these elements or their positioning [9].

Further information about the use of wooden chains is provided by another document concerning the reconstruction works of a *domus magna* (a palace), located behind the Santa Elisabetta monastery in Palermo, designed by architect Giovanni Battista Cascione. Among the various actions listed for the consolidation, including the adoption of an iron chain, the document prescribes the use of two chains in *ruvolo* of five *palmi* (a typical unit of measure used in Sicily at that time, corresponding to 26 cm approx.) to repair the façade of the building [10].

Moreover, we know that wooden chains had been inserted at the main floor of the Merendino-Costantino palace along via Maqueda after the seismic sequence that struck Palermo between the 13th of July and the 23rd of August 1751. They had been placed on all the four sides of the interior courtyard, functioning as tie rods and reinforcing the courtyard walls. To overcome the considerable length of the internal façades, many wood connections and a considerable number of iron tie rods were introduced, joining the different parts of the wooden beams and maximize tensile strength [11].

Wood as an Important Fellow: Anti-seismic Support Systems and Light Frameworks

Technicians, aware of the important seismic role offered by floors and roofs, often intervened in the consolidation and improvement of existing buildings essentially by lightening structures, abolishing possible mobile masses, and introducing more solid connections among construction wooden elements. They also proceeded to the dismantlement of all those floors presenting a secondary joist structure and overlying dry mortar, with poor gravel (*tercisato*), which showed to be a risk because of its considerable weight and the presence of dissolved materials that could have collapsed in case of horizontal actions, as a result of a dangerous dynamic load.

The connection and the fixing of the extremities of trusses and beams to vertical structures became more frequent, by applying iron terminals nailed down to the wooden parts, provided with loop and iron planks useful to constitute a block as a tension system. This practice, besides creating a connection in several points between parallel walls, also decreased the risk of cracks due to hammering in the wooden beams, as the principal elements of the floor had been blocked and it could not move to a different direction than the supporting walls [12].

The elimination of wooden or vaulted floors with different heights at the same elevation (a usual effect of expansions and consolidations occurred over time in a historical building) was also considered advisable. This also had an anti-seismic function: frequently, this possibility appeared in the occasion of heavy refurbishments, after extensive damages. Contextually, the floor impost band was improved with extended brick insertions, or with a wooden beam used for an equal load distribution on the wall where it was placed—that is, under the wooden beams of the floor, especially in the case of ordinary masonry, often inconsistent and chaotic.

With regard to the trusses, the post-seismic devices were essentially the bracing of the trusses themselves, in order to limit the risk of mutual displacement and their overturning as an effect of horizontal forces. Some beams in the roof were fixed to trusses with nails and specific flat metallic bars, positioned over the trusses; these bars were probably connected with nails to the extremities of the beams, rather than extended for the whole span.

Local constructive practices employed wood as a structural element and for the realization of accessory artefacts too. Expertise of the material and the carpenters' technical knowledge derived from centuries of application, favoured the widespread diffusion of wooden ceilings and light partition walls (the last ones were called *intilarate*). Their use continued beyond the 18th century; it was adopted throughout the 19th and until the first decades of the 20th century. While wooden ceilings constituted a self-supporting structural system, flat and vaulted ceilings were not unfrequently connected to the beams in the upper floor or to the chains of the trusses (especially in the last floor). In this case, the structural failure or the collapses produced by an earthquake on the load-bearing structures had serious repercussions on the wooden ceiling, which suffered both of its inherent fragility arising from the poor materials used (poplar wood, reeds, chestnut slats), and the precarious connections with the riveting parts.

Together with cracks and partial collapses in the structural parts, damages also included the irremediable deterioration of stuccoes and other painted decorations that embellished the ceiling. To overcome these problems, an effective consolidation system resulted in the insertion of a new wooden beam (where a gap between the ceiling and the upper floor was present). The beam was independent from the floor, useful to support the most risky portions of the ceiling (especially the larger, diagonal wooden ribs), through wooden or metal slats (called *appenditori*). The local anti-seismic building experience used large wooden beams during reconstructions, set in correspondence of the rib base (*bordoni* or *custanoni*), which was placed around the walls. Therefore, it also had a function of enchainment and load distribution, as we have previously illustrated [13] (Figs. 3 and 4). The use of

Fig. 3 Wooden beam supporting the ribs of a wooden ceiling and constituting a long and continuous lintel

Fig. 4 Wooden joints and planks between a wooden ceiling and an upper reinforcement structure

wooden vaulted systems with a structural function, instead, is quite different. At a regional level, this constructive solution characterized post-earthquake reconstructions of churches, since stone domes were particularly vulnerable to disastrous collapses during a seismic event. We know that in Palermo, after the earthquake in 1751, constructive choices were not oriented by a philological reconstruction of a cupola in the same material, but preferred to use a lighter double-shell system, in which the exterior shell was made by ashlars, and the interior one, also with a load-bearing function, was made with a wooden structure. Wooden circular chains (*rondelli*), located at the base and at the top of the dome, joined the two shells, in addition to a wooden support put in correspondence of the interspace between the shells, thus promoting load transmission and enchainment [14].

As for the widely used timber-frame walls construction, the structure of the wooden frame was built using chestnut wood (we refer to the typologies known as *mezze costane* or *costane*), completed in its structural definition and finishing elements in a second time [15]. The subdivision of larger rooms was often assigned to wooden frame walls, with a function of enchainment for unconnected walls. They had also the function to regularize room planimetry, and were able to support frameworks or completion structures such as vaulted or flat ceilings, which often rested precisely on this type of wooden partitions (this construction technique could guarantee elasticity and the optimal absorption of deformations between the two described wooden structures). If located at the last elevations, this light wooden walls represented an effective system to build external thin walls, which were often discontinuous vertical walls, used to support loggias, balconies, and accessible roof terraces).

The frame structure (*graticola*, as sometimes we can read in the coeval building yard documents) made with wooden joists (*costane*) determined squares whose sides measured approximately two Sicilian *palmi* (corresponding to about 52 cm).

Vertical and horizontal joists were connected by nails, made coplanar thanks to special slots producing the appropriate mutual coupling and cohesion. Both the insertion of bracings (diagonal or saltire, useful to reinforce and increase rigidity in the partition wall) and a suitable connection to the adjacent masonry walls were necessary, because of the height and the scant thickness of the wooden wall. The effectiveness of unions and joints were guaranteed by appropriate nail connections: the so-called *cinquantini* nails were used for the connection of the joint frame (*cinquanta* in Italian means fifty, and fifty nails made up to one *rotolo*, which corresponded to a weight of 0.793 kg). *Novantini* nails (*novanta* means ninety, and ninety nails weighed one *rotolo*, too), instead, were applied for fixing reed mats. Technicians often had opted for nails longer than the joining parts, in order to suitably fold a part of the nail shaft on the face of the vertical or horizontal wooden joints, thus preventing them to pull out. Similarly, nails having wider heads were especially employed to ensure adhesion of the reeds woven mat to the partition wall, also through the interposition of barked chestnut twigs (called *circhi bianchi*) at the points of riveting.

In order to improve the connection of the wooden wall to its bottom and upper sides, this was anchored both to the floor beams on which it rested and to the intrados of the upper floor, especially when the partition had a considerable height, if compared to its thickness.

It was a good constructive practice to realize perimetral, larger vertical and horizontal joints, so that they would create a cornice-frame, nailed to the adjacent structures. It was not uncommon to find special supports (called *scironi*), too, which prevented warping and deformation due to the weight of the structure.

The empty areas of the frame could be filled using whole reeds, or crushed reeds woven into mats, if the partition had to be light and was positioned inside. However, if the partition was an external perimeter wall, or it had to carry an excessive load to increase its structural capacity, empty areas of the wooden frame could be filled with thin ashlars (*terzalori*), or thin bricks (*pantofali*) [16]. The arrangement of the woven reed mat could be differently disposed and oriented in various boxes, to give a greater rigidity and prevent the transmission of any micro-cracks in the overlapping layers of plaster, put above the reeds. In mountain or wooded areas, these empty spaces could be saturated by structural panels composed by a conglomerate of stone, mortar, and cork [17].

Well-built wooden walls, especially if reinforced by the presence of stone fills, could assume precisely the function of bracing, when seismic actions occurred (Figs. 5 and 6).

In case of frameworks having a larger mesh, with the same partition function, this mesh could be reinforced by the application of wooden boards, on one side only, or on both sides of the partition wall. A continuous layer of chestnut rafters having a semi-circular section (*mezzi ginelli*) was seldom used.

A special mention must be reserved to overhangs and staircases: in the first case, strict rules and notices prescribed the elimination of protruding and heavy stone elements, due to the collapse of many balconies and cornices, preferring instead, in

Fig. 5 Wooden partition walls: empty spaces are filled with reed mats

Fig. 6 Wooden partition walls: with thin ashlars

the immediate reconstruction and in modest buildings, the use of overhangs with a wooden nucleus, protected by oil, tar, or beeswax, left in view or then plastered.

In the second case, rebuilding was often associated with the desire to not overburden the existing damaged masonry structures, thus orienting designers towards a reconstruction of large and monumental staircases using wood for the entire main structure, especially in noble buildings. Even if it was considered as a "temporary" structure to be replaced with stone—more magnificent and long-lasting—, this choice is an important example of those building systems using wood as an emergency measure.

The long permanence of these structures, the executive attention, and the technical expertise of all the operators involved are well testified in the building yard documents [18].

References

1. On the earthquake in Palermo in 1726, see Scibilia F (2015) Terremoto e architettura storica. Palermo e il sisma del 1726. Caracol, Palermo
2. The document is in the State Archive of Palermo (ASPa), Notai defunti, Michelangelo Lopresti, vol. 4751, cc. 825 r-827v., cfr. Nobile MR (2004) Cupole e calotte "finte" nel XVIII secolo. In: Gambardella A (ed) Ferdinando Sanfelice. Napoli e l'Europa. Napoli, Edizioni Scientifiche Italiane, pp 151–16 (Scibilia F 2015 op cit)
3. On the "false" vaults and their seismic aspects, see Tobriner S (2009) Safety and the reconstruction after the Sicilian Earthquake of 1693, the 18th century context. In: Casamento A, Guidoni E (eds) Le città ricostruite dopo il terremoto siciliano del 1693. Tecniche e significati delle progettazioni urbane, Storia dell'Urbanistica/Sicilia II, proceedings (Roma, 20–21 Marzo 1995). Edizioni Kappa, Roma, pp 26–41. Nobile MR, Bares MM (2015) The 'false vaults' in the architecture of Sicily. In: Construction History. International Journal of Construction History Society, vol 30, No. 1, pp 53–70
4. Bares MM (2013) L'architetto e la costruzione. In: Nobile MR, Bares MM (eds) Rosario Gagliardi (1690 ca.-1762), exhibition catalog (Noto, 22 marzo-21 giugno 2013). Caracol, Palermo, pp 61–91
5. Nifosì P (1988) Scicli. Una via tardobarocca. Comune di Scicli, Scicli (RG), pp 37–38
6. Lo Faro A, Salemi A (2009) Cultura tecnica e sisma nella Sicilia orientale: il terremoto del 1818. In: Fiandaca O, Lione R (eds) Il Sisma. Ricordare prevenire progettare. Alinea, Città di Castello (PG), pp 109–122
7. On this topic see Patetta L (2000) Le «catene» come scelta progettuale negli edifici tra XIII e XV secolo. In: Scritti sull'architettura del Rinascimento. Guerini, Milano, pp 205–217. In relation to the anti-seismic techniques in historical architecture see also Arce García I (1996) Elementos y sistemas constructivos antisísmicos en la antigüedad. Aplición a la restauración de estructuras históricas. In: de las Casas A, Huerta S, Rabasa E (eds) Actas del Primer Congreso Nacional de Historia de la Construcción (Madrid, 19–21 septiembre 1996). Instituto Juan de Herrera, Madrid, pp 39–47
8. On the use of wooden elements as strengthening devices in the architecture of the modern era in Sicily see Montana S (2014) L'uso dei dispositivi di rinforzo ligneo nell'architettura siciliana di età moderna. In: Lexicon. Storie e architettura in Sicilia e nel Mediterraneo, vol 18, pp 32–39
9. ASPa (2015) Notai defunti, Giuseppe Magliocco, vol. 5223, cc. 156r–157r (cfr., Scibilia F, op. cit)
10. ASPa (2015) Notai defunti, Giuseppe Magliocco, vol. 5222, cc. 380r-v, c. 380r (cfr., Scibilia F, op. cit)
11. Cfr. Campisi T, Saeli M (2015) Aristocratic palaces in the XVIII century in Palermo. The construction site of Merendino-Costantino palace. In: Bowen B, Friedman D, Leslie T, Ochsendorf J (eds) Proceedings of the 5th international congress on construction history, Chicago, 3–7 June 2015, 3 voll., s.l. Construction History Society of America, I, pp 333–340
12. Campisi T, Fatta G (2009) I terribili tremuoti nel XVIII secolo a Palermo. Dai danni alle nuove esperienze costruttive, in Il Sisma. Ricordare…, op.cit., pp 19–33
13. Campisi T (2014) I soffitti in legno e canne: conoscenza e tecniche di intervento per una manutenzione e restauro sostenibili, 2014. In: Proceedings of the congress "Quale sostenibilità per il restauro?", XXX-2014, Bressanone 1–4 luglio 2014, Arcadia ricerche ed., pp 659–668
14. Campisi T (2013) Wooden domes, vaulted and flat ceilings, partition walls in the Sicilian constructive tradition. Knowledge for the safeguard of architecture and technic. In: Proceedings of the 6th international congress "science and technology for the safeguard of cultural heritage in the Mediterranean Basin", vol I, Athens, Greece, 22–25 Oct 2013, pp 78–85
15. Campisi T (2001) I soffitti in legno e canne nella tradizione costruttiva palermitana del XVII e XVIII secolo. In Recupero & Conservazione, no 41, pp 38–44; no 42, pp 62–66

16. Campisi T, Mutolo S (2003) Palermo pietra su pietra. Apparecchi murari dell'edilizia settecentesca, ed. Ila Palma
17. Fatta G, Campisi T, Li Castri M, Lo Piccolo S, Vinci C (2007) La tradizione costruttiva nell'area delle Madonie. In: Aymerich C, Dell'Acqua AC, Fatta G, Pastore P, Tagliaventi G, Zordan L (eds) Architettura di base. Alinea, Firenze, pp 431–472
18. All the informations inserted in this paragraph are deduced by specific archival researches conducted at the State Archive of Palermo (*Fondo Notai Defunti* and *Congregazioni religiose soppresse*)

Urban Planning and Building Reconstruction of Southern Italy after the 1783 Earthquake: The Case of Mileto

Stefania Stellacci, Francesca Geremia, Enrico Pagano, Nicola Ruggieri, Ginevra Salerno and Michele Zampilli

Abstract The seismic period, that started in 1783 and lasted three years in the southern Calabria (South of Italy), caused massive damages: over 35,000 victims and relevant upheavals in the hydrogeological asset of the site. The *Borbone* government immediately dealt with the reconstruction of backward provinces, attempting to reorganize the economic and administrative systems by innovative criteria and by the construction with anti-seismic techniques. In this paper a preliminary reviewing of the historical sources is analysed to frame the reconstruction, highlighting the differences compared with the initial models. By the regional to the urban scale, it was assumed as case study the historical centre of Mileto, one of the five cities entirely rebuilt on another site after the earthquake. Mileto preserves the scheme of the urban planning conceived in the end of the Eighteenth century. Also in Mileto, there is the Episcopal Palace, characterized by several architectural and construction features to increase the seismic response. Despite the advanced state of deterioration, the

S. Stellacci (✉)
ISCTE—Instituto Universitário de Lisboa, DINÂMIA'CET-IUL,
Av. das Forças Armadas, 1649-026 Lisbon, Portugal
e-mail: stestellacci@gmail.com

F. Geremia · E. Pagano · G. Salerno · M. Zampilli
Dipartimento di Architettura Universitá Roma Tre,
Via della Madonna dei Monti 40, 00184 Rome, Italy
e-mail: francesca.geremia@uniroma3.it

E. Pagano
e-mail: enrico.pagano.89@gmail.com

G. Salerno
e-mail: ginevra.salerno@uniroma3.it

M. Zampilli
e-mail: michele.zampilli@uniroma3.it

N. Ruggieri
DiBEST, Università della Calabria, Via Pietro Bucci,
87036 Arcavacata di Rende, Italy
e-mail: nicola.ruggieri@unical.it

H. Cruz et al. (eds.), *Historical Earthquake-Resistant Timber Framing in the Mediterranean Area*, Lecture Notes in Civil Engineering,
DOI 10.1007/978-3-319-39492-3_10

original structure is still perfectly recognizable and allows to deepen the knowledge regarding the technology, that is unquestionably a cultural heritage to preserve, enhance and possibly to be reused. Accordingly, the authors prefigure the importance of carrying out a census of still standing buildings characterized by the *Borbone* anti-seismic system by suggesting possible research methodologies.

Keywords 1783 earthquake · *Borbone* anti-seismic code · Masonry reinforced with timber frames · Seismic vulnerability

Historical and Territorial Overview

The 1783 macro-earthquake struck down more the southern zone of Calabria, so-called *Ultra* or *Greca*, then the northern area (*Citra* or *Latina*), part of *Regno di Napoli* (Fig. 1).

After the first period of the emergency response, in which the government was more interested in the construction of temporary houses, in the implementation of infrastructure and in the draining of flooded areas, the reconstruction works were undertaken from March 1784. These activities, initiated with great diligence and dedication, had to undergo ups and downs, setbacks and accelerations.

The conflict between the enlightenment culture of the planners and the local culture (traditionalist and focused on economic and feudal relations) ended with spotty results.

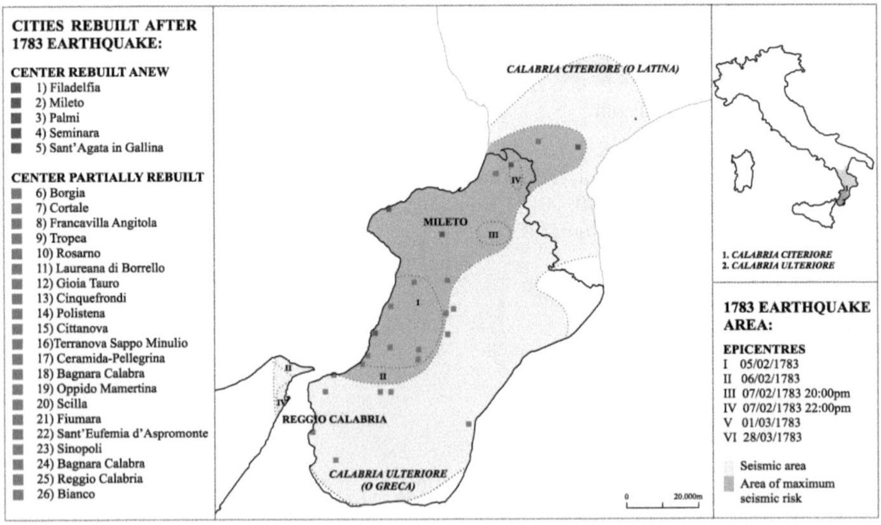

Fig. 1 Calabria map: the isoseismal map overlap on the program of the cities rebuilt after 1783 earthquake (data processed by Stellacci from Baratta [1])

Among the 33 cities to rebuild on another site indicated by early reports, just 5 were actually built anew (Fig. 1). The others were the scope of differentiated urban or building renovations.

The choice of new sites, historically attributed to Winspeare and La Vega, gave origin to the movement of population from morphologically diverse areas (spurs, hills, ridges) to flat areas, with better hygienic and hydraulic conditions and near the main roads.

Mileto: The Anew City from the Original Design to the Implementation Plan

The royal architect Vincenzo Ferraresi, Milizia's student in Rome and Naples, elaborated the urban solutions for six cities to be rebuilt in Calabria, including Mileto [2].

The plan was executed on the basis of the instructions issued by the engineers Winspeare, La Vega and Frangipane and set on a rectangular grid of 50 quarters of variable sizes for a total area of approximately 22 ha (Fig. 2).

The Ferraresi's plan is schematic and does not contain, unlike other contemporary projects, information on the socio-typological distribution of the buildings in the blocks, except for the houses of farmers. The location of the site, a high meadow along the ridge near to the old town, is justified by the *Strada Regia* and by another existing building, the *Baracca Baronale*.

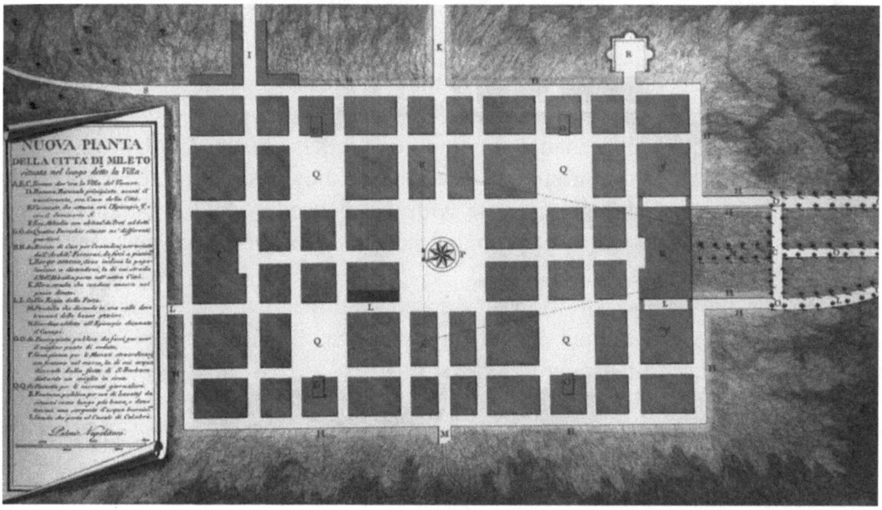

Fig. 2 Mileto map [2]

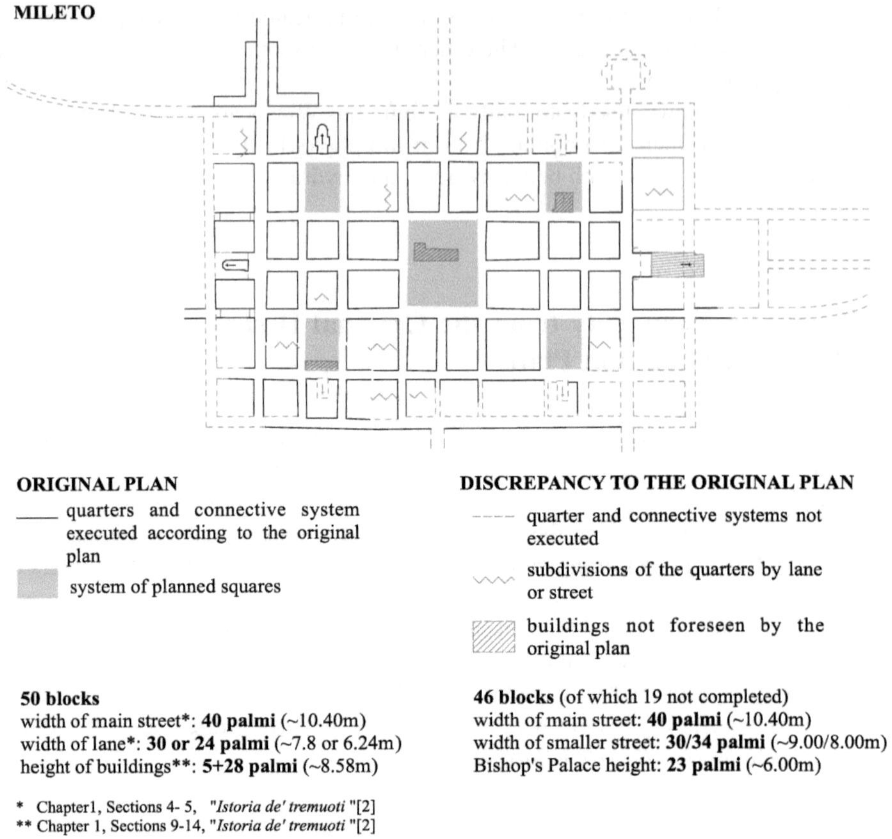

Fig. 3 Comparative analysis between the anew planned city and the executed plan

The layout planned partially respected the previous connective tissue, but did not take account of the landowners system. Accordingly, the multiple demands of the population, few inclined to change the tradition models of urban settlement and the housing, influenced the environment changes and the construction techniques.

The comparison of the current state with the planned system highlights several discrepancies [3, 4] (Fig. 3) that probably depended on the active interests, as the royal power and the clergy.

Moreover, the land subdivision on the cadastral map dated 1907 as well as the actual condition of buildings do not allow an easy recognition of the type of construction related to the rebuilding first phase.

A rather substantial fractioning process, occurred to the original quarters, was executed with the aim of increasing the number of building units and the building fronts facing the street. The subdivision of the quarters by narrow streets did not respect the standards for seismic safety reasons (Fig. 3).

The average height of the buildings (28 *palmi*) in addition to the basement (5 *palmi*) was reasonably respected in that era, according to the indication provided in the Chap. 1, Sects. 4 and 5 [2]; rather, the urban uniformity and modularity were recently lost by the urban unidirectional expansion, along the East-West direction [3].

Case Study: Bishop's Palace of Mileto (Vv)

Prolegomena to the Post-earthquake Reconstruction

The Treatise "*Istoria e teoria de' tremuoti*" by the physic of the court Vivenzio is crucial in order to focus on the issues addressed by the process of reconstruction [2].

It was published a few months after the earthquake and included the local seismic history, a comprehensive report on the recent damages and also a set of urban plans with architectural drawings. The latter became known as the prototype of the "*casa baraccata*", designed by Ferraresi and represented examples of the technological model proposed for the post-earthquake reconstruction.

In this explanatory text of the drawing, regarding the choice of associating the timber frame to the masonry, there is a reference to a specific case, the *Palazzo del Conte di Nocera* in Filogaso (Vv) that after the earthquake had remained unscathed, unlike the other buildings. As suggested in the publication edited by Rubino [2], the *casa baraccata* seems to be inspired mainly in the contemporary European experience, especially, from the North of Europe with timber frame and coated to maintain the prevalent stylistic features and also to protect the timber structure. Ferraresi's drawing does not appear in the next edition of Vivenzio's Treatise dated 1788; however it's possible that he have contributed in some way to the drafting included in the *Istruzioni Regie*, whose authorship is mainly attributed to La Vega and Winspeare.

The Treatise included a briefly description regarding the activities of analysis and interpretation of the damage mechanisms suffered by the buildings in the *Calabrian* historic centers and it was proposed an alternative system to take greater account of local technical traditions.

In conclusion, the inspiration of the lawful construction technique probably derived from Treatise of Vivenzio (1783) and it is configured as a completely new system, which combines the characteristics of the half-timbered structures with features derived from the local building tradition.

On the Scientific debate concerning the new earthquake-resistant construction set up in those few years it would be interesting to examine further based on the archival research.

Historical, Architectural and Construction Features

The construction system of the Bishop's Palace, characterized by the materials and elements from the ancient Mileto, establishes an historical continuity of memory for the community, while the project enshrines the positivist momentum *post* quake, characteristic of the *Borbone* approach, comparable to the *Pombalino* attitude.

The project, probably attributed to an architect from Campania, dated back to 1784; however, the works were undertaken after a decade and intermittently, followed by the bishop Minutolo (1792–1824) and later by the bishop Mincione (1847–1882) [5].

The wide front of the building is on two levels, on a high ashlar base and marked by an order of pilasters, simple in the spans central and coupled to the outside courts. The main facade is punctuated by vertical openings with alternating gables in cotto (curved and triangular) and at the bottom by rectangular windows in the basement. The sandstone and rusticated portal recovered from the ruins of Mileto is recurrent in contemporary buildings of the old town. Despite the regular drawing, the ground floor plan is dissymmetric, not centered with respect to the "*non double square*" [3].

At the entrance the staircase, tripartite in large ramps, is framed by a timber lodge that divides symmetrically the space, where the depth of the staircase corresponds exactly to the dimension of the adjacent room, probably used as "*cubitum salutatorium*".

Despite the incompleteness (Fig. 4), the neoclassical building is of exquisite workmanship as a whole and in the architectural details (e.g. the band in smooth ashlar, the elegant interior lodge, the grooves of the pilasters, the shape of the timber lintels, the surface finishes, the rustic capitals made with by the vertically bricks) (Fig. 5).

The construction system provides a clear hierarchy and a firmly connection of the structural components. A timber framework on the inner side connected to the

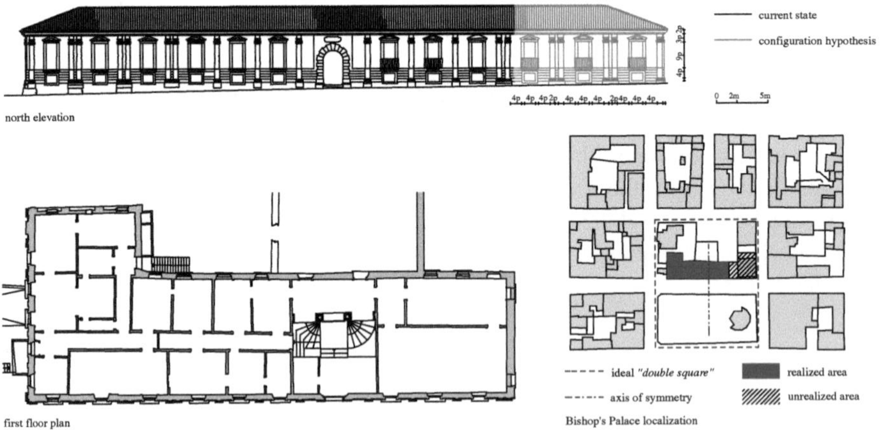

Fig. 4 Bishop's Palace in Mileto (Vv): configuration hypothesis versus realisation (drawing by S. Stellacci)

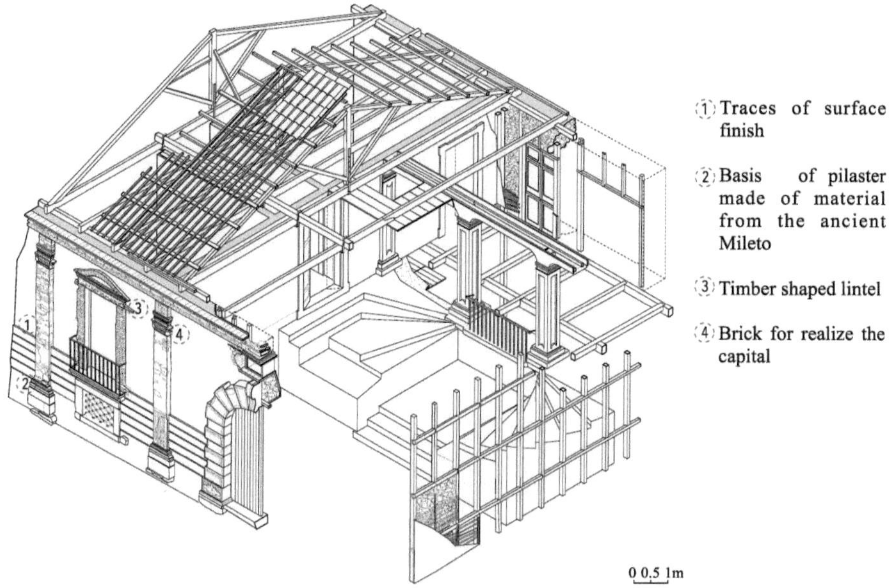

Fig. 5 Isometric view of the Bishop's Palace in Mileto (Vv) (drawing by S. Stellacci)

floors and the roof structure reinforced the external walls that in the previous tradition could be barely made in *Brest*, similar to the adobe; the infill was irregular stone with few transversal bond-stones ("*diatoni*").

Vertical main and secondary uprights, stiffened horizontally by cross-members constituted the timber frame on the perimeter walls (Figs. 5 and 6) [6].

In order to ensure greater stability and structural continuity, some timber upright reach the half-basement area, in contrast with the concern relating to the contact between woods with wetted parts materials. Despite its simplification compared to the Vivenzio' description, the main structural principles were respected (e.g. continuity of the uprights, connection of the components, protection of the framework by plaster) [6].

Nowadays, several volumetric accretions and especially a general state of decay have greatly compromised the monumental and representative character of the Palace.

Criteria for the Census of *Case Baraccate*

The census of the Calabrian *case baraccate* (datable between 1784 and the first decade of the twentieth century) should be a complex process, due to the rights of lawful owners, the limited visibility of internal structures and to the relevant adulteration that occurred over time even in the absence of specific protection rules.

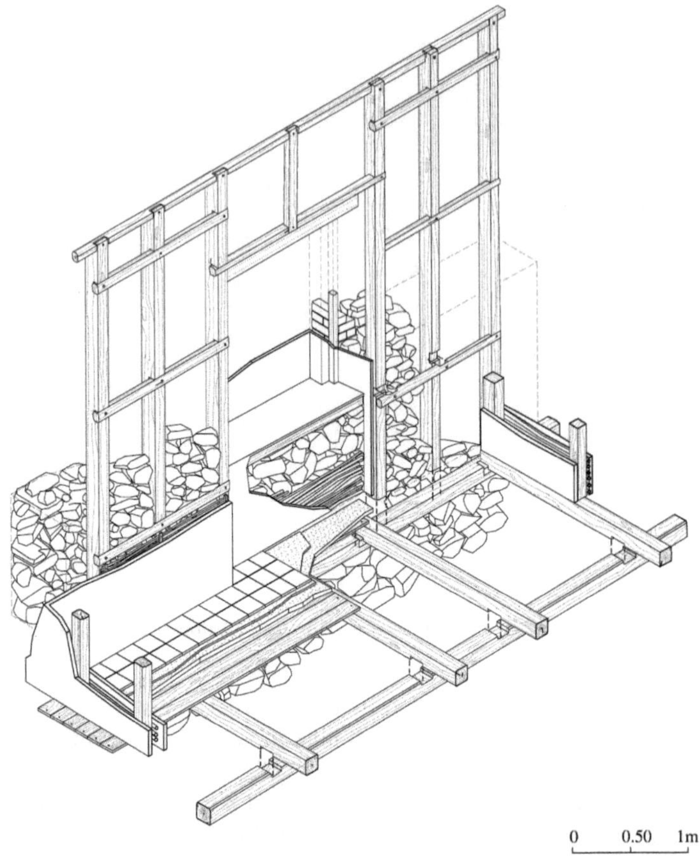

Fig. 6 Isometric view of the external wall of the Bishop's Palace in Mileto (Vv) [6]

In addition, further damages by the earthquake in the early twentieth century hampered and slowed Mileto rebuilding, as suggested by Maretto [3].

Unlike other cities (e.g. Reggio Calabria), the combined use of timber and masonry remained unchanged in Mileto until the twentieth century, possibly due to the backwardness, the difficulty or the excessive cost (e.g. appropriate sand and clay) in supplying the materials for the concrete system [7], as witnessed by the *Palace Borgia* on the corner of Via *Real Badia* and Via *Seminario*. As resulting from a postcard (dated 1920), the *Palace Borgia* construction system was still "*baraccato*".

By broadening the scope to a territorial scale, the study regarding the consistency of the residual anti-seismic heritage derives from a critical and multidisciplinary reading of historical and cadastral data, and surveys in situ. The following procedure should be carried out within the steps: (a) study of bibliographic sources; (b) interpretation and comparison of historical maps, namely the cadastral maps, to check the preservation status of the original subdivisions of land and the building units for the first building up; (c) morphological, typological and architectural in situ analysis and compilation of technical sheet; (d) non-invasive studies (i.e. infrared camera, resistometer, endoscopy, etc.), carried on the units previously selected would enable to assess the degree of authenticity of the structural system [8].

The assessment of the existing *case baraccate* in Calabria, relating the consistency and the state of conservation, allows to highlight the civic and identity values of this construction system, expressed by the local culture in an important historical phase.

Conclusions

Since the second decade of the Eighteenth century the seismic issue becomes central in the debate of the scientific community, as well the earthquake-resistant models and the regulations are fundamental in the process of rebuilding, as suggested by Barucci [7].

In Calabria, the post-earthquake program covers innovators principles within the planning and construction features compared with the previous local models.

The new chessboard plan of Mileto (Vv) replaced the articulated medieval urban tissue and established relations with the territory through the previous road system.

The lack of attention given to the landowners system and the slow reconstruction activities (worsened due to the 1905 and 1908 earthquakes) implies several divergences to the original program conceived by the royalist engineers. The construction system achieved a simplification, keeping, however, a significant seismic resistance, as exemplified by the case study, the Bishop's Palace in Mileto.

The Palace is an interesting example for the central location (in place of the hypothetical "*double-square*"), the architectural features (spatial and in details) and the construction techniques (timber framed structure), as gathered from this report.

Accordingly, the documentation of the traditional construction systems and the methodology to inventory the *case baraccate* represent the key strategic aims in order to aware the Institutions responsible for the Built Heritage Conservation.

Acknowledgments The authors acknowledge *Portuguese Foundation for Science and Technology* (FCT) for a financial support through grant SFRH/BD/94980/2013, ISCTE-IUL/DINÂMIA'CET—IUL(Lisbon) and the Municipality of Mileto (Vv).

References

1. Placanica (1985) Il filosofo e la catastrofe. Un terremoto nel Settecento. Storica Einaudi, Torino
2. Vivenzio G (1993) Istoria de' tremuoti avvenuti nella provincia della Calabria Ulteriore, e nella città di Messina nell'anno 1783 e di quanto nella Calabria fu fatto per lo suo risorgimento fino al 1787 di Giovanni Vivenzio: Atlante/premessa, saggio introduttivo e schede edited by Gregorio E. Rubino; and in annex Corso di Architettura civile di Vincenzo Ferrarese, 2. ed. Mario Giuditta, Catanzaro

3. Maretto P (1975) Edificazioni tardo-settecentesche nella Calabria meridionale. Teorema, Firenze
4. Principe I (1976) Città nuove in Calabria nel tardo Settecento. Effe Emme, Chiaravalle centrale
5. Valentia CSCV (1982) Beni culturali a Mileto di Calabria. Barbaro editore, Villa S. Giovanni (Rc), pp 117–119
6. Stellacci S, Ruggieri N, Rato V (2015) *Gaiola vs Borbone* system: a comparison between 18th C anti-seismic case studies. Int J Arch Heritage. doi:10.1080/15583058.2015.1086840
7. Barucci C (1990) La casa antisismica, Prototipi e brevetti, Materiali per una storia delle tecniche e del cantiere. Gangemi editore, Roma
8. Ceradini V, Bianco A (2008) The wooden structures of "case baraccate" after 1908 earthquake in Reggio Calabria: investigation methods to knowledge of the techniques and the evaluation of the conservative conditions. In: Proceedings of the conference SaCoMaTis2008, international RILEM conference. RILEM Publications, Bagneux-FRANCE 2008, pp 1209–1218

The Masonry Timber Framed Load Bearing Structure of the *Palazzo Vescovile* in Mileto (Italy)

Nicola Ruggieri, Francesca Geremia, Enrico Pagano, Ginevra Salerno, Stefania Stellacci and Michele Zampilli

Abstract The *Palazzo Vescovile* in Mileto (Southern Italy), erected shortly after the catastrophic earthquake that struck the Calabria region, emblematically represents the application of principles and technical indications recommended in the Borbone seismic code *Istruzioni* and included in treatises of Eighteenth century seismic engineering. The building, characterised by various constructive events that caused a construction completion delay, presents devices aimed at consciously mitigating its seismic vulnerability, including the building geometry (i.e. in plan bi-axial symmetry, a front elevation regular development and a height limited to only one storey) as well as constructive solutions regarding the nature and the organisation of materials and resistant elements forming the load bearing structure. The vertical structural system is represented by masonry reinforced with timber frames whose main function is to bond the building intersecting walls with the aim of obtaining a box behaviour of the *Palazzo Vescovile* under earthquake excitations.

N. Ruggieri (✉)
DiBEST, Università della Calabria, Via Pietro Bucci, 87036 Arcavacata di Rende, Italy
e-mail: nicola.ruggieri@unical.it

F. Geremia · E. Pagano · G. Salerno · M. Zampilli
Dipartimento di Architettura Università Roma Tre, Via della Madonna dei Monti 40, 00184 Rome, Italy
e-mail: francesca.geremia@uniroma3.it

E. Pagano
e-mail: enrico.pagano.89@gmail.com

G. Salerno
e-mail: ginevra.salerno@uniroma3.it

M. Zampilli
e-mail: michele.zampilli@uniroma3.it

S. Stellacci
ISCTE—Instituto Universitário de Lisboa,
Av. das Forças Armadas, 1649-026 Lisbon, Portugal
e-mail: stestellacci@gmail.com

H. Cruz et al. (eds.), *Historical Earthquake-Resistant Timber Framing in the Mediterranean Area*, Lecture Notes in Civil Engineering,
DOI 10.1007/978-3-319-39492-3_11

On that purpose the particular arrangement and morphology of the members nodes, often with the addition of metallic devices, contributes to transfer stresses to a three-dimensional system of orthogonal walls. In addition to the vertical load bearing system the inter-storey floor also has a key role in influencing the building seismic response. In fact, the presence of notches strengthening the node beam to the wooden ring, improves, together with the latter, the resistance to out-of-plan actions. Furthermore, the floor is characterized by a moderate deformability, thanks to the narrow spacing and geometric features of the beams, useful for redistributing horizontal forces to the shear walls. The roof structure, constituted by king post trusses ensuring the absence of outward thrusts, benefits from the presence of saint Andrew's crosses arranged orthogonally to the structural unit with the purpose of preventing the trusses from stacking under horizontal loads and, generally speaking, of ensuring a roof longitudinal stiffness. Moreover, the document devotes its attention to the behaviour of the *Palazzo Vescovile* in Mileto under gravitational actions, analysing the wooden frame contribution to the compression resistance of the entire constructive system.

Keywords 1783 Earthquake · Borbone seismic code · Masonry reinforced with timber frames · Seismic vulnerability

Introduction

The villages reconstruction in the Calabria region (Southern Italy) after 1783 earthquake was characterised by a systematic use of compound framed system, masonry reinforced with timber frames, both in the current and monumental buildings, except for churches. As for the latter, the Borbone code recommends the use of seismic measures as an alternative to timber framing because of difficulties in supplying wooden members of appropriate length.[1]

That kind of constructive system could represent the implementation of the first European seismic code (*Istruzioni Generali per la riedificazione dè paesi diruti della Calabria*) [1].

This paper aims at studying a late 18th century building, the *Palazzo Vescovile* in Mileto (Vibo Valentia, Calabria) (Fig. 1), analysing the various structural systems constituting the building and emphasising, in particular, the aware attempt, among other seismic principles, to execute a load bearing structure characterised by a three-dimensional response to inertial seismic actions, hence, a suitable *box like* behaviour [2].

[1]Moreover, that indication is also justified in the mechanical and geometrical features of the posts. In fact, timber elements of such size, if continuous, may be slender and imply instability especially in the direction in which the frame is not confined by the masonry.

Fig. 1 The main façade of the *Palazzo Vescovile* in Mileto

State of the Art

References to the technical and technological features of the buildings erected during the Borbone Age in the Calabria region have been published by Tobriner [3]. The author related about a private house in Filadelfia (nowadays unfortunately demolished) and several examples characterised by frames braced with Saint Andrew's crosses, e.g. Seminara (Rc), Palmi (Rc) and Reggio Calabria, the latter dated back to the 19th century.

Barucci [4] comprised in her work the Vivenzio's seismic prototype, by focusing on a historical analyses regarding constructive typologies developed after the 1908 earthquake.

Ruggieri [1] addressed extensive studies on the seismic knowledge in the *Regno di Napoli* and the Borbone system application in eighteenth century *Calabrian* buildings.

Furthermore, the first author conducted an experimental campaign on the masonry wall of the *Palazzo Vescovile* in Mileto [5].

General Features of the Building

The case study, which was a seminary and Bishop's Palace until the 19th century, belongs to the first core built of *Nuova Mileto* (Calabria), a village entirely reconstructed after the 1783 earthquake.

The building is situated in a strategic position, overlooking the village central square and along a street prospectively included between two churches.

The plan development is quite regular: a main façade with an extension of 50 m and lateral fronts of about 13 m. Although close to rectangular plan and characterised by a symmetrical distribution of the structural members and masses, the elongated plan shape could result in a non-uniform global response of the building; even if the described geometry results within the *regularity* limit according to Eurocode 8, *Design of structure for earthquake resistance.*[2]

The one storey structure, excluding the basement, stands 0.70 m above street level, according to the *Istruzioni Borboniche* that allowed a masonry building base devoid of wooden frame up to a height of 5 *piedi* (about 130 cm).

The total building height is 5.5 m with a limited inter-story floor of 3.5 m.

Vertical Structural System

The load bearing system includes masonry, about 50 cm thick, reinforced with regular spacing timber frames, localised in the inner part of the wall. Other horizontal wooden members are present inside the panel arranged in a random way.

That configuration emphasises a dimensional hierarchy with the apex beam (i.e. ring beam) of 30 × 20 cm, whilst studs and horizontal elements measures respectively 12 × 10 cm and 7 × 7 cm. The frame bears on the upper edge of the floor beam and is fastened to the latter by means of a notch. Some vertical members, posts, continue beyond the floor up to and including the whole basement.

All the nodes among the resistant elements, characterised by the stiffening presence of a unique metallic nail, are half-lap joints, in which the indentation is executed, in general, only in one of the two members connected. Such an arrangement entails a wooden structure installation simplification even if the resistant section, to ensure an effective junction, is significantly reduced. The inner walls are entirely constituted by the wooden framing filled by masonry; the posts belonging to those panels are simply juxtaposed to the orthogonal walls, while a constraining element is represented by the apex horizontal member running beyond the wall up to bond, by means of a notch, the perpendicular masonry panel.

The masonry, in plane stabilizing element for the frame devoid of saint Andrew crosses, relating back to the *opus mixtum,* is constituted basically by roughly squared stones, whose origin of the rock can be classified as *calcilitute*.

A certain fabric arrangement regularity is observed, given by the presence of variable size pseudo-ashlars, wedged by means of pieces of bricks, elements probably

[2]The building is in plan regular, according to Eurocode 8, if the slenderness λ = Lmax/Lmin is not higher than 4, where Lmax and Lmin are respectively the larger and smaller in plan dimension of the building, measured in orthogonal directions.

originated from buildings collapsed after the 1783 earthquake. Furthermore, the masonry is sometimes traversed by irregular horizontal beds of bricks.

The mortar, organised in layers of approximately of 2–3 cm, is composed by lime and quartzose-granitic aggregates and, relied on a macroscopic analysis, presents a discrete degree of rigidity, a cohesive aspect and the absence of evident decay phenomena.

Horizontal Structural Systems, Floor and Roof

The inter-storey floor is characterised by roughly squared and not defects[3] devoid beams, whose dimensions are 25 × 20 cm spacing about 80 cm and carrying a wooden boarding. The latter is fastened to the beams through pyramidal nails. The beams, arranged with north-west direction, hence perpendicular to the main façade, are constrained to the wooden ring by means of a half lap junction.

An interesting constructive peculiarity is represented by an inclined timber element[4] localised at the lower edge of the floor, in proximity of the angle area (Fig. 2), generating, in addition to the beams narrow spacing, a reduction (even if low) of the in plan deformability of the floor, useful for redistributing horizontal forces to the shear walls (Fig. 3).

The roof carpentry is constituted by king post trusses, a secondary structure of purlins and the covering constituted by the Calabrian typical organization of wooden slats carrying the shingles. Such an arrangement ensures a certain ventilation of the roof structure and therefore generates an adverse hygrometric condition for biotic attacks.

The king post trusses, including struts and king post,[5] emphasise notches and nails at the member extremities aimed at improving the cooperation among timbers. The king post is higher dimensions than other truss members. A geometry that does not find a mechanical justification[6]; however it can be assumed a reason in having a sufficient surface to allow the knots execution for the struts slot.

---

[3]In particular high dimensions knots are present in the timber traction area. Furthermore a macroscopic analysis highlighted significant checks due to the shrinkage and some cases of spiral grain.

[4]A similar device was used in the Lefkas Island after the 1826 earthquake relative to an anti-seismic system characterized, at the ground floor, by independent wooden frames from the masonry, whilst at the first level the presence of compound walls [6].

[5]The material features described above relate to the frames members are valid for the roof structure: members not of excellent quality, in particular emphasizing the presence of inclined checks irrefutable symptom of timber spiral grain.

[6]It is worth observing that in Mediaeval Age, for example in the San Marco roof carpentry in Florence [7], a common element is represented by the choice of a wooden specie with better mechanical properties for the king post compared to the other truss members.

Fig. 2 Inclined floor in plane member

Fig. 3 Roof Saint Andrew's crosses

A stirrup is present at the foot of the king post aimed at holding the structural unit in the vertical plan. The node tie-beam to rafter, with an appropriate dimension of the heel, shifted few centimetres[7] from the wall section, is stiffened by the presence of a metal plate and nails (Fig. 4).

The installation of metal-ware on trusses nodes will become a widespread procedure only in the 19th century thanks to the improvement of the productive process and a consequent large scale production of metallic elements.

[7]However the shear onset is limited resulting the node almost centered on the timber ring.

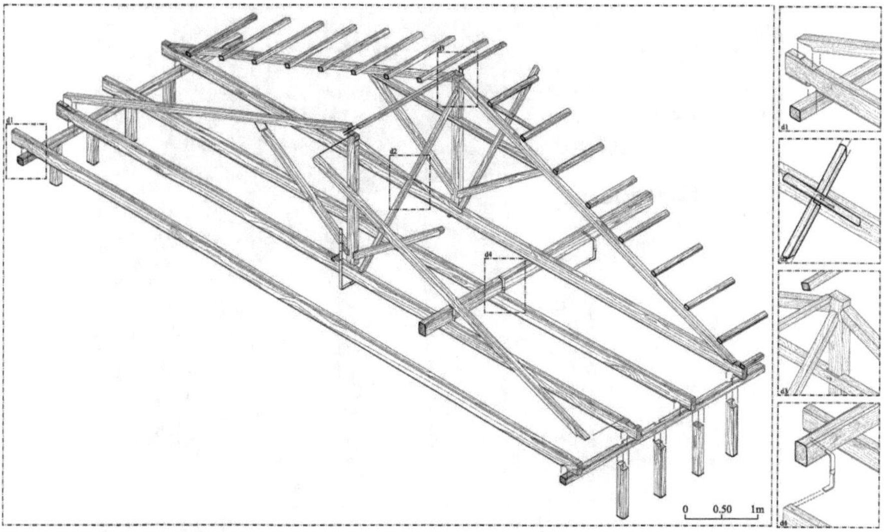

Fig. 4 Isometric view and details of roof structure of the *Palazzo Vescovile* in Mileto (drawing by S. Stellacci)

Hence we are in presence of "gravity junctions" characterized, for moderate accelerations, by an appropriate behaviour under cyclic action.[8] It is worth noting that a more effective answer is obtained if the potential vertical component of the earthquake is counteracted by means of metallic elements, as in the connection rafter to tie-beam of the Mileto trusses.

The roof presents some devices aimed at transversally stiffening the structural system.[9]

In fact bracings, i.e. saint Andrew's crosses (Fig. 3), are arranged perpendicularly to the trusses with the evident purpose of improving the longitudinal stability of the roof, avoiding the trusses stacking. A further contribution to that objective is provided by a continuous wooden member arranged orthogonal to the trusses, in the proximity of the structural unit middle, above the tie-beam and to this secured by means of a metallic stirrup.

[8]On this purpose studies have been conducted by Branco et al. [8] and by Piazza and Parisi [9] in which, by means of an experimental campaign including monotonic and cyclic tests, have been observed that the carpentry joints through the various inaccurate surfaces contact given by the handmade production can generate frictions, hence can dissipate seismic energy. Furthermore, it is possible a more appropriate earthquake answer if those nodes present metallic fasteners such as stirrup, bolt and binding strip.

[9]The roof transversal stiffness research is widespread in the North European carpentry, where an elevated snow charge entails a relevant covering slope and consequently more surfaces opposed, in both directions, to possible horizontal forces (in the specific case, in general, wind force). In the Mediterranean area such a structural organization represents a uniqueness and finds a reason only in the conscious improvement of the earthquake resistance properties of the building.

Fig. 5 Timber frame emphasizing alternated notches at the ring level

Out of Plane Behaviour

The connection among studs, posts and the ring, stiffened thanks to the presence of a pyramidal nail,[10] presents technical artifices aimed at avoiding both the inward and outward wall overturning; in fact the notches are approximately regular alternated in the external and internal part of the wooden ring (Fig. 5).

Such a device, although it is a limited benefit[11] under horizontal action perpendicular to the wall, highlights a clear comprehension of a crucial damage mode for masonry building, prelude to the modern discussion about the wall overturning begun by Rondelet and continued with the current calculation models.

In case of seismic excitation also the horizontal timbers, localised in the inner masonry, contribute, through the elements deflection, to counteracting the outward panel rotation.

[10]It is defined by Italian carpenters "a lasciare" (leaving) since has not traction effectiveness; however it can work, although with low values, under shear stresses and bending moment.

Furthermore the seismic action can provoke the nail complete detachment even after few charge cycles. Anyway it is worth emphasizing that, at least relatively the wall localised at the building West side, such metallic devices result clenched in the frame back [5], with a consequent mechanical behaviour that can be assimilated to the bolts one.

[11]In fact it is improbable the inwards building wall overturning, given the presence of boundary constraints such as orthogonal walls, floors, etc.

Such members can play, together the clay or calcilitute rock courses, an important role in the building seismic response. In fact those devices, under horizontal action perpendicular to the panel, can "direct" the damage modality toward a cylindrical hinge arising and the generation of macro-blocks instead of entailing the wall crumbling.

The floors, arranged orthogonal to the main façade, with beams and/or tie-beam constrained to the wooden ring through half lap nodes, limit the wall overturning tendency.

In Plane Behaviour

The behaviour of a wall module copy of the *Palazzo Vescovile* in Mileto has been analysed by means of cyclic tests performed in the CNR Ivalsa laboratory (Trento, Italy) in 2013 by one of the authors [5]. Such an experimental campaign has recorded an extremely ductile panel behaviour with a value about of 7 (Vu/Vy). Hence the tested sample is classifiable among the structures that present a *high ductility*, according to Eurocode 8 *Design of structures for Earthquakes resistance.* The Mileto masonry reinforced panel has dissipated a discrete energy quantity thanks in particular to the masonry features. In fact, the fragile mortar beds have facilitated the slip and in some areas the expulsion of stones, with high friction values and leaving the timber frame without or at least with limited damage.

The wooden frame, in general, remained in elastic domain for the entire duration of the test.

However, also the half lap joints among timber members have counteracted the rotation imposed by the actuator horizontal force, generating, in particular, compression perpendicular to the grain. The latter stress type can generate one of the few possible plastic damage modalities with regard to timber structures.

Peculiar Aspects: Behaviour Under Gravitational Actions

The Borbone system is characterized by a load bearing structure represented, in static field, almost exclusively by the masonry. In fact the high difference between the elastic modulus of the latter compared to the wooden elements entails a greater burden of the masonry under gravitational loads.

Nevertheless, the inner frame can represent a transversal dilatation limitation, at least in one direction, of the masonry fabric.

A similar behaviour has been addressed by Vintzileou [10] through an experimental campaign on historical Greek walls, characterised by the presence of an horizontal timber frame placed in the masonry section. The tests have proved a mechanical properties improvement, in the specific the compression resistance, of the masonry reinforced compared to a panel without timbers in the three directions.

Fig. 6 The soil settlement relative to a portion of wall

Moreover, the frame can influence the deformability of the system absorbing or limiting the cracks onset due to soil settlements. An example of such behaviour is showed in the Mileto wall localised in the South of the building, where the vertical translation, with a rotational component, probable result of a central soil settlement, has been in part counteracted by means of the timber deflection (Fig. 6).

Conclusion

The theoretical and practical reconstruction process of the Calabria region villages, after the 1783 earthquake, beyond the well known framed constructive system, included a more complex set of technical and technological seismic measures.

The research of a *box-like* behaviour, mass reduction, symmetry in plan and in front elevation, limit of the height and increase of the inter-storey floors and roof stiffness, are fundamental principles and criteria to pursue, in accordance with the Italian 18th century treatises hence with the *Borbone Istruzioni* [1].

A combination of constructive devices, analysed through the case study of the *Palazzo Vescovile,* were perhaps applied for the first time in a such systematic, complex and aware mode, in order to reduce the building seismic vulnerability.

References

1. Ruggieri N (2015) L'ingegneria antisismica nel Regno di Napoli. Aracne Editrice, Roma
2. Ruggieri N (2005) La casa antisismica. In: Proceedings of the international conference on the conservation of historic wooden structures. Tampone Ed., Firenze

3. Tobriner S (1983) La Casa Baraccata: earthquake-resistant construction in 18th-century Calabria. J Soc Arch Historians 42(2):131–138
4. Barucci C (1990) La casa antisismica: prototipi e brevetti: materiali per una storia delle tecniche e del cantiere. Gangemi editore, Roma
5. Ruggieri N, Tampone G, Zinno R (2015) In-plane vs out-of-plane "Behaviour" of an Italian Timber Framed System: the Borbone constructive system. Historical analysis and experimental evaluation. Int J Cult Arch Heritage 9:1–16
6. Vintzeleou E, Touliatous P, Zeris C, Repanis K, Zagotsis A, Leonardos E, Palieraki V, Adami C (2007) Earthquake planning and protection organization. Public library of Lefkada
7. Tampone G (1996) Il restauro delle strutture di legno. Hoepli, Milano
8. Branco J, Cruz P, Piazza M, Humberto V (2006) Experimental analysis of original and strengthened traditional timber connections. In: WCTE2006, IX world conference on timber engineering, Portland, Oregon, USA, 6–10 Aug 2006
9. Parisi MA, Piazza M (2000) Mechanics of plain and retrofitted traditional timber connections. J Struct Eng ASCE 126(12):1395–1403
10. Vintzileou E (2008) Effect of timber ties on the behavior of historic masonry. ASCE J Struct Eng 134(6):961–972

A Multidisciplinary Approach to the Analysis of the Traditional Lefkada Houses

Sandra Tonna and Claudio Chesi

Abstract In the pre-modern age, the concept of preservation was mainly related to the possibility of achieving profits; the decision about restoring derived from the willing of people to keep living their houses and any configuration change was done only if unavoidable. In addition, within a context marked by strong and frequent destructive events, it is interesting to note how conservation may give the way to empirical attempts to improve the construction system. In the present work, the problem is discussed with reference to the particular case of the Lefkada Greek island which, being characterized by a high seismic hazard with a short return period in the context of a poor economy, has experienced a number of different foreign governments in history. Despite a long sequence of critical events, inhabitants decided to stay, developing an empirical resilient structural system able to control simultaneously several criticalities. The main purpose of this paper is to characterize, through a multidisciplinary analysis, this traditional construction system, the only one still surviving in the Ionian Islands. The preservation of this particular situation is a difficult issue, which can not be planned without the assumption of the transformation process which unavoidably affects both things and local people.

Keywords Lefkada · Timber · Masonry · Seismic resistance · Multidisciplinary · Resilient

S. Tonna (✉) · C. Chesi
Politecnico di Milano, Piazza Leonardo da Vinci, 32, 20133 Milan, Italy
e-mail: sandra.tonna@yahoo.it

C. Chesi
e-mail: claudio.chesi@polimi.it

H. Cruz et al. (eds.), *Historical Earthquake-Resistant Timber Framing in the Mediterranean Area*, Lecture Notes in Civil Engineering,
DOI 10.1007/978-3-319-39492-3_12

Introduction

The combined use of wood and masonry in building structures is connected to a widely spread ancient tradition. From the beginning, builders and local populations were induced to a careful analysis of the earthquake phenomenon through the basic means at their disposal: observation and experience. Earthquakes, due to the nature of recurring phenomena, feed experience and play a somehow educational role, highlighting the most effective constructive solutions and misconceptions.

Especially in the southern Mediterranean regions, characterized by a strong seismic hazard, buildings and structures which survived through the time testify the effectiveness of local building traditions, rarely described in manuals, but empirically applied in both construction and restoration works. Interesting examples still exist in Italy, Greece, Portugal and Turkey [1, 2], all referring to the combined use of timber and masonry. Within the specific context of vernacular building, this tradition has been expressed in a variety of solutions and details, at different levels of sophistication.

In some cases, wood elements were used inside masonry with the role of reinforcement, to provide resistance towards tensile and shear actions; this corresponds to the specific technique of wood reinforced masonry, a different concept with respect to timber framed masonry.

Both solutions are finalized to the improvement of the seismic behavior of masonry walls, although different in nature. In the first and simpler case, the material behavior is improved through the insertion of individual wooden elements arranged within the stone structure; this was more a construction practice, rather than a structural intervention. In the second case, a real timber structure is realized, which may be inside the wall, or parallel, or also above it. The wood framed structure, obviously, represents the evolution of the first technique, requiring a larger amount of materials and a better carpentry expertise.

Special interest comes from the specific use of timber and masonry which can be found in the typical traditional houses of the Lefkada island. This research aims to offer an analysis of this system through a qualitative description and simplified resistance considerations. Attention has been focused onto this subject because the corresponding global design is borderline if compared to other examples of the "Mediterranean timber frame family", like the Turkish *Himis*, the Italian *Casa Baraccata* and the Portuguese *Gaiola Pombalina*.

The Study Case

Lefkada belongs to the Mediterranean Ionian Islands and is separated from the Greek mainland by an isthmus. Differently from the other islands, it has always held the role of outpost or "border", falling under the control of different administrations:

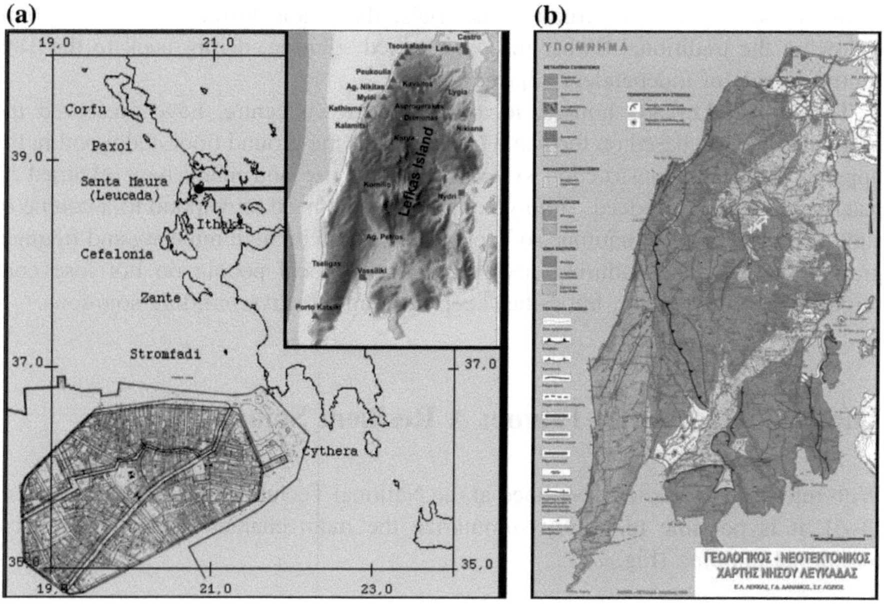

Fig. 1 **a** Lefkada, Ionian Island. Greece [11], **b** geological map, 1:100,000 (http://www.elekkas.gr/en/research/mapping.html)

the Venetians (1684–1797), the French (1797), the Turks (1798–1807) and, last, the British (1810–1864). Finally, the island became part of the Greek Nation in 1864.

Undoubtedly, Lefkada offers a meaningful case study, being characterized by a couple of critical issues: frequent and strong earthquakes and critical soil conditions. Studies on historical seismicity have estimated that an earthquake with Magnitude around 6 occurs in the island about every 30 years. As to the second criticism, Lefkada is characterized by an almost inaccessible western coast due to unstable slope conditions, by a mountainous backcountry not suitable for agriculture, and by a marshy eastern coast (Fig. 1a, b).

In the capital city, Lefkada town, the soil is an alluvial deposit generated by a reclaim project dating back to the Venetian administration. According to Eurocode 8 [3], it could be classified as a C class soil ("deep deposits of dense or medium-dense sand, gravel or stiff clay with thickness from several tens to many hundreds of metres"), prone to ground subsidence and liquefaction phenomena.

Due to the permanent need to face critical environmental conditions, the building tradition has developed efficient solutions for both the urban setting and the building design, in relation to the different realities over the Island.

While in the backcountry squat masonry buildings are in use interconnected in structural aggregates, along the coast, and mainly in Lefkada town, a specific system was developed. The historical centre was planned so as to allow the drainage and natural flow of rain water to the sea and to take advantage of a frequent wind condition, blowing from N-NW, which helps reducing humidity. From the point of view of the construction materials, the whole historical center is built following the traditional timber-masonry mixed solution, dating back to the 18th century, made of independent adjacent units.

The typical Lefkada houses, located in the town centre, have preserved the original structure, based on the use of masonry for the ground floor and wood in the upper levels. This construction system is still in use and many have claimed its paternity. Its origin, indeed, simply comes from the need to respond to a couple of simultaneous critical situations: soft soil, associated with high intensity and frequent ground motion. The enduring need implies that local people do not lose consciousness of the danger, but rather keep refining effective building solutions.

The Lefkada Timber Frame: A Resilient Solution

With reference to studies developed at the National Technical University of Athens [4–6], it is possible to briefly summarize the main characteristics of a typical traditional building (Fig. 2).

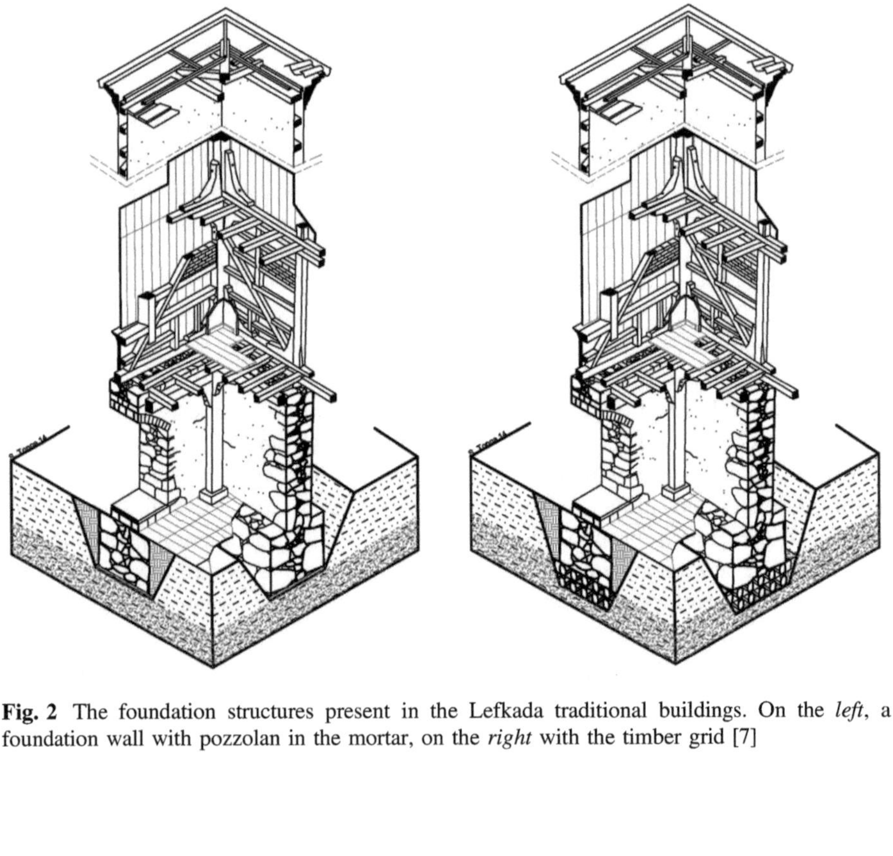

Fig. 2 The foundation structures present in the Lefkada traditional buildings. On the *left*, a foundation wall with pozzolan in the mortar, on the *right* with the timber grid [7]

This normally has from one single floor to a maximum of three; the ground floor presents a stone masonry structure, while the upper levels present a frame structure, based on the mixed use of timber and clay bricks. All the floor slabs are timber structures and the roof is covered by tiles. Openings are arranged symmetrically in plan and, in order to protect timber structures at upper levels from humidity, the outer façade was covered with wood boards.

The ground floor has a maximum height of 3 m, and is made of stone masonry walls with thickness in the range 60–100 cm, with the external leaf composed by roughly shaped elements, whereas those used in the corners are well shaped; stones on the internal leaf are plastered. The space between the external leaves is filled with mortar and small pieces of stones and bricks.

In parallel to the masonry structure, a secondary structure is built, consisting of timber columns with a centre to centre distance of 2–3 m. The slab timber beams provide connection of these columns to the masonry walls. Due to the low horizontal stiffness, this secondary system does not contribute to the seismic resistance; however, due to the strong connection to the slab, it plays a fundamental role in case of partial or total collapse of the masonry walls, providing support to vertical loads. It is therefore a passive protection system, which is activated in case of damage to masonry, the primary resistance system; in addition, it would allow masonry repair without interfering with the rest of the building.

Timber floors play a double role, resisting to both vertical loads coming from upper levels and to those directly acting on them; such loads are transferred not only to the perimeter walls, but to the secondary timber structure as well. All the connections in the timber structure are extremely accurate and well detailed.

Considering that economic resources have always been very limited in Lefkada, it can be assumed that the secondary system was conceived also to the purpose of increasing the structure seismic performance in the sense of allowing an easier and cheaper repair of the damaged building.

This system requires, for maximum effectiveness, that the columns rest on a stable supporting surface, not allowing for differential settlements; it is also required that masonry at the ground level is compact and well connected to the foundation walls.

The Foundation System

The high level of soil deformability is a permanent condition for the urban area, and it might affect also the behavior of the sophisticated structural system conceived to resist earthquakes. In the absence of a rigid base, indeed, the passive secondary system might lose effectiveness due to relative settlements and the consequent modification in the stress distribution.

From the analysis of literature, it appears that the Lefkada foundation system has only been studied to a limited extent. However, thanks to a recent study [7], it is now possible to formulate considerations on the foundation system, assuming that a couple of different solutions were systematically adopted, depending on the availability of time and economic resources (Fig. 2).

In order to explain the reasons and the modalities of this foundation system, a clear understanding of the local soil condition is required. The foundation excavation was about 80–100 cm deep, in order to reach the original lagoon bottom. In the past, neither efficient digging machines nor drainage pumps were available; because of this, builders had to set the foundation lower surface relatively high in the ground, at a maximum depth of 150 cm.

They were therefore forced to develop techniques which enabled the mortar to set in the presence of water or air, whichever was the situation; from this need, two different solutions were generated. Where the use of pozzolan or volcanic materials could be allowed, a hydraulic mortar was used for the foundation walls, which could then be directly in contact with the ground. Where, instead, the use of such materials could not be afforded, a timber grid was used, providing a flat surface for the erection of walls; any contact with water was so avoided. The grid extended homogeneously under the entire house perimeter; in the case of a higher water table level, additional timber layers were inserted, normal to underlying ones.

From the analysis of data recorded during a recent earthquake occurred in 2003 [8, 9], satisfactory results have been observed with both the construction techniques. In the first case, indeed, the foundation system is a hard and monolithic structure, almost a "concrete box" directly connected to the ground floor; in the second case, a hard wall is present, resting on a grid filled with sand, rocks and hydraulic mortar, which plays the role of a stiff surface for an effective support of the upper structure.

From the above description, clearly results that the traditional Lefkada house provides an unconventional system, with the building divided into two parts: a lower masonry structure and an upper timber part, directly connected to the foundation through the secondary timber system. This last passive structure, in addition to the double foundation solution and to the general structural design explain the previously given definition of a "borderline" timber frame.

This can be considered, at any effect, a proper resilient solution; the relative potential is discussed, in the following, through the analysis of the seismic history and some simplified numerical evaluations.

The Seismic History of the Area Around Lefkada Town

Thanks to the availability of relevant data for the frequency of strong seismic events and archival reports of their damage [10, 11], on the basis of a concept expressed by P. Touliatos in 2003 [12], the diagram in Fig. 3 has been obtained, which is representative of a specific period in the history of Lefkada, ending at the beginning of the modern age [11].

The diagram relates the magnitude value to the level of damage; this last is indicated in Table 1 for all the considered seismic events. In the abscissas,

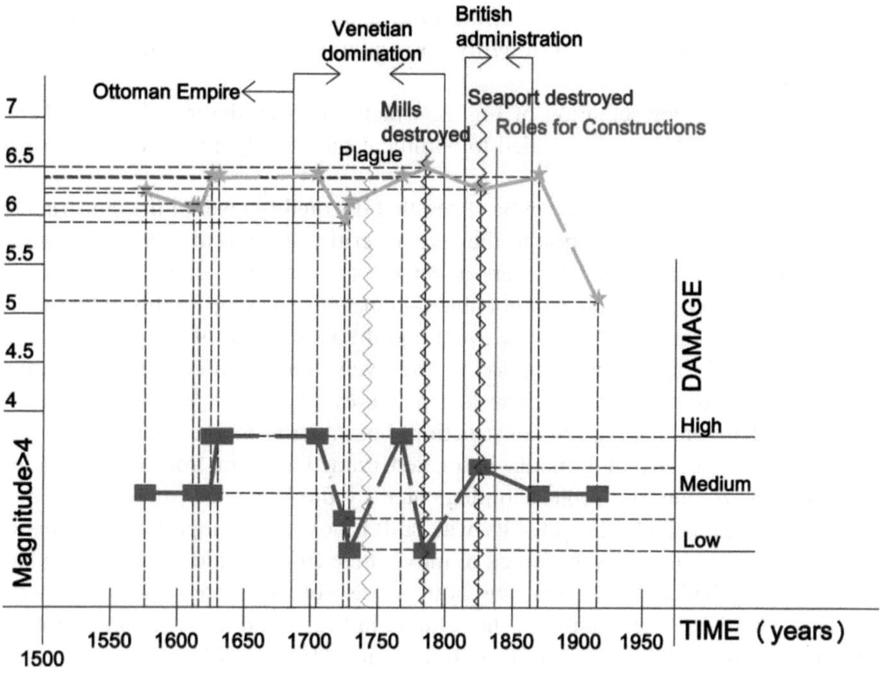

Fig. 3 Graphical representation of Lefkada earthquakes between the XVI and the XX century [11]

Table 1 The M value and the damage level for the selected earthquakes

Y	M	D
1577	6.2	2
1612	6.1	2
1613	6.1	2
1625	6.4	3
1630	6.4	3
1704	6.4	3
1722	5.9	1
1723	6.1	1
1769	6.4	3
1783	6.5	1
1825	6.25	2.5
1869	6.4	2
1914	5.1	2

earthquake frequencies can be read; the time schedule proposed by Touliatos has been kept: a fifty-year interval, roughly corresponding to a generation in terms of work experience [13].

The graphical representation of Fig. 3, which has been described in detail in a previous study [11], is here recalled to comment on the origin and effectiveness of the Lefkada system.

As discussed in [11, 12], it can be stated that the quality level of the construction techniques is often in proportion to the level of criticality typical of the area: where severe and frequent natural disasters affect the area, specific solutions had to be developed in order to survive and not abandon the site. In order for this to happen, it is necessary that the memory of the event and the knowledge of the damage that it caused are not lost; the return period of the event must then coincide, generally, with the length of a generation. The fear of the event induces local society not to forget nor abandon that type of construction, but rather to maintain and improve it.

Nowadays, the city of Lefkada is located immediately in front of the ancient fortified *Castro*, due to a project implemented by the Venetians shortly after their settlement in the island. The new city was built following an important reclamation work and the inhabitants were forced to move there from the old location. The land on which they had to build their new houses was not any more rock but, rather, a soft soil. Therefore, the reading of the scheme in Fig. 3 and the first considerations will start from the Venetian administration period (1684–1797).

The 1704 earthquake hit the island and caused severe damage. As said, the nature of the soil had changed and the inhabitants had to adapt their knowledge; this earthquake ($M = 6.4$) was a test for the new buildings. A marked improvement in the building structural behavior, indeed, can be observed 18 years later: the 1722 and 1723 earthquakes, slightly less severe than the previous one ($M = 5.9$ and $M = 6.1$, respectively), produced medium-to-low damage, with only some minor cracks in the stone churches. The 1769 earthquake ($M = 6.4$) tested those solutions and during the subsequent reconstruction works people could improve the building structure. Almost 20 years later, during the 1783 earthquake, people could benefit of the previous improvement and, although presenting the same magnitude ($M = 6.5$), the earthquake produced minimal damage to the houses.

During the British administration, an earthquake occurred in 1825; due to a careful analysis of the most affected buildings and of the construction materials, well defined rules were imposed for new construction and repair works. The following earthquakes occurred in 1869 and 1914 ($M = 6.4$ and $M = 5.1$, respectively); the damage trend was homogeneous, and can be estimated as average.

As a conclusion, although earthquakes are not predictable and many aspects influence the affected area, the knowledge of the damage mechanisms and the adoption of all the necessary precautions, in a relatively quiet political and economical situation, allow to keep under control the consequences they can produce on buildings. In Lefkada, the frequent recurrence of earthquakes has allowed the development of resilient buildings.

The earthquake which took place on August 14, 2003 ($M = 6.2$) provided a further proof of this. From the report "*Learning from Earthquakes*" [8] it is possible to know that, out of 3165 inspected houses, access was forbidden to 126 only; the remaining 1544 were immediately usable and 1495 in need of repair. An outstanding result, also considering the modifications suffered by the traditional system owed to the use of new materials and to demolition works with the purpose of upgrading, expanding and renovating.

The Seismic Resistance Level

Finally, it is of interest to assess the level of global shear resistance, in order to define the corresponding level for the seismic resistance. This can be done starting from a visual evaluation of the quality of masonry with reference to standard parameters and criteria.

In a previous study [10] it has been shown that the maximum value for the ground acceleration (a_{SLV}) can be approximated as:

$$a_{SLV} = \frac{1}{2} \cdot \frac{\tau_d}{\gamma \cdot H}$$

on the basis of the masonry shear resistance τ_d, density γ and total height H.

With reference to the problem under discussion, a single house has been selected, two stories high, built according to tradition, with the ground floor in stone (wall thickness: 60 cm) and the upper floors in timber framed masonry (thickness of walls: 12 cm); floors and roof are timber structures.

Therefore:

$$a_{SLV} = \frac{1}{2} \cdot \frac{\tau_d}{\gamma \cdot H} = \frac{1}{2} \cdot \frac{68.6}{10.4 \times 6.6} = 0.50g$$

where the following values have been be adopted:

- shear resistance at the base of masonry walls: $\tau_d = 68.6$ N/cm^2
- total building height: $H = 3.0 + 2.7 + 1.8/2 = 6.6$ m
- average density for the walls over the entire building height: $\gamma = 10.4$ kN/m^3.

This last value for the density comes from a weighted combination of the values relative to the ground floor masonry ($\gamma = 20$ kN/m^3) and the upper floor timber walls ($\gamma = 12$ kN/m^3—50 % clay bricks and 50 % wood).

The value of 0.50 (g units) for the maximum ground acceleration is in line with the expected values for the region (in the range 0.4–0.5g), as indicated in the 2013 European Seismic Hazard Model (ESHM13) [14]. In addition, it is worth noting that during the 2003 earthquake, the maximum acceleration recorded in the city of Lefkada was 0.43g.

The above calculations, although approximate, lead to a simple explanation for the system seismic effectiveness: the high value of resistance is related to the strong mass reduction in the upper levels. Such a reduction has a twofold reason: from the first to second story, the thickness of the panels is reduced to 1/5 (from 60 to 12 cm), while the material changes from masonry to timber (the density, correspondingly, varies from 20 to 12 kN/m^3).

It should also be considered that, as shown in [1, 6], that the fundamental period of these buildings does not exceed the value of 0.2 s; the structure, therefore, as it can be seen in the response spectra of the 2003 earthquake, is not affected by the high spectral amplification related to the deformability of the soil surface layer.

The above simple yet significant quantitative evaluation confirms the intuition in the conceptual design of the Lefkada typical house, suggesting that large safety margins are available, despite the high seismic hazard.

Conclusions

This study provides a multidisciplinary framework to a very complex issue: vernacular architecture in seismic areas. From both on site and archive analyses, it was possible to provide a heterogeneous and original state of the art.

The Lefkada system exhibits a peculiarity, at the limit of the usual classifications; at the same time, it shows a remarkable behavior during earthquakes. In relation to these buildings, which provide the only surviving example of the building tradition in the Ionian Islands, the preservation need is accompanied by the interest to acquire a deep understanding of the conceptual design, as well as of the real resources of the building system.

For this reason, once established an overall description of the structural system, the research next step will consist in a more detailed analysis. The approximate procedure, in fact, has the advantage of giving full evidence to the role of the average density of the masonry, which results in a direct increase of the ground acceleration. The peculiarity of the structural configuration, however, characterized by a high variation of mass and stiffness from the ground floor to the upper levels, suggests the importance of combining the simple analysis with evaluations based on more refined structural models.

A further research step would deal with a model taking into account the interaction between the complex structure-foundation system and the ground. This phase of the study should highlight the real potential of the traditional building system, with reference to a precise description of the earthquake source and site effects. It should also help to understand whether the original conceptual design was simply based on socio-economic reasons, or it was rather linked to an insight of the system dynamic behavior.

In general, this research should lead to a significant improvement of the knowledge attained for the Lefkada traditional construction techniques and for the local seismic hazard, thus allowing for conservation interventions of the architectural heritage fully compatible with the surrounding environment, characterized by high uniqueness and remarkable historical interest.

References

1. Vintzileou E (2008, June) Effect of timber ties on the behavior of historic masonry. J Struct Eng © ASCE
2. Vintzileou E (2009) Timber-reinforced structures in Greece: 2500 BC–1900 AD. In: Proceedings of the Institution of Civil Engineers Structures and Buildings, June 2011 Issue SB3. ICE Publishing, pp 167–180
3. CEN—European Committee for Standardisation (2004, Dec) EN 1998-1—Eurocode 8: design of structures for earthquake resistance part 1: general rules, seismic actions and rules for buildings
4. Touliatos P, Gante D (1995) Local historical antiseismic constructions. The example of Lefkada. Research Report, National Technical University of Athens (NTUA), Athens
5. Vintzileou E et al (2004) Assessment of and recommendations for interventions to buildings in the nucleus of the city of Lefkada, Research Report. Laboratory of RC, NTUA, Athens
6. Vintzileou E et al (2007) Seismic behaviour of the historical structural system of the island of Lefkada, Greece. Elsevier Constr Build Mater 21:225–236
7. Tonna S et al (2014) Traditional building criteria in the Lefkada island: peculiarities of the foundation system. In: Peña F, Chávez M (eds) SAHC2014—9th international conference on structural analysis of historical constructions, Mexico City, Mexico
8. Learning from Earthquakes Preliminary Observations on the August 14, 2003, Lefkada Island (Western Greece) Earthquake, EERI Special Earthquake Report, November 2003
9. Kassaras I et al (2014) Correlation between damage distribution and soil characteristics deduced from ambient vibrations in the old town of Lefkada (W. Greece). In: Second European conference on earthquake engineering and seismology, Istanbul 25–29 Aug 2014
10. Tonna S (2014) Earthquake resistant systems: experience based construction techniques in vernacular architecture—two study cases: the village of Castelvecchio Calvisio (L'Aquila, Italy) and the town of Lefkada (Lefkada Island, Greece). Ph.D. thesis, Doctorate course in Preservation of the Architectural Heritage, Politecnico di Milano, Italy
11. Tonna S, Chesi C (2015) Implications of earthquake return periods on the building quality. In: Proceedings of the XII international forum "Le vie dei Mercanti", Best practice in Heritage conservation and management, Aversa and Capri, Italy
12. Touliatos P (2003) Approach, identification and intervention in historic constructions. The technological point of view. Athens, NTUA
13. La Longa F, Crescimbene M (2011) Edurisk 2002–2011 Dieci anni di progetti di educazione al rischio. Istituto Nazionale di Geofisica e Vulcanologia, Dipartimento della Protezione Civile, Istituto Nazionale di Oceanografia e di Geofisica Sperimentale. Roma 30 Novembre
14. Woessner J et al (2015) The 2013 European seismic hazard model: key components and results. Bull Earthq Eng (Springer)

Part II
Numerical Analysis

The "Gothic House" in Arquata Scrivia (Italy): Constructive Details and Seismic Assessment of a Timber Framed Masonry Structure

S. Cattari, S. Lagomarsino, R. Vecchiattini and G. Stagno

Abstract The paper presents the analysis of the structural and constructive system together with that of the seismic response of the "Gothic house" in ArquataScrivia, that constitutes a rare example of timber framed masonry structure in Piedmont (Italy). Besides a certain diffusion around Europe, which is documented in the 19th century chronicles, in Italy there are only a few timber-framed buildings left. The timber frame system of the "Gothic house" presents some analogies with the French and German one (area in which it is more widespread than Italy) although it is characterized exclusively by the presence of horizontal and vertical timber elements. In the paper, the analysis of the building has been supported by an extensive investigation campaign that included the execution of several non-destructive tests on both masonry and timber elements. Preliminary step for setting-up the investigation plan was a sensitivity analysis carried out on an equivalent frame model of the building by performing nonlinear static analyses. Aims of such sensitivity analysis were, firstly, to address the as-built information phase and, then, to improve the reliability of the final model to be adopted for the seismic assessment.

Keywords Timber framed masonry · Sensitivity analysis · Seismic response

---

S. Cattari (✉) · S. Lagomarsino
Department of Civil, Chemical and Environmental Engineering,
University of Genoa, Genoa, Italy
e-mail: serena.cattari@unige.it

S. Lagomarsino
e-mail: sergio.lagomarsino@unige.it

R. Vecchiattini
Department of Sciences for Architecture, University of Genoa, Genoa, Italy
e-mail: rvecchiattini@arch.unige.it

G. Stagno
STD Stagno, Private Company, Genoa, Italy
e-mail: stdstagno@libero.it

H. Cruz et al. (eds.), *Historical Earthquake-Resistant Timber Framing in the Mediterranean Area*, Lecture Notes in Civil Engineering,
DOI 10.1007/978-3-319-39492-3_13

Introduction

The paper presents the analysis of the structural and constructive system, as supported by in situ investigations, survey and numerical analyses, of the "Gothic house" in Arquata Scrivia (Alessandria) that constitutes one rare example of timber framed masonry structure in Piedmont (Italy).

As known, timber-framed buildings with infill masonry walls can be observed in many European countries like Turkey, Sweden, Romania and Portugal. They are essentially part of a traditional architectural culture that exploited a renewable raw material, the wood, which was once more widespread in the territory. Timber-framed buildings were already documented during the High Middle Ages but it is between the 17th and 18th centuries that these buildings gained (more or less explicitly) their anti-seismic function, especially in high-risk countries, i.e. Greece, Turkey, Portugal, Italy and Romania. In these countries the method of timber-frame construction, which was adopted because the effective resistance provided to particularly violent earthquakes, shows more architectural complexity: diagonal elements, joint trenches and raft foundations, which are explicitly linked to their function. In some European areas timber-framed houses became successful after observing how these buildings resisted to earthquakes and in some cases timber-framing was codified as a construction method (as documented in Dos Santos 1758 [1], Vivenzio 1783 [2], Milizia 1785 [3]). Some treatises provide indications on timber-framed buildings: "The seismic prevention requires building timber houses, by connecting all structural elements in order to have a global (box type) behaviour. The inertial masses do not have to be fixed to the ground by rigid foundations but it is better to build the house over a basement of dry stones. […] The building height must be lower then its length or wideness. In this way, the centre of gravity remains within the plan of the building. The seismic actions will produce vibrations but no overturning or collapse, as in the case of masonry buildings: this is a box" [3, pp. 190].

Besides a certain diffusion around Europe, which is documented in the 19th century chronicles, in Italy there are only a few timber-framed buildings left. One of these is the "gothic house" in Arquata Scrivia which is located within the medieval centre of the town on the Ligurian-Piedmontese Apennines. The timber frame system presents some analogies with the French and German one (area in which it is more widespread than Italy) although it is characterized exclusively by the presence of horizontal and vertical timber elements for which a specific role in relation to the seismic response could be questionable.

In the paper, the understanding and analysis of the structural system have been supported by an extensive investigation campaign that included the execution of several non-destructive tests on both masonry and timber elements. Preliminary step for setting-up the investigation plan was a sensitivity analysis carried out on an equivalent frame model of the building realized with 3Muri program (distributed by S.T.A. Data [4]) by performing nonlinear static analyses. Aims of such sensitivity analysis were, firstly, to address the as-built information phase and, then, to improve the reliability of the final model to be adopted for the seismic assessment.

Description of the "Gothic House"

The so-called "Gothic House" is built on Via Interiore, which was constructed as part of the new fortified village of Arquata in the second half of the 13th century. The timber-framed structure and its infill walls are associated with old medieval buildings and they are not common construction methods in the area anymore. In fact, due to its particularity in the setting it has become known as the "gothic house". Alfredo D'Andrade, painter, archaeologist and restorer from the 19th century, studied the house of Arquata in his research on Italian medieval architecture. It was a preliminary study before the construction of the Medieval Village realised for the occasion of the Italian General Exhibition in Turin in 1884. In 1882, he sketched a perspective view of the house in his notepad as well as one lateral side and a series of architectural details with useful comments to understand the construction of the house, which he identified as one of the many timber-framed houses that still existed in the village at the end of the 19th century.

The house is built on a rectangular area of land, at right angles to the main road (Via Interiore), attached to another building at its posterior side and separated lengthwise from the other houses by an alley to the south (Vico del Gelsomino) and by a narrow *ambitus* to the north, which (to this day) collects the drainage water from the roof and where some vents can be observed. The house has three above-ground floors, each one of which protrudes out from the beam of the floor underneath, in accordance with the methods of medieval architecture (in timber as well as stone). The double timber roof truss is covered with terracotta tiles while the ridge roof has chestnut-tree A-frames and is covered with shingles.

The walls of the ground floor and those that separate the *ambitus* are made of split local sandstones and bricks, while the front and the side of the two higher floors, besides the internal partition walls, are built with a timber-framed structure and infill walls. The infill wall is mainly made up of full terracotta bricks placed edgeways. The bricks on the main side are thicker and are a mixture of split stones and fragments of sandstones and limestone with full terracotta bricks. Today most infill walls have been covered with plaster. The vertical connection between floors is provided by stairs with monolithic steps in sandstone.

The frame structure is made of struts (sometimes vertically aligned on the two floors) that lay on the beams of the wooden floors, thanks to the rafter interposition, and are connected with each other by horizontal beams. There are no diagonal elements present (Fig. 1). In the lower part, "mortise and tenon" joints are used for the timber elements, while in the higher part timber joints with forged iron nails were used.

Some elements are particularly detailed in their aesthetic and decorative aspect, such as: the serrated pigeonhole in the gable; the infill wall of full terracotta bricks

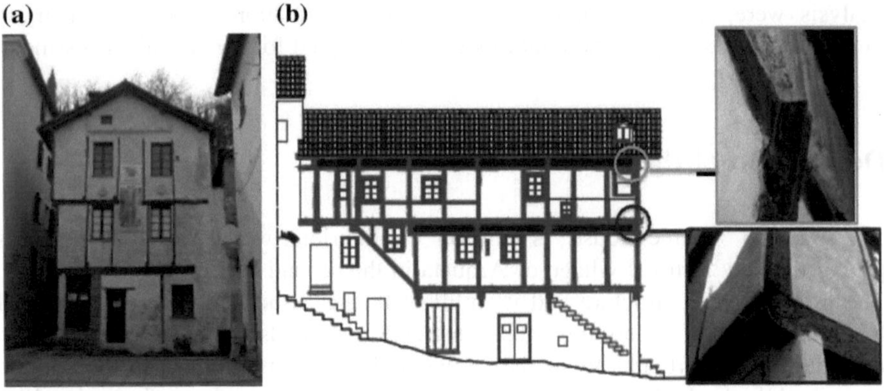

Fig. 1 **a** View of the main façade of "Gothic House"; **b** Façade in Vico Gelsomino and details of timber elements connections

that shows a fishbone pattern between the top section of the main beams of the timber floors; the timber floors of the main rooms on Via Interiano that are completed with butt straps, "zembini" (finishing slabs) and frames (common in medieval times), that might have once been painted.

To date the house two methods of absolute dating were applied on different parts of the house: dendrochronology and mensiochronology of bricks. Previously the house was generically thought to be medieval by Alfredo D'Andrade and it was more recently dated by local historians to the 14th century (this was based on the dating of other buildings in the village, i.e. the Hospital from 1325). On the basis of the dendrochronological curve of the chestnut tree in the Canton Ticino region, dendrochronology was used to date a number of beams of the timber floor. According to the data collected, the section was dated between the end of the 16th century and the beginning of the 17th century. On the basis of the mensiochronological curve of Genoa, mensiochronology was used to date some wall portions. The resulting time period was between the 15th and 16th centuries, which was coherent with the timber beams. A number of internal partition walls made of bricks were dated back to the 19th century: they therefore are probably part of subsequent additions. Thanks to the coherence of the dendrochronological and mensiochronological dating, the "gothic house" can be dated to the late Middle Ages, a period of development for the village of Arquata, which was located on an important route connecting the sea with the hinterland.

Between 1999 and 2005, the house, owned by the city council, underwent two restoration interventions. These were part of the same project (Studio Fera - Genoa) aiming to ensure the house stability for the possibility of its public use. Both interventions sought to preserve its original structure as much as possible, maintaining its function and characteristics. For this reason, a number of steel bars were inserted in some timber beams and the top sections of some damaged beams were reconstructed (mostly in the *ambitus* section) and a beam cage with threaded rebars has been added to stabilise the timber-timber and timber-brick joints.

Uncertain Factors and Sensitivity Analysis to Optimize the Investigation Plan

First step for a reliable assessment of existing buildings and the proper understanding of their structural system is constituted by the knowledge phase. Thus, it becomes crucial to have, on one hand, effective procedures to optimize the investigation protocol in term of cost-invasiveness-benefit and, on the other one, reliable approaches to account for the residual uncertainties in the final assessment. The common approach adopted by standards or guidelines for the seismic assessment of existing buildings (e.g. at international levels, Eurocode 8-3 [5]) is in the framework of a deterministic procedure (at least concerning the capacity). It is based on the definition of a discrete number of *Knowledge Levels* (KL), achievable as a function of information gathered to overcome the incomplete knowledge, and on the application of a *Confidence Factor* (CF) to one specific parameter, assumed a priori by the code as being the most critical in affecting the outcome of the assessment (usually a strength parameter). Several critical issues have been raised by various authors on the current approach proposed by codes, as discussed in [6, 7] with reference to masonry structures.

Focusing the attention on the as-built information phase, codes recommend to address investigation on three main aspects, related to geometry, constructive details and mechanical properties by proposing criteria in terms of number and type of tests to be performed that are more or less detailed as a function of the structural type (if r.c., steel or masonry). In particular, three levels of completeness associated to the three aforementioned levels are classified as *limited*, *extended* and *comprehensive*. In most cases, a given KL is assigned to the whole building, thus implicitly assuming that the sensitivity to all groups of parameters is equivalent (geometry, constructive details and mechanical properties): in other words, codes in general do not introduce the concept of deepening the knowledge on the factors that most affect the structural response. To this aim the codified use of sensitivity analysis is an emerging tool that allows to optimize the investigation plan and strengthen the link between knowledge and assessment, as discussed in [7] and also explicitly recommended in the Instruction DT 212 recently issued by the Italian National Council of Research [8].

In the case of "Gothic House", the sensitivity analysis has been adopted to investigate the role of the following uncertainties: the mechanical properties of the different masonry types present in the building; the stiffness of diaphragms; the stiffness of timber elements (by distinguishing the role of vertical and horizontal elements). Following the general principles illustrated in [7], the sensitivity has been assessed with respect to a selected Structural Performance Indicator, chosen as the maximum Intensity Measure (IM_{PLi}) compatible with the fulfilment of selected performance levels: in this case the Peak Ground Accelerelation has been adopted as IM and the Life Safety has been assumed as reference as Ultimate Limit State. To this aim nonlinear static analyses have been performed on a 3D model of the building set on basis on the preliminary data available enriched by those from the

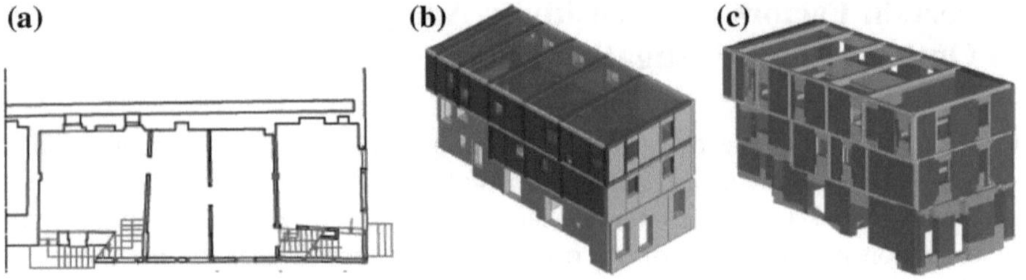

Fig. 2 **a** Plan view of the first floor; 3D view of the geometrical model (**b**) and equivalent frame idealisation of the building (**c**, the spandrels, piers and rigid nodes are marked in *green*, *orange* and *light blue*, respectively)

two previous restoration interventions which were comprehensive also of detailed drawings, photos and technical reports (Fig. 2).

In particular, the equivalent frame model has been adopted as modelling approach by using the software 3Muri [4]. Masonry panels have been modelled as nonlinear beams (with a bilinear constitutive law coherent with the requirement of Italian Structural Code 2008 [9] and Eurocode 8-3 [5]), while diaphragms as equivalent orthotropic membranes that allow to account for the actual stiffness of floor, quite flexible in this case. Also the timber beams have been modelled as nonlinear beams. Further details on the hypotheses which the model is based on are illustrated in [10]. Non linear static analyses have been performed varying: the two main directions of the building footprint (X, Y), the positive or negative sense, the load pattern (proportional to mass and to the mass and height product), the accidental eccentricity (assumed as proposed by codes as the 5 % of the maximum length in the direction orthogonal to that examined). For each of these possible combinations, 2N + 1 analyses (with N the total number of aleatory variables considered) have been performed: a first one by adopting as reference for all the parameters the plausible mean value (defined on basis of the initial knowledge state); a set of 2N analyses in which each parameter is changed one by one according to the lower ($x_{k,low}$) or higher ($x_{k,up}$) bound of the rational range defined.

Final aim of the sensitivity analysis is to attribute a sensitivity class (High, Medium and Low) to each of parameters investigated and, then, on basis of such result address the in situ investigations to be performed. The sensitivity is attributed through a quantitative indicator computed on basis of nonlinear static analyses performed [7]. Taking into account the results of all analyses performed and the different behaviour highlighted by the structure in the two directions, Table 1 summarizes the main results obtained in terms of sensitivity class and tests carried out to improve the knowledge.

Indeed in terms of value of IM_{PL} the building satisfies the seismic safety criteria proposed in the Italian Structural Code, being higher than the PGA expected for Arquata Scrivia (equal to 0.092 m/s^2 with a return period of 475 years) that is characterized by a low seismicity level.

Table 1 Summary of results of sensitivity analysis and plan of investigation

Parameter	Sensitivity class	Test performed
Mechanical parameters of stone masonry	High in X dir.	Pulse sonic test, hardness test (pendulum hammer for mortar and rebound schmidt hammer for stone and brick), chemical and physical analyses on mortar samples, endoscopy, thermography
Mechanical parameters of brick masonry	Medium/High in Y dir.	
Mechanical parameters of mixed stone-brick masonry	Low–high only for few analysis in X dir.	
Stiffness of diaphragms	High in X dir.–medium in Y dir.	Dynamic identification
Stiffness of standing out diaphragms (on Vico Gelsomino)	High in X dir.	–
Horizontal timber beams	Low	Electric instrumental hammer, electrical resistivity, samples, hardness and sclerometrictest, resistographic and ultrasonic test
Vertical timber elements	High in Y dir.–medium for few analysis in X dir.	

Main Results of Investigations and Tests Performed

Figure 3 identifies the position of tests performed which involve masonry walls, timber diaphragms, elements and connections, while Fig. 4 shows the execution of some of them.

As far as the masonry concerns (Fig. 5), the integrated use of the different NDT tests adopted was addressed to confirm the rational range of values adopted for the sensitivity analysis. Indeed, the latter has been defined on basis of values suggested by the Italian Code [11] as a function of various masonry types; moreover, the code proposes also some corrective coefficient to be applied to such basic values as a function of the quality of constructive details (e.g. mortar quality, transversal connection, etc.). Thus the tests performed were very useful to assess the latter ones for example: the endoscopic surveys to examine the stratigraphy of the walls; or the laboratory tests on mortar samples to assess its quality. Moreover the use of thermography has allowed to properly attribute the correct masonry type to various panels covered by plaster; thermography was used with a passive system pre-heating with gas devices. Quality of mortar revealed quite poor. Pulse sonic tests highlighted a high dispersion of velocity values as a function of different masonry patterns that characterize panels (e.g. if member of a timber frame or not). If a significant residual uncertainty still remained on some parameters, the minimum value of the corresponding rational range defined has been adopted in the final assessment: that is the case for example of mechanical properties of masonry types associated to an high sensitivity for which it was not possible to execute tests capable to provide a direct characterization of stiffness and strength properties. In this case to limit the invasiveness and on basis of the economic resources available, the execution and integrated use of non destructive tests have been favoured.

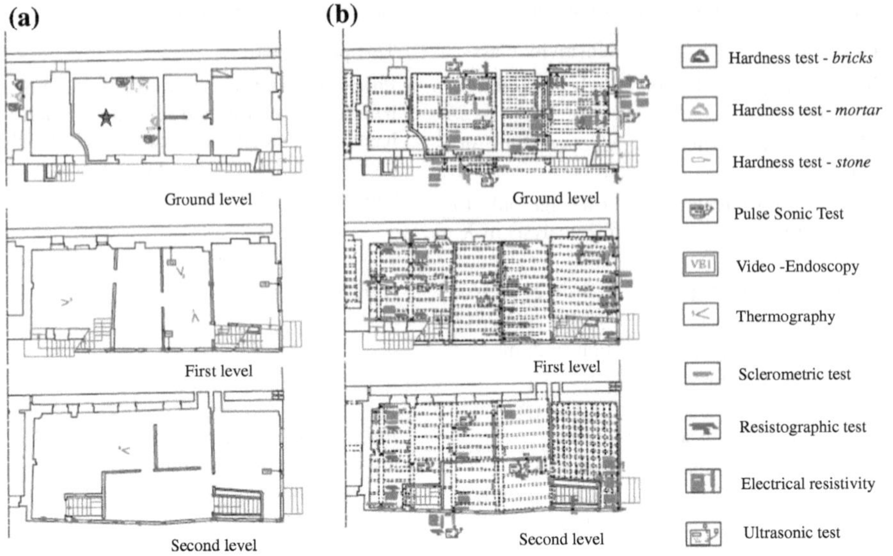

Fig. 3 In-situ investigations and tests performed on masonry (**a**) and timber (**b**) elements; the *black star* indicates the diaphragm investigated by structural dynamic identification

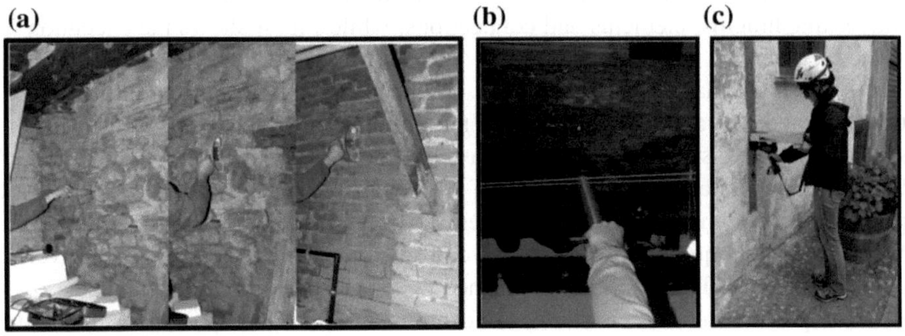

Fig. 4 Examples of tests performed: **a** hardness tests on mortar and blocks; **b** timber hardness test; **c** resistographic test

Of course, the optimization of the investigation plan requires to balance needs related to costs, conservation and reliability of obtainable data.

As far as the timber elements concern, protocols for technological inspections of UNI 11119 [12] have been applied which require firstly the classification of wood in structural categories based on the identification of the type (microscopic method with samples and laboratory analyses), the humidity level (applied on the spot for electrical resistivity), the defects and the alterations of the wood. Moreover, the sclerometric test has been used to identify any superficial decay and the density of each section. A total of 29 elements have been analysed through the entire or partial set of

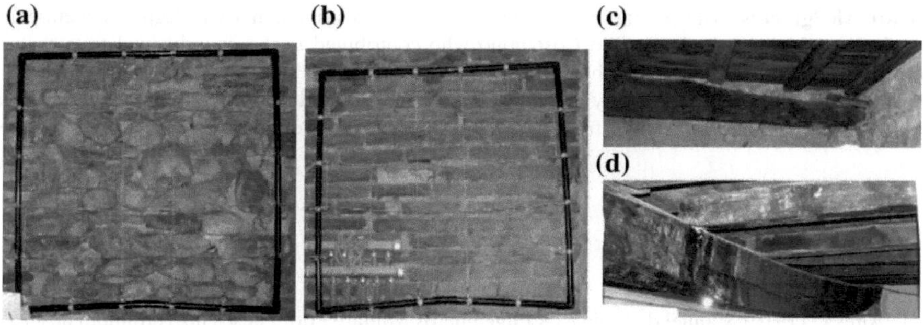

Fig. 5 Two of main masonry types that characterize the “Gothic House” (stone **a**) and brick (**b**); example of strong reduction of the transversal section (**c**) and cracks (**d**) in timber beams

tests listed in Table 1; for example, the electric hammer that drills into the entire section of the material didn’t carried out on elements strengthened by the steel bars. In some cases, also significant cracks or section reduction have been noticed (Fig. 5c, d). Tests performed allowed to identify 9 critical elements for which a reduction of the initial values of the mechanical properties ranging from 30 to 50 % has been assumed in the final assessment. Great attention was paid to the quality of connection. Ultrasonic tests were carried out to examine the continuity of the timber joints between the horizontal elements in the floor and the vertical supports; such technique proved to be extremely efficient and fast in the wave propagation.

The dynamic identification of diaphragms based on ambient vibration tests suggested a quite good quality of the connection between different timber elements that compose them. Thus, the contributions of nails in defining the equivalent shear modulus of diaphragms has been taken into account according the analytical expressions proposed in [13]. Such contribution revealed not negligible confirming how the assumption of a completely flexible floor would not have been appropriate. Being the sensitivity to such parameter high, this investigation is relevant in improving the reliability of the model, with particular reference to the capability of transferring seismic actions among walls.

Final Remarks

In the paper, the potential of the sensitivity analysis, firstly, to address the as-built information phase and, then, to improve the reliability of the final model to be adopted for the seismic assessment has been illustrated. Thanks also to the results achieved through the sensitivity analysis, a quite comprehensive investigation plan has been outlined focused on both masonry and timber elements. Although for the “Gothic house” the outcome of the seismic assessment has not be detected as crucial, results of tests allowed to identify some elements characterized by critical conditions that could locally affect the response of the building to ordinary actions and for which specific strengthening interventions could be designed.

Acknowledgments The authors acknowledge: Lucia Macario, a master degree student who worked on this case study; Emilio Costantino who contributed to the experimental tests carried out; Severino Fossati who contributed to the dendrochronological analysis; the municipality of Arquata Scrivia for the availability given to examine the Gothic House; the Technical Study Fera for the data provided as resulting from the previous restoration works carried out on the structure.

References

1. Aires C (1910) Manuel da Maia e os Engenhieros Militare Portugueses no Terramoto de 1755, Lisboa
2. Vivenzio G (1783) Istoria e teoria de' tremuoti in generale ed in particolare di quelli della Calabria e di Messina del MDCCLXXXIII, Naples
3. Milizia F (1785) Principj di architettura civile, III, Bassano, Venice
4. 3Muri, release 10.0.1.2, 2015, S.T.A. Data s.r.l
5. CEN (2005) Eurocode 8—design of structures for earthquake resistance. Part 3: assessment and retrofitting of buildings, Brussels, Belgium
6. Tondelli M, Rota M, Penna A, Magenes G (2012) Evaluation of uncertainties in the seismic assessment of existing masonry buildings. J Earthq Eng 16(S1):36–64
7. Cattari S, Lagomarsino S, Bosiljkov V, D'Ayala D (2015) Sensitivity analysis for setting up the investigation protocol and defining proper confidence factors for masonry buildings. Bull Earthq Eng 13(1):129–151
8. CNR-DT212 (2013) Recommendations for the probabilistic seismic assessment of existing buildings. ConsiglioNazionaledelleRicerche, Rome (in Italian)
9. NTC 2008 (2008) DecretoMinisteriale 14/1/2008. Normetecniche per le costruzioni.Ministry of Infrastructures and Transportations. G.U. S.O. n.30 on 4/2/2008 (in Italian)
10. Lagomarsino S, Penna A, Galasco A, Cattari S (2013) TREMURI program: an equivalent frame model for the nonlinear seismic analysis of masonry buildings. Eng Struct 56: 1787–1799
11. Circolare del Ministero delle Infrastrutture e dei Trasporti (2009) No. 617,: "Istruzioni per l'applicazione delle nuove norme tecniche per le costruzioni di cui al Decreto Ministeriale 14 Gennaio 2008," Consiglio superiore dei lavori pubblici. S.O. n.27 alla G.U. del 26.02.2009, No. 47
12. UNI 11119:2004 - Beni culturali - Manufatti lignei - Strutture portanti degli edifici - Ispezione in situ per la diagnosi degli elementi in opera (In Italian)
13. Brignola A, Pampanin S, Podesta' S (2012) Experimental evaluation of the in-plane stiffness of timber diahgrams. Earthq Spectra 28(4):1687–1909

Seismic Assessment of *Pombalino* Buildings

Rita Bento, Ana Paula Falcão, Ricardo Catulo and Jelena Milosevic

Abstract Portuguese *Pombalino* masonry buildings were built mainly in Lisbon after the 1755s Lisbon Earthquake. These buildings are characterized by a mixed timber-masonry structure, known as the *gaiola pombalina* (cage), and were built aiming to withstand the horizontal seismic loads. The cage is composed of timber floors and improved mixed timber-masonry shear walls ("frontal" walls). It was arguably the first case in history of an entire town built with the purpose of providing seismic resistance to its buildings. In this paper, a description of the *Pombalino* buildings is presented, GIS mapping tool was developed to identify the main features of *Pombalino* buildings in Lisbon downtown area and the results of the seismic assessment of existing building are briefly discussed. A three-dimensional model was developed in Tremuri program based on the equivalent frame model approach. The building was modelled using non-linear beams for masonry panels. The "frontal" walls were modelled by a macro-element model developed and calibrated according to an experimental campaign. The seismic assessment of the global response of the building was determined through non-linear static (pushover) analyses. Due to the fact that building does not fulfil the safety requirements for the ultimate limit state, a strengthening solution was proposed in order to improve the seismic behavior of the building.

Keywords *Pombalino* buildings · Cyclic static shear testing · Macro-element modelling · Performance-based assessment · Non-linear static analyses · Arcgis · BIM

R. Bento (✉) · A.P. Falcão · R. Catulo · J. Milosevic
CERIS, Instituto Superior Técnico, Universidade de Lisboa,
Av. Rovisco Pais 1, 1049-001 Lisbon, Portugal
e-mail: rita.bento@tecnico.ulisboa.pt

A.P. Falcão
e-mail: anapmfflor@gmail.com

R. Catulo
e-mail: rdcatulo@hotmail.com

J. Milosevic
e-mail: jelena.milosevic@tecnico.ulisboa.pt

H. Cruz et al. (eds.), *Historical Earthquake-Resistant Timber Framing in the Mediterranean Area*, Lecture Notes in Civil Engineering,
DOI 10.1007/978-3-319-39492-3_14

Introduction

Modern societies have been emphasizing the need of preserving the built heritage. It is estimated that old masonry buildings, which were built before the introduction of seismic code provisions, compose half of the existing building stock in Lisbon County. The structural assessment of these buildings is a crucial issue, in particular in areas with a moderate to high seismic risk. It was estimated that during the 1755 earthquake, which was followed by a tsunami, 75 % of the buildings collapsed or became uninhabitable [1]. The affected area was after rebuilt based on rigorous planning and construction techniques, which are generally referred to as *Pombalino* construction.

The *Pombalino* buildings were composed by exterior stone masonry walls and a three-dimensional timber structure named *gaiola pombalina* enclosed on the interior masonry walls above the first storey. The *gaiola pombalina* which was made by the timber floors and mixed timber-masonry shear walls, (called *frontal* walls), provided resistance to horizontal forces in any direction of the building structure. Due to the standardization of the *Pombalino* building's construction process, it was the first time in history that an entire city was built with anti-seismic provisions.

Focusing on these types of buildings, and taking into account research developed until now, more information and investigation should be provided focusing on *Pombalino* buildings. Namely, only few scientific studies can be found in the literature related to: (i) the experimental tests on frontal walls [2–7], as well as (ii) the seismic assessment of the global response of the *Pombalino* building [8]. Due to this, and to the fact that these buildings still represent an important part of the Lisbon building stock and have an undeniable historical and cultural value, developments of the studies aimed to assess their seismic behaviour and afterwards to propose strengthen solutions are crucial.

The main goal of the work herein presented is to describe the main features of the *Pombalino* structures and to present the seismic vulnerability of a typical *Pombalino* building. A GIS mapping tool was developed to identify the main features of *Pombalino* buildings in Lisbon downtown area. The performance-based assessment was carried out to evaluate the global seismic response of this building typology. A three-dimensional model was developed in Tremuri Program based on the equivalent frame model approach. The building was modelled using non-linear beams for masonry panels, whereas the "frontal" walls were modelled by a macro-element model calibrated according to the previous experimental campaign [8].

The seismic assessment of the global response of the building was determined through non-linear static (pushover) analyses. Regarding the obtained results it was concluded that building does not fulfil the safety requirements for the ultimate limit state. Due to this fact, different strengthening techniques were analysed in order to improve the seismic behaviour of the building, but only one of them will be discussed in the presented paper.

The Downtown of Lisbon

Based on the statistical data collected by Sequeira [9] and taking into account the construction period in Lisbon, it was estimated that *Pombalino* buildings correspond to 27.2 % of Lisbon existing building stock.

This study involved a complete identification of the old masonry buildings located in the downtown of Lisbon. The data gathered from the City Hall of Lisbon were processed and a database management system integrated into a GIS application was developed to manage, compare and analyse the information. Currently, identification of old masonry buildings was carried out on four different domains: (a) architectural typologies and drawings; (b) structural and non-structural features; (c) structural changes or retrofitting solutions and (d) state of preservation (structural damage). Figure 1 present the downtown area of Lisbon placed in the GIS format, including the existing buildings and roads in the study area. It should be mentioned that this study should be considered as a starting point for future work, due to the fact that the use of the information gathered in the database is of particular importance to the seismic vulnerability assessment of old masonry buildings at city scale. The CityEngine software [10], which is a 3D modelling application specialized in the automatic generation of urban environments based on procedural modelling and shape grammars, was used in this work. This application combines geo-referenced data of the building location and parameter information. In this case, each polygon (which corresponds to a building) is associated with the features and attributes mentioned above (four different domains), allowing their visualization, selection and search.

Pombalino buildings are usually located in the rectangular quarters, where each consists of an average of 10 buildings with four to five stories (Fig. 2).

The *Pombalino* buildings were supported on a timber grid laid on short timber piles with no more than 0.15 m diameter and 1–6 m length (Fig. 3a), which hardly achieved the resistant soil. It should be mentioned that, based on the experience, it was noticed that the length of timber piles are closer to 1 m rather than 6 m.

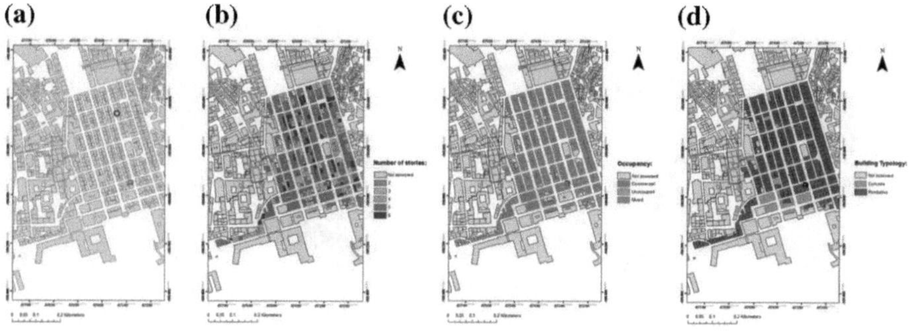

Fig. 1 Identification of old masonry buildings according: **a** location of a *Pombalino* building studied in detail [8]; **b** number of stories; **c** occupancy and **d** building typology

Fig. 2 Collection of *Pombalino* buildings, from a Lisbon downtown street, modeled in 3D-GIS [10]

This support system intended to stiffen the alluvial soil and to create a working platform above the water level [11]. The general foundation system consisted in continuous masonry walls or caissons filled with rubble masonry covered by a group of arches (Fig. 3b). In Fig. 4 the typical example of the foundation system below the partition interior walls is depicted, where timber pillars support the walls.

(a) **(b)**

Fig. 3 **a** Timber pile from *Pombalino* building; **b** foundation system: masonry arches and caissons supported in timber piles [12]

Fig. 4 Foundation system for the interior walls

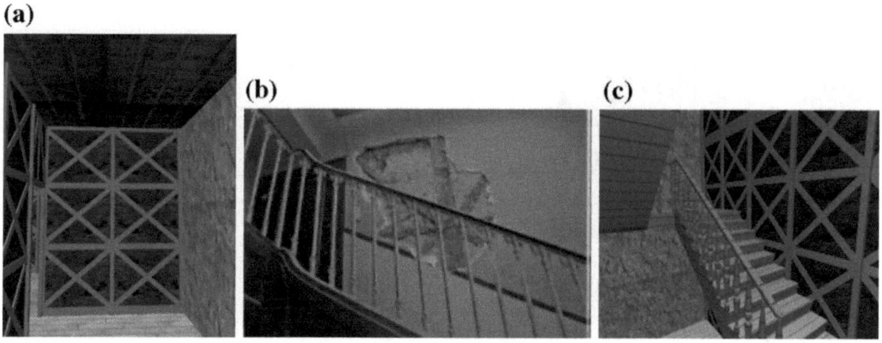

Fig. 5 **a** *Frontal* walls; **b**, **c** stairs in *Pombalino* buildings

The ground floor of the building consists of solid stonewalls and piers linked by a system of masonry arches. In more elaborate cases, thick-groined vaults spanned between the arches are used to protect the upper floors from the spread of any fire that might start at ground floor level. The *Pombalino* buildings are characterized by the design of a mixed timber-masonry structure (above the ground floor), known as *gaiola pombalina* (cage) intended to withstand the horizontal seismic loads [13, 14]. The *gaiola* is composed of traditional timber floors and improved mixed timber-masonry shear walls (*frontal* walls). They are made of a wooden truss system filled with a weak masonry in the empty spaces (Fig. 5a). Some interior walls (called *tabiques*), used only as partition walls, made of one or two sets of boards or small wood bars, are thinner and have much less resistance to horizontal loads than *frontal* walls. The façade and side (gable) walls are made of stone and rubble stone masonry with air lime mortar (with poor strength capacity) with decreased thickness in height. The staircase on the ground floor was made of masonry (like the other structural elements), while on the upper floors it was made of wood and supported on timber truss walls (Fig. 5). The staircase also has an important contribution to the *gaiola* structure, working as a resistant core throughout the building height, providing interior support to the floor beams.

Case Study

In order to assess the seismic performance of the *Pombalino* buildings, one representative building was chosen and analyzed. The selection of the building was based on previous works carried out by Meireles et al. [14] where the seismic behavior of a *Pombalino* building based on pushover analyses, before and after the application retrofitting schemes, was studied. This existing building is located in the downtown of Lisbon (marked in Fig. 1a) and it was subjected of some structural modifications, with respect to the original layout. As it is common in the *Pombalino* buildings of downtown, in this particular case one floor has been added to the original layout of

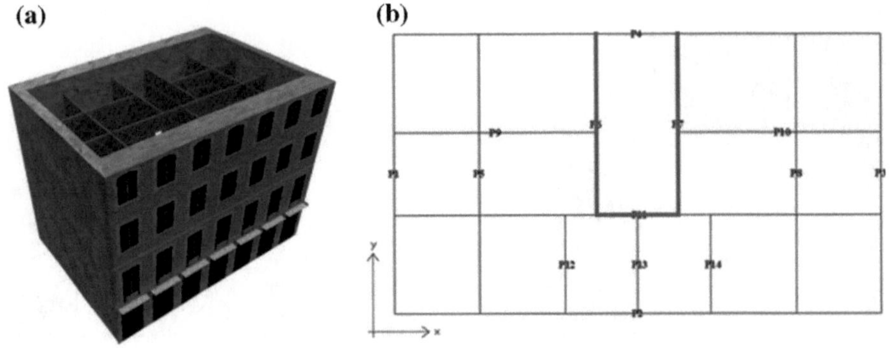

Fig. 6 **a** 3D view of the *Pombalino* building; **b** strengthening solution: Introduction of RC walls

4 floors plus roof, making a total number of 5 floors plus attic. In the present study, as it is intended to be representative of a typical *Pombalino* building, only 4 floors plus roof was considered in the layout. The building under study has a rectangular plane shape with 18 m × 11 m and a total height of 15 m (Fig. 6). The ground floor has 5 internal piers with dimensions of 0.7 × 0.7 m^2. There are stairs in the middle of the building facing towards the back façade. Only on the ground floor stairs were made of brick masonry with the thickness of 0.24 m, whereas at the upper floors the staircases are "frontal walls". On the ground floor, the front and back façade as well as the interior walls are made of stone masonry with decreased thickness up the height, whereas the front and back façades of the upper floors including the gable walls are constituted of rubble masonry. Two alignments of *frontal* walls parallel to the façades and five alignments (including the staircase) of *frontal* walls parallel to the gable walls were observed. Connecting the *frontal* walls there are openings (doors) of 0.8 m wide. The actions considered on the structure are the self-weight loads given by the weights of the roof, floors, ceilings, partition walls and *frontal* walls themselves combined with the live loads respectively given by the Eurocode 1 [15].

The mechanical masonry parameters in this study were defined based on the average values proposed in Italian code [16] for appropriate masonry types and by experimental campaign [5]. All adopted parameters, as well as gravity and live loads considered in the case study are presented in the Table 1. Regarding the *frontal* walls, a comprehensive experimental study for the characterization of the hysteretic behavior of these walls was performed [7] based on static cyclic shear tests with imposed displacements. These experimental results were after used to calibrate a macro-element model developed and implemented in Tremuri program, and considered in this work. In order to assess the seismic global response, the modelling of the *Pombalino* building was performed by means of the equivalent frame method based on the solutions adopted in Tremuri software [17]. Thus, each wall is discretized by a set of masonry panels (*piers* and *spandrels*), in which the

Table 1 Mechanical parameters

Mechanical and geometrical properties	Thickness (m)	Young modulus (GPa)	Shear modulus G (GPa)	Comp. strength f_m (MPa)	Shear strength τ_0 (MPa)	Specific weight γ (kN/m^3)	Gravity loads G (variable loads Q) (kN/m^2)
Regular stone masonry	0.90	2.80[a]	0.86[a]	7	0.105	22	Floors 0.7 (2.0)
Rubble stone masonry	0.85–0.75	1.23	0.41	2.5	0.043	20	Staircase 0.7 (4.0)
Solid clay brick masonry	0.24	1.5[a]	0.5[a]	3.2	0.076	18	Roof 1.4 (2.0)

[a]cracked stiffness assumed, 50 % of the value used

non-linear response is concentrated, connected by a rigid area (*nodes*). More detailed about modeling can be found in [8].

Strengthening solution Pushover analyses performed for the basic configuration are discussed in detail in [8] and the main conclusions are: stiffness and strength is much higher in the direction of the gabble walls (Y direction) than in the direction of the main façades (X direction), due to the fact that there are no openings in two perimeter walls of the Y direction. On the other hand, the ductility of the system is much higher on the X direction and is practically non-existing in the Y direction. Due to the configuration of ground floor, a soft story failure mechanism has been stressed. Furthermore, due to the fact that the safety requirements for the ultimate state were not fulfilled for any of two directions, strengthening technique was proposed in order to improve the seismic behavior of the building.

Namely, taking into account the above mentioned damage pattern which appear in the building (soft story mechanism), the strengthening solution proposed and presented in the paper is the addition of the RC shear walls around the staircase (Fig. 6b), in longitudinal and transverse direction in the entire height of the building, keeping the in-plane stiffening of the floors. These walls were modelled with the thickness of 0.25 m, fulfilling the minimum thickness required by EC8 [18]. Reinforcement in the walls was adopted to respect the minimum (R_{min}) and maximum (R_{max}) requirements of longitudinal reinforcement recommended by EC2 [19] and EC8 [18]. Similar work was done in [20, 21]. Nonlinear static (pushover) analysis and N2 method were performed for both directions and for two load patterns, as proposed in EC8 [18]: (i) uniform, proportional to the mass; and (ii) pseudo-triangular, proportional to the product between the mass and height. In order to have a better comparison of the benefits of the strengthening technique, pushover curves in function of the average displacement (d) of the nodes located at the roof level and the base shear force (V_b) were plotted and presented in Fig. 7. The analyses were stopped for 20 % decay of the maximum V_b (Fig. 7) following the criterion proposed in [16, 18]. This condition was assumed as reference in order to define the ultimate displacement (d_u) capacity of the structure. Comparing the obtained results between the original and reinforced building (Fig. 7), can be seen that this strengthen solution increase effectively the stiffness and strength of the building, especially for the Y direction, as the area of the RC walls is more representative in this Y direction.

In Table 2 values for the maximum load F_y, ultimate and target displacements (d_u and d_t) are presented. The latter were obtained by means N2 method and for the response spectrum that represents the seismic action in Lisbon. Comparing the structure in original and strengthened stage in terms of d_u a huge improvement can be noticed, whereas the d_t values show some reduction for strengthened structure for both, minimum and maximum reinforcement. It is clear from the results obtained that the original building does not withstands the seismic demand defined on the Eurocode 8 for Lisbon as d_t is significantly smaller that d_u, for both directions and both distribution of load.

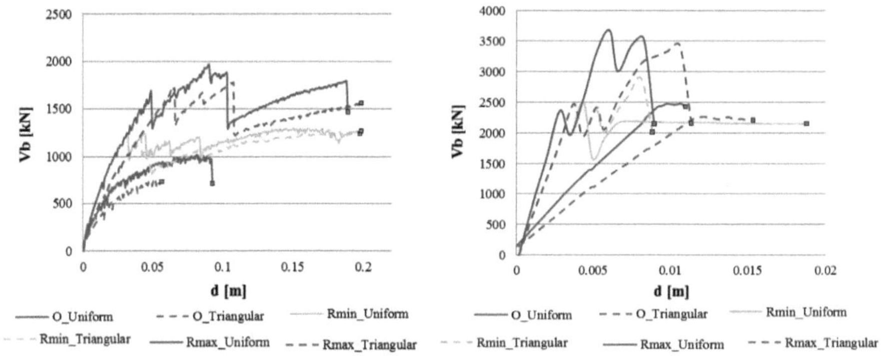

Fig. 7 Pushover curves: X (*left*) and Y (*right*) direction

Table 2 Values for maximum load, ultimate and target displacement for analyzed building

Load pattern		O	R_{max}	Δ O-R_{max} (%)	R_{min}	Δ O-R_{min} (%)
X Uniform	Fy (kN)	882.91	1670.16	89.2	1193.19	35.1
	du (m)	**0.093**	**0.189**	103.2	**0.198**	113.7
	dt (m)	0.09881	0.095	−4.0	0.081	−17.7
X Triangular	Fy (kN)	646.63	1507.42	133.1	1131.36	75.0
	du (m)	**0.057**	**0.199**	247.9	**0.198**	246.6
	dt (m)	0.110	0.107	−2.8	0.098	−11.2
Y Uniform	Fy (kN)	2464.82	3651.99	49.1	2135.63	−12.8
	du (m)	**0.011**	**0.009**	−18.4	**0.019**	70.8
	dt (m)	0.028	0.011	−59.9	0.016	−41.5
Y Triangular	Fy (kN)	1934.51	2765.81	43.0	2535.64	31.1
	du (m)	**0.015**	**0.011**	−25.9	**0.009**	−42.4
	dt (m)	0.036	0.015	−58.5	0.016	−54.3

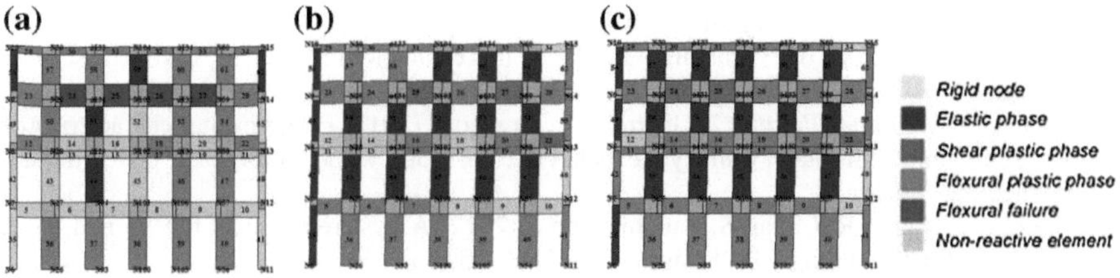

Fig. 8 Damage pattern for front façade, triangular load, X direction: **a** Original; **b** R_{max}; **c** R_{min}

In Fig. 8 damage patterns, corresponding to d_u, for front façade and triangular load pattern are presented. For the original building the damage is less well distributed throughout the structure (most of the beams of the third floor collapse), while for the strengthen configurations it is clear the existence of a greater number of elements in the nonlinear phase. Damage pattern is characterized by flexural failure of spandrel beams and piers. This type of behavior derives from the presence of slender piers and from the moderate coupling provided by spandrels.

Conclusions

This work describes the main features of the *Pombalino* buildings, and a study to evaluate the seismic resistance of a typical *Pombalino* building was performed. It was considered the *Pombalino* building as-built and including different proposed strengthening techniques. The one considering the floors stiffener in-plan and the addition of RC walls around the staircase is herein presented. This strengthening solution leads to an improvement in terms of strength, stiffness and ductility, thus leading to the increase of the seismic resistance of the building. This work is only a preliminary study, essential to propose effective strengthen solutions, economic feasibly, contributing for the preservation of *Pombalino* buildings; in fact, instead of RC walls could be considered alternatively wood or steel walls around the staircase.

References

1. LNEC (2005) Evolução das Tipologias Construtivas em Portugal. Núcleo de Engenharia Sísmica e Dinâmica de Estruturas, Laboratório Nacional de Engenharia Civil (LNEC), Lisbon, Portugal. Available at http://www.lnec.pt/LNEC/NE/NESDE (in Portuguese)
2. Teixeira MJ (2010) Rehabilitation of pombalino buildings. Experimental analysis of frontal walls. M.Sc dissertation, IST (in Portuguese)
3. Carmo L (2013) Wood connections in historic structures: study on Pombalino cage. M.Sc. dissertation, University of Tras-os-Montes e Alto Douro, Portugal (in Portuguese)
4. Vasconcelos G, Poletti E, Salavessa E, Lourenço P, Pilaon P (2013) In-plane shear behavior of traditional timber walls. Eng Struct 1028–1048
5. Santos S (1997) Tests of "Pombalino" walls. Nota Técnica Mº15/97-NCE, LNEC, Lisbon (in Portuguese)
6. Gonçalves A, Gomes Ferreira J, Guerreiro L, Branco F (2012) Experimental characterization of "Pombalino" "frontal" wall cyclic behavior. In: 15th world conference on earthquake engineering, Lisbon
7. Meireles H, Bento R, Cattari S, Lagomarsino S (2012) A hysteretic model for "frontal" walls in "Pombalino" buildings. Bull Earthq Eng 1481–1502
8. Meireles HA (2012) Seismic Vulnerability of "Pombalino" buildings. Ph.D. Thesis, IST, Universidade de Lisboa
9. Sequeira A (1999) Characterization and assessment of the market related to maintenance and conservation of the architectural heritage. Report GEcORPA—Grémio do Património, Lisbon, Portugal (in Portuguese)
10. Ildefonso S, Falcão AP, Rua H (2014) An application of CIM Models in Urban design and Land management: the case of Lisbon city. In: Proceedings of BIM international conference: challenges to overcome, Lisboa

11. Appleton J (2003) *Reabilitação de Edifícios Antigos. Patologias e Técnicas de Intervenção.* Orion Edition, 1st Edition, Lisbon, Portugal (in Portuguese)
12. Silva VC (2007) Reabilitação Estrutural de Edifícios Antigos. Alvenaria e Madeira. Técnicas Pouco Intrusivas. Argumentum, Grémio das Empresas de Conservação e Restauro do Património Arquitectónico (GECoRPA), 2nd Edn. Lisbon, Portugal (in Portuguese)
13. Lopes M, Meireles H, Cattari S, Bento R, Lagomarsino S (2013) Pombalino constructions: description and seismic assessment. In: Costa A, Guedes J, Varum H (eds) Structural rehabilitation of old buildings, vol 2. Springer, pp 187–234
14. Meireles H, Bento R, Cattari S, Lagomarsino S (2014) Seismic assessment and retrofitting of "Pombalino" buildings by pushover analyses. Earthq Struct 57–82. doi:10.12989/eas.2014.7.1.057
15. European Committee for Standardization (CEN) (2002) Eurocode 1: Actions on structures—part 1-1: general actions—densities, self-weight, imposed loads for buildings (EC1)
16. Italian Code for Structural Design, (Norme Tecniche per le Costruzioni—NTC) (2008) D.M. 14/1/2008, Official Bulletin N° 29 of February 4, 2008 (in Italian)
17. Lagomarsino S, Penna A, Galasco A, Cattari S (2013) TREMURI program: an equivalent frame model for the nonlinear seismic analysis of masonry buildings. Eng Struct 1787–1799. doi:10.1016/j.engstruct.2013.08.002
18. European Committee for Standardization (CEN) (2004) Eurocode 8: Design of structures for earthquake resistance—part 1: general rules, seismic actions and rules for buildings (EC8-1)
19. European Committee for Standardization (CEN) (2004) Eurocode 2: Design of concrete structures—part 1-1: general rules and rules for buildings (EC2-1)
20. Costa A, Oliveira CS, Arêde A, Varum H, Costa A, Rodrigues H, Neves F, Furtado A (2013) Reforço estrutural de edifícios: casos concretos e custos. Seminário Vulnerabilidade Sísmica no Contexto da Reabilitação Urbana, Ordem dos Arquitetos, Lisboa
21. Branco M, Guerreiro L (2006) Comparação de métodos de reforço sísmico de edifícios "Gaioleiros", JPEE2006

Numerical Approaches for the Analysis of Timber Frame Walls

Elisa Poletti, Graça Vasconcelos, Paulo B. Lourenço and Maria Pia Ciocci

Abstract Timber frame structures constitute an important cultural heritage of many countries, since they represent a typical anti-seismic construction adopted worldwide and are worth preserving. Recent experimental results have shown that the seismic response of traditional timber frame walls varies greatly with the type of infill, the type of connection and wall geometry. Two simplified 2D numerical models were created and in-plane monotonic tests were simulated and calibrated using recent experimental results. Material non-linearity was considered for all materials. Considering the numerous advantages of the use of simplified models, parametric analyses were carried out in order to evaluate the influence of parameters such as type of infill, infill-frame connectivity, timber type, connection type. The greater influence on the behaviour of the walls is given by the quality of the joints as well as the connectivity between timber and masonry, behaviour that is in accordance with experimental results. Additionally, the geometry of the wall influences its stiffness and load capacity.

Keywords Timber frame · Traditional connections · Numerical analysis · Calibration · Parametric analysis

E. Poletti (✉) · G. Vasconcelos · P.B. Lourenço · M.P. Ciocci
Institute for Sustainability and Innovation in Structural Engineering (ISISE), University of Minho, Guimarães, Portugal
e-mail: elisapoletti@gmail.com

G. Vasconcelos
e-mail: graca@civil.uminho.pt

P.B. Lourenço
e-mail: pbl@civil.uminho.pt

M.P. Ciocci
e-mail: mariapiaciocci@gmail.com

H. Cruz et al. (eds.), *Historical Earthquake-Resistant Timber Framing in the Mediterranean Area*, Lecture Notes in Civil Engineering,
DOI 10.1007/978-3-319-39492-3_15

Introduction

Numerical modelling is an important tool for the analysis of the behaviour of structural elements when appropriately validated on experimental results and can be used to predict the response of a structure after interventions are made. A number of numerical studies analysing the seismic performance of modern timber frame shear walls exist, while only recently the attention has focused on traditional timber frame walls, which represent an important historic seismic-resistant construction typology, particularly in Mediterranean countries.

Timber frame walls have proven to be effective when subjected to seismic loads, as testified by real case studies [1] and by experimental observations [2–5]. The origin and diffusion of these structures has been thoroughly described in these works and it will not be discussed here. Nevertheless, the numerical tool would allow to complement these findings and help in the preservation of existing structures.

Different approaches can be taken when modelling this type of elements, varying from complicated 3D modelling, accurately representing connections, pegs and nails, to simplified 2D modelling, adopting linear and springs elements to represent timber members and joints.

Kouris and Kappos [6] performed non-linear numerical analyses on traditional half-timbered walls. Based on the test results of Santos [7], the authors first performed a detailed 2-D modelling of a wall, adopting plane-stress elements and considering non-linear orthotropic properties for wood, while the contacts in timber elements was simulated adopting interface elements. Infill was not included in the model. The model was able to appropriately simulate experimental results, being able to capture the gradual softening of the walls as well as their capacity. To be able to model whole buildings in a simplified way, the authors created a simplified model using beam and link elements, while non-linear axial hinges were used for the diagonals. The application of this model to a building façade proved to be satisfactory. The model was later expanded to take into consideration masonry infill by adopting a plasticity model for masonry and it can provide predictions of local quantities such as penetration of elements and relative sliding [8].

Quinn and D'Ayala [9] modelled traditional Peruvian timber frame walls (quincha) using 2-D beam elements and adopting semi-rigid spring elements to model the mortise and tenon connections of the wall. The springs were calibrated based on experimental results and using the components method and the preliminary results appear to be reasonably good.

Another approach to the modelling of timber frame walls is the use of analytical hysteretic models calibrated on experimental results to a simplified numerical wall model. Ceccotti and Sandhaas [10] performed such an analysis on a traditional timber frame wall creating a numerical model consisting of lumped masses and rotational springs, where the rigid elements represent the timber elements, while the global cyclic behaviour of the wall is given only by the springs. The springs could model both rotations and vertical uplifting. This model was then extended to model a whole building. A similar approach was used by Meireles [11] introducing an analytical hysteretic model calibrated on experimental results in the finite element software, modelling a traditional Pombalino wall representing the connections by means of spring elements. A macro-element was developed to represent the Portuguese frontal walls inside a masonry building. The hysteretic model was also implemented in an existing finite element model for timber frame structures [8] providing acceptable results. Hicyilmaz et al. [12] developed a numerical model of a dhajji dewari building, taking into account parameters such as vertical load, shortening of the diagonal bracings and removal of nails from the wall.

Numerical models are available even for traditional timber connections. Hong and Barrett [13] developed a 3-D finite element model for Japanese post and beam connections (basically a mortise and tenon joint), modelling the nails and steel plates of the connection and the crushing of timber that occurred at the mortise-tenon interface. Branco et al. [14] modelled the behaviour of traditional timber roof connections using nonlinear moment-rotation laws and hysteretic rules in order to represent the cyclic response of the joint. Considering modern connections, Tannert et al. [15] studied the strength prediction of rounded dovetail connections considering size effects.

In this paper, two distinct numerical models have been validated to simulate the mechanical behaviour observed in the experimental results of cyclic tests on traditional timber frame walls. A parametric study is then presented considering some variables such as post continuity, type of infill and wall geometry.

Brief Overview of Experimental Observations

To study the seismic response of traditional Portuguese timber frame walls, quasi-static in-plane cyclic tests were performed on real scale specimens [2]. Half lap joints were used for the connections between the elements of the main frame, while the diagonals were simply nailed to the frame (Fig. 1). In general the walls showed a good capacity and ductility. Results greatly varied depending on the level of vertical pre-compression and on the presence of infill, which could alter the response of the wall from a shear one to a flexural one (Fig. 1). Damages were concentrated at the connections and uplifting of the non-continuous lower connections was severe particularly for infill walls. For a full description of the experimental results, see [2].

Numerical Modelling of Walls

In order to simulate numerically the results obtained experimentally, the walls were modelled adopting 2-D finite elements, specifically plane stress elements and beam elements. The models were calibrated based on the experimental envelope curves.

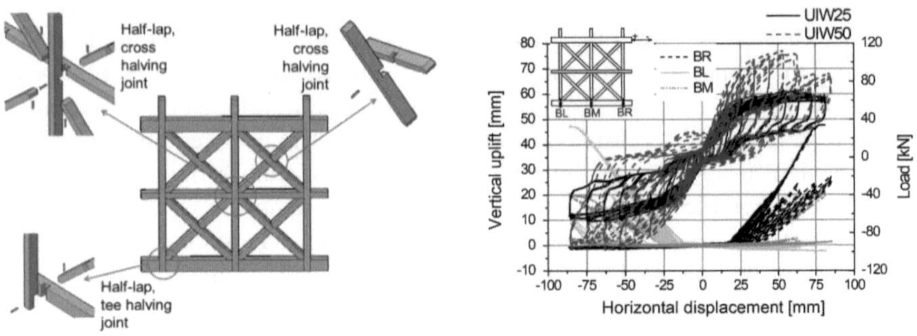

Fig. 1 Wall geometry (*left*) and flexural behaviour of timber frame wall with infill (*right*) [2]

2-D model with plane stress elements. A first model of the wall was created adopting plane stress elements and interface elements for the connections using the finite element code DIANA [16]. The elements used for timber were eight-node quadrilateral isoparametric plane stress element CQ16M, whereas the interfaces used were line based elements CL12I with 3 + 3 nodes.

To take into account the uplift observed experimentally [2], interface elements were introduced between the posts and the bottom beam. The horizontal interface elements should simulate the bottom half-lap joint. Moreover, interface elements were introduced between the diagonals and the main frame to represent the nailed joints (Fig. 2). These joints were crucial during the experimental tests, particularly when infill is not present, as they would penetrate the main frame. Moreover, using a rigid connection would have stiffened the wall in an unrealistic way.

To simplify the calibration of the model, it was assumed that the half-lap connections of the diagonals and of the middle and top beam of the wall were rigid. To simulate the crushing observed in these connections, mainly at the central post, different materials properties were considered in this region. Infill was modelled adopting six-node triangular isoparametric plane stress elements (CT12M). Masonry infill was assumed as a continuous and isotropic material. Interface

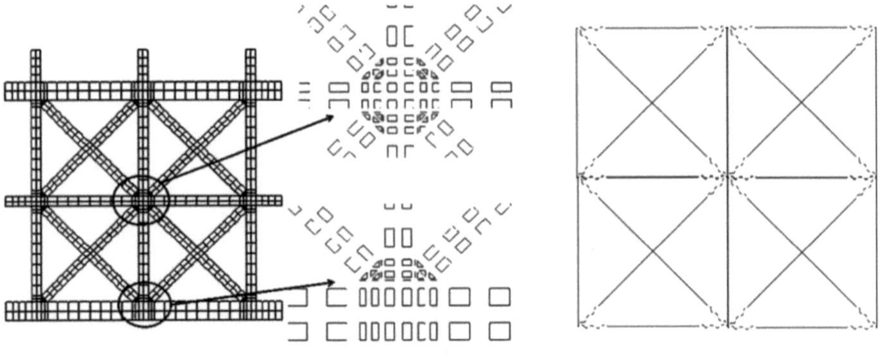

Fig. 2 Model geometry: plane stress elements (*left*) and beam elements (*right*)

elements (CL12I) were introduced between the masonry infill and the timber frame to represent the connectivity between the two materials. Boundary conditions and loading protocols adopted reproduced the ones of the experimental tests [2]; note that two vertical load levels were considered, namely 25 kN/post and 50 kN/post. The solution algorithm used for the non-linear analyses performed was the full Newton-Raphson method and the convergence criterion was energy based.

Non-linear properties were assumed for timber and for the timber-to-timber and infill-to-timber interfaces. The orthotropic behaviour of timber is taken into account by using the Rankine-Hill Anisotropic model available in DIANA [16]. To model the anisotropic behaviour of the material, the model considers a Hill yield criterion in compression and a Rankine yield criterion in tension. For interface elements, a Coulomb Friction model was used, combining the friction criterion with a gap criterion, in order to allow the detachment of the diagonals, and assuming a Mode-II constant shear retention model after gap appearance [17]. The material properties were obtained from the results of the material characterization carried out in [2]. The additional information necessary was obtained using the JCSS probabilistic model code [18] (Table 1). For the half-lap joints which were not directly modelled, timber properties were reduced by 40 %. For interface elements representing the bottom joints, a normal stiffness of 1.25 N/mm^3 was assumed, while tangential stiffness was 5 N/mm^3. For the interfaces at the diagonals, a lower cohesion and friction angle were assumed in relation to half-lap joints, since no interlocking was present. Normal and tangential stiffness were assumed as 1 N/mm^3. Infill properties were obtained by Poletti and Vasconcelos [2], while infill-to-timber interfaces were assumed as weak, with a normal and tangential stiffness of 1 and 0.5 N/mm^3 respectively.

Assuming such values of material properties, the numerical model was calibrated based on experimental results with no infill material; results were compared to the envelope curves obtained from the experimental quasi-static in plane cyclic tests. It is observed that a reasonable fitting between experimental and numerical results was obtained both in terms of stiffness, lateral resistance and maximum displacement (Fig. 3). In terms of initial stiffness, the numerical model exhibits a value of

Table 1 Material properties used for timber [18]

$f_{t,0}$	25 MPa	G_v	700 MPa	γ	9
$f_{t,90}$	9 MPa	ν	0.3	$G_{ft,0}$	70 Nmm/mm^2
E_0	11,000 MPa	ρ	590 kg/m^3	$G_{ft,90}$	50 Nmm/mm^2
E_{90}	900 MPa	α_τ	1	$G_{fc,0}$	130 Nmm/mm^2
$f_{c,0}$	35 MPa	α_h	1	$G_{fc,90}$	70 Nmm/mm^2
$f_{c,90}$	5 MPa	β	−1	κ_p	0.001

ν is Poisson's ratio; G_v is the shear modulus; γ is the material density; E_0 and E_{90} are the moduli of elasticity parallel and perpendicular to the grain; $f_{c,0}$, $f_{c,90}$, $f_{t,0}$ and $f_{t,90}$ are the compressive and tensile strength parallel and perpendicular to the grain; $G_{fc,0}$; $G_{fc,90}$; $G_{ft,0}$; $G_{ft,90}$ are the corresponding fracture energies in compression and tension parallel and perpendicular to the grain; α_τ is a factor that determines the shear stress contribution to tensile failure; α_h, is a factor that relates the equivalent length of the element to the area of the finite element; factor β couples the normal compressive stresses; factor γ controls shear stress contribution to compressive failure; κ_p is the equivalent plastic strain corresponding to the peak compressive stress

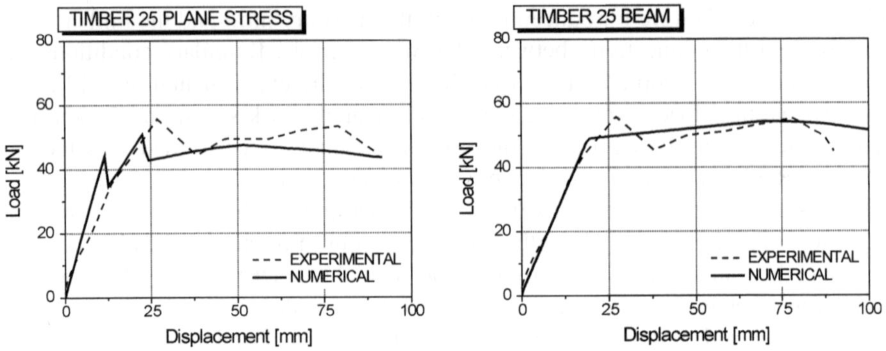

Fig. 3 Model calibration: plane stress elements (*left*) and beam elements (*right*) model results

3.8 kN/mm for both load levels, which is in good accordance with the experimental results for the higher vertical load, with an increase of only 6 %. For the lower vertical load the error is higher, with an increase of stiffness of approximately 45 %. In terms of maximum load, both load levels exhibited a good approximation, with a decrease of 9 and 15 % respectively for the lower and higher vertical load [17]. In terms of deformation patterns and damages, the uplift of the lateral connection for the lower vertical load level reached 65 mm, a value comparable with what observed experimentally. In terms of stresses distribution, the models presented important stress concentrations at the central joints, with detachment of the diagonals, replicating what was observed experimentally. Similarly, a good approximation was obtained for the model with infill.

2-D model with beam elements. Considering that the plane stress model still presents some complications, particularly in terms of material properties, since non-linearities are assumed for all elements, and having as a goal the modelling of a whole structure, for which such a model would be computationally expensive, a simplified 2-D model was created using beam elements on the finite element software SAP2000 [19]. Replicating the real geometry of the walls (Fig. 1), each timber element of the main frame was modelled using a frame element with a linear behaviour. No infill was considered and the linear elastic timber properties of Table 1 were assumed for this model. All non-linearities were concentrated in the semi–rigid connections that were modelled using link elements with multi–linear elasticity property. Each link element was assumed to be composed of six separate springs, one for each deformational degree of freedom. In particular, two link elements were considered: link 1 and link 2; link 1 was located at each nailed connection between the main frame and the diagonal elements, while link 2 at each half–lap connection between the elements of the main frame. Because of the planar nature of the problem, only the axial u1, shear u2 and rotational r3 stiffness were considered for the element deformational DOFs. While the axial and the shear stiffness were specified for link 1, the rotational one was defined for link 2. Non-linear force-deformation relationships were used for all link elements (Fig. 4) [20]. For link 2, a tri-linear behaviour was assumed based on in-plane cyclic tests carried out

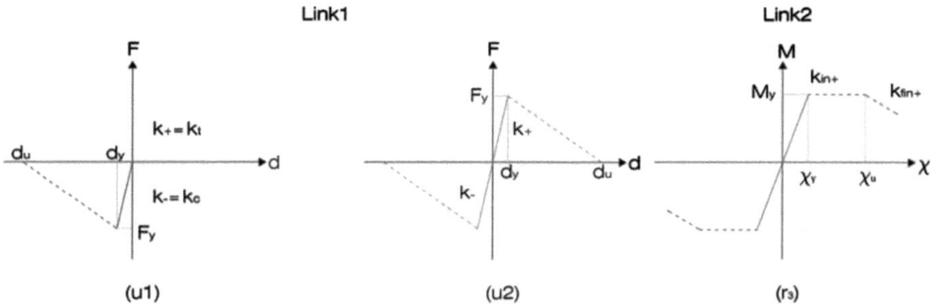

Fig. 4 Force-displacement and moment-curvature relationships assumed for link [20]

on the half-lap connections of the wall [21]. In particular, the mean values of the initial and the final rotational stiffness were calculated as k_{in} = 171 kNm/rad and k_{fin} = 47 kNm/rad, respectively. For link 1, after reaching the maximum capacity, the load progressively decreased until a null value until the ultimate displacement. The value of 10,000 kN/m obtained by applying the inverse fitting method in the elastic analysis was assumed as the initial stiffness of the axial and shear stiffness. The capacity of the connection was calculated by considering the compressive strength perpendicular to the grain and the contact areas with the presence of initial clearances. The ultimate displacement was assumed to be the same of that of half–lap connections: when the failure occurred for the half–lap connection, the connection between the diagonal and the main frame no longer worked [20].

Comparing the numerical results thus obtained with the experimental envelope curve of the cyclic tests performed on timber frame walls [2] (Fig. 3), a good approximation was obtained with this simplified model. The consideration of all the connections of the wall and the calibration of the model with the inverse fitting method allowed for a better prediction of the stiffness of the wall. Moreover, the capacity and softening response of the timber frame wall were satisfactory predicted.

Parametric analyses. Parametric analyses were performed for the first model to better understand the influence of different parameters on the lateral response of the wall. Parametric analyses are an important tool in numerical modelling as they allow to study the influence of different parameters on a calibrated model, a feat which would otherwise require extensive experimental testing.

During the experimental campaign [2], the partially rocking response of the walls was caused by the uplifting of the bottom joints, due to the fact that the half-lap was not complete. To better capture the importance of post continuity, the normal stiffness of the interface elements linking the bottom beam to the posts was varied, considering values ranging from 25 to 500 % of the calibrated value. The tangential stiffness was kept constant, as it does not influence the uplifting of the post.

Both initial stiffness and maximum load are influenced by the post continuity. For an increase of normal stiffness k_n higher than 500 % the parameter loses its influence, as the connection becomes rigid (Fig. 5). For high normal stiffness values, the tensile stresses in the joints of the central beam reached the maximum

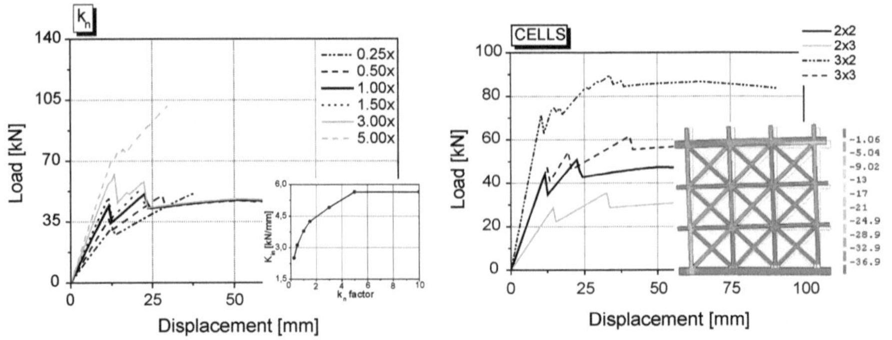

Fig. 5 Parametric analyses: influence of post continuity (*left*) and of length-to-height ratio (*right*)

tensile strength value assumed for those elements, as the response of the wall was predominantly in shear. In terms of initial stiffness K_{in}, it increases with increasing values of normal stiffness of the connection until a constant value is reached, after which other parameters, such as timber strength and connection to the diagonals should have more relevance [17].

Other parameters influencing the behaviour of the wall are represented by material properties, namely timber and infill properties. They influence mainly the capacity and the post peak behaviour of the wall, while timber-to-timber and infill-to-timber connection properties affect the wall stiffness [17]. The same was observed in the analyses with the simplified beam model [20].

Another parameter to be taken into consideration is the wall geometry. Timber frame walls have a great variability in terms of geometry, namely height to length ratio of the wall, height to length ratio of a single cell, sectional dimensions of timber elements, positioning of diagonals. In this study, only the first issue will be discussed, keeping as a future development the analysis of the other parameters. Three wall geometries were analysed, namely a wall with three cells along the length and two cells along the height (3 × 2), a wall with two cells along the length and three cells along the height (2 × 3) and a wall with three cells both along the length and along the height (3 × 3).

The general behaviour of the models in terms of stress paths and deformation was similar to what observed for the original model (2 × 2 configuration) (Fig. 5), with similar stresses concentrations in the connections and at the compressed post. The response of the walls changed in terms of initial stiffness and load capacity (Fig. 5). For the same height, a higher width of the wall, corresponding to a lower height to length ratio, leads to higher initial stiffness and load capacity. For example, model 3 × 2 has an initial stiffness 80 % higher than that of model 2 × 2, while model 3 × 3 has an increase of 92 % on the stiffness when compared to model 2 × 3. In terms of maximum load, the increase is of 76 % for two cells in height and of 98 % for three cells in height. On the contrary, for the same width, a higher height (i.e. an higher height to length ratio) leads to smaller values of initial stiffness and load capacity, as the higher slenderness of the wall leads to a lower resistance.

Further Developments

The numerical analyses carried out in this work are only a small part of what could be analysed when studying traditional timber frame walls. Among the future developments that could be mentioned, the most interesting ones are: (1) the implementation of a hysteretic model derived from the available experimental results [17]; (2) the simulation of the seismic behaviour of a whole timber frame structures through the execution of pushover and dynamic analyses on a simplified model. Additionally, the beam model can be modified in order to capture the rocking response of the wall, as in its current state the bottom connections cannot uplift. Other studies that could be considered are the assumption of different types of timber frame walls, with different geometries, configurations and materials than the Portuguese ones assumed in this study.

Conclusions

In order to numerically represent a traditional timber frame wall, two 2-D models were created taking into account material and connection non-linearities, one adopting plane stress elements and one beam elements. Both models were calibrated on experimental results. For the plane stress model, a good correlation was found between the experimental results and the calibrated numerical model, for both timber frame and infill walls. Material non-linearity was considered for both timber and masonry as well as for timber-to-timber and timber-to-masonry interfaces.

The beam model was able to capture more accurately the stiffness of the wall, thanks to a more complete representation of the wall connections. The capacity was overpredicted and further parametric studies are necessary for this model.

From the parametric analyses it was observed that the greater influence on the behaviour of the wall is given by the continuity of the post as well as the connectivity between timber and masonry. This seems in accordance with experimental results as the response of the wall was governed by its connections. Additionally, the geometry of the wall influences its stiffness and load capacity, as a higher height to length ratio led to lower values of these parameters.

References

1. Tsakanika-Theohari E, Mouzakis H (2010) A post-Byzantine mansion in Athens. Restoration project of the timber structural elements. In: Proceedings of WCTE World conference on timber engineering. Riva Del Garda, Trento, Italy, 20–24 June 2010
2. Poletti E, Vasconcelos G (2015) Seismic behaviour of traditional timber frame walls: experimental results on unreinforced walls. Bull Earthq Eng 13:885–916

3. Ruggieri N, Tampone G, Zinno R (2015) In-plane vs out-of-plane "behaviour" of an italian timber framed system: the Borbone constructive system. Hist Anal Exp Eval Int J Architectural Heritage. doi:10.1080/15583058.2015.1041189
4. Vieux-Champagne F, Sieffert Y, Grange S, Polastri A, Ceccoti A, Daudeville L (2014) Experimental analysis of seismic resistance of timber-framed structures with stones and earth infill. Eng Struct 69:102–115
5. Dutu A, Sakata H, Yamazaki Y, ShindoT (2015) In-plane behavior of timber frames with masonry infills under static cyclic loading. J Struct Eng. doi:10.1061/(ASCE)ST.1943-541X.0001405, 04015140
6. Kouris L, Kappos AJ (2012) Detailed and simplified non-linear models for timber-framed masonry structures. Journal of Cultural Heritage 13(1):47–58
7. Santos S (1997) Tests of Pombalino walls. Nota Técnica No 15/97-NCE, LNEC, Lisbon (in Portuguese)
8. Kouris LAS, Meireles H, Bento R, Kappos AJ (2014) Simple and complex modelling of timber-framed masonry walls in Pombalino buildings. Bull Earthq Eng 12:1777–1803
9. Quinn N, D'Ayala D (2013) Assessment of the realistic stiffness and capacity of the connections in quincha frames to develop numerical models. Adv Mater Res 778:526–533
10. Ceccotti A, Sandhaas C (2010) A proposal for a standard procedure to establish the seismic behaviour factor q of timber buildings. In: Proceedings of WCTE world conference on timber engineering. Riva Del Garda, Trento, Italy, 20–24 June 2010
11. Meireles H, Bento R, Cattari S, Lagomarsino S (2012) A hysteretic model for "frontal" walls in Pombalino buildings. Bull Earthq Eng 10(5):1481–1502
12. Hicyilmaz KMO, Wilcock T, Izatt C, da Silva J, Langenbach R (2012) Seismic performance of Dhajji Dewari. In: Proceedings of the 15th world conference on earthquake engineering (15WCEE). Lisbon, 24–28 Sept
13. Hong JP, Barrett D (2008) Three-dimensional finite solid element model for Japanese post and beam connection. In: Proceeding of 10th world conference on timber engineering (WCTE). Miyazaki, Japan, 2–5 June
14. Branco J, Cruz P, Piazza M, Varum H (2006) Modelling of timber joints in traditional structures. In: Proceedings of international workshop on "earthquake engineering on timber structures". Coimbra, Portugal
15. Tannert T, Lam F, Vallée T (2010) Strength prediction for rounded dovetail connections considering size effects. J Eng Mech 136(3):358–366
16. DIANA (2009) Diana Finite Element Code, Version 9.4.3. TNO Building and Construction Research, The Netherlands
17. Poletti E (2013) Characterization of the seismic behaviour of traditional timber frame walls, Ph.D. Thesis, University of Minho, Guimarães, Portugal
18. JCSS (2006) Joint committee on structural safety (JCSS) probabilistic model code—part 3: Resistance models—3.5 properties of timber
19. CSI (2014) Analysis reference manual for SAP 2000, ETABS, SAFE and CSiBrisge. Berkeley, California, USA: Computers & Structures
20. Ciocci MP (2015) Structural analysis of the timber structure of Ica Cathedral, Peru, M.Sc. Thesis, structural analysis and historical constructions (SAHC) international master course. Guimarães, Portugal
21. Poletti E, Vasconcelos G, Branco JM (2015) Experimental analysis of the cyclic response of traditional timber joints and their influence on the seismic capacity of timber frame structures. In: 3rd international conference on structural health assessment of timber structures. Wroclaw—Poland, 9–11 Sept 2015

Seismic Vulnerability of Borbone Masonry Reinforced with Timber Frames

Nicola Ruggieri, Carmen Sandhaas and Ario Ceccotti

Abstract After the 1783 earthquake, under the Borbone kingdom, new building regulations were introduced in the Calabria region (Italy) with the main scope of reducing the seismic vulnerability of the building asset in these highly seismic areas. The Borbone system consisted of masonry walls reinforced with timber frames and prescribed a symmetrical development in plan and elevation as well as height limitations. These measures aimed at increasing the seismic resistance of buildings and hence, in general, at reducing the socio-economic impact of future seismic events. In order to maintain, repair or retrofit these buildings, it is indispensable not only to understand their qualitative performance under an earthquake loading, but also to gain quantitative information on stiffness, load bearing and energy dissipation capacity and failure modes. This paper presents first investigations on the seismic vulnerability of simple Borbone structures where modelling assumptions such regularity in plan and elevation hold and only one shear wall system was used. Re-built shear wall specimens mirroring as exactly as possible a wall of the Palazzo Vescovile in Mileto (Italy), built according to the Borbone *Istruzioni* after the 1783 earthquake, were tested under quasi-static cyclic in-plane loading. By means of these tests, lateral properties and damage patterns of Borbone shear walls under horizontal loading were assessed and, even if limited to a single panel, earthquake behaviour patterns were evaluated. The test results on the shear walls in terms of hysteretic loops were used to develop a non-linear dynamic lumped mass model of a case study Borbone building which was subjected to various earthquake accelerograms. The single earthquake's peak ground acceleration (PGA) values were increased until a previously defined near-collapse state in

N. Ruggieri (✉)
DIBEST, Università della Calabria, Cosenza, Italy
e-mail: nicola.ruggieri@unical.it

C. Sandhaas
Karlsruhe Institute of Technology, Karlsruhe, Germany
e-mail: carmen.sandhaas@kit.edu

A. Ceccotti
IUAV, Università di Venezia, Venice, Italy
e-mail: ario@iuav.it

H. Cruz et al. (eds.), *Historical Earthquake-Resistant Timber Framing in the Mediterranean Area*, Lecture Notes in Civil Engineering,
DOI 10.1007/978-3-319-39492-3_16

terms of maximum interstorey drift of the modelled building was reached. The thus obtained ultimate values for the PGA represent the sustainable level of seismic action of the investigated Borbone structure and ranged from 0.25*g* (North American earthquake) up to 1.52*g*. The effectivity of Borbone building regulations could be shown.

Keywords Borbone system · Masonry reinforced with timber frames · Seismic engineering · Quasi-static cyclic tests · Non-linear dynamic modelling

Introduction

The town reconstructions in the Calabria region following the 1783 terrible earthquake were executed under the building regulations established by the Borbone seismic ordinance. These regulations show the scientific progress at the end of the 18th century concerning European seismic engineering. This seismic knowledge characterising the *Regno di Napoli* during the 18th century has been documented by Ruggieri [1, 2] who reports and comments many treatises of the time containing principles and technical indications for improving the building answer during an earthquake.

Building height limitation, symmetry in plan and elevation and above all proper connections between the load bearing elements represent innovative and modern rules for the best practice of the time for buildings in a zone with high seismicity. The Borbone measures concerning improved earthquake behaviour of buildings are described in the *Istruzioni per la ricostruzione dè paesi diruti della Calabria* [1].

In this paper, a two-storeys model building, applying Borbone construction criteria and technical regulations, has been developed to assess the seismic behaviour of the Borbone building prescriptions by means of non-linear dynamic analyses using the DRAIN2DX software [3]. The numerical modelling has been based on the results provided by an experimental campaign performed in the CNR-IVALSA laboratories in 2013 [4].

State of the Art of Numerical Modelling of Historical Structures

The scientific literature presents a complete lack of studies relative to the seismic vulnerability analysis of entire Borbone buildings by means of numerical approaches.

However, a preliminary mechanical interpretation relative to a simple wall module deriving from a Borbone building was conducted by Galassi et al. [5] using the software *BrikWORK.*[1] In Galassi et al., the masonry infill was modelled by rigid blocks connected with unilateral elastic contact constraints. The timber frame was modelled by using finite elements with elastic and bilateral behaviour. The contact interfaces between elements in wood and masonry blocks were represented by unilateral contact constraints which took into account friction.

Several authors investigated other typologies of timber framestructures by means of non-linear analyses using various finite element software packages. Kouris and Kappos [6] carried out a numerical analysis in ANSYS on masonry walls reinforced with timber elements found in Greek traditional edifices. Their numerical approach modelled the horizontal and vertical members through a linear-elastic beam element whilst the diagonals of the timber frame were modelled with a link element pinned at its ends and characterised by the presence of a plastic axial spring. The model was validated using cyclic loading tests on timber framed masonry, showing a good match between the results of the numerical analyses and those of the experimental proof.

Ramos and Lourenço [7] used DIANA to model traditional Pombalino buildings. Ramos and Lourenço aimed at assessing the contribution of internal Pombalino walls to the overall building behaviour under seismic actions. The numerical analyses were confirmed by means of quasi-static cyclic tests on three specimens removed from existing Pombalino buildings. The seismic analysis showed a safety factor not in compliance with the value required by the Portuguese code and a collapse induced by the out-of-plane failure of the external masonry walls.

Historical Building Regulations

In the 18th century, timber has been widely used to build in the Calabria region. In fact, wooden structures are recommended by several writers of treatises [4] to connect orthogonal walls constituting the building in order to achieve a box-type behaviour of the whole construction. One of the most important sections included in the *Istruzioni sul metodo da tenersi nella riedificazione dei paesi diruti della Calabria,* the Borbone Construction Practice Code, confirms this: «*le case si costruiranno…con ossatura di grossi travi … legati con altri travi trasversali…*».[2]

[1]*BrikWork* is a specifically updated version of a previously defined one in the context of research on the behaviour of masonry structures (Galassi, Paradiso, Tempesta.Dep. of Architecture DiDa, Section Materials and Structures, Università di Firenze).

[2]«…the buildings will be built…with trusses made of big members…connected to other transversal members…».

Furthermore, the Borbone Construction Practice Code, on which all reconstructions in the Calabria region were based, suggested to limit the overhang of eaves and balconies and a maximum permissible elevation height of new buildings of 28 *palmireggini.*[3]

The dependence of the seismic vulnerability on the building height was described by Sarti [8] who observed different earthquake damages that provoked the building collapse «*dallalorocima*».[4] On thisissue, alsoPoleni [9] states: «...*son più che le basse fabbriche, le alte soggette al pericolo del danno*...».[5] The recommendation by Mori [10], a Neapolitan engineer of the Enlightenment Age, «...*di renderetuttigliabitatori in case a pianterreno*...»,[6] represents the ultimate opinion of Borbone engineers on how to protect human life from earthquakes. Furthermore, the strong instruction to pursue a symmetric shape both in plan and elevation of the buildings is reported by many authors of the 18th century. Vivenzio [11], for example, described an anti-seismic building prototype characterised by a biaxial symmetrical development in plan, regularity in elevation and reinforcement of the walls by, in this case, double timber framing.[7]

Moreover, Neapolitan engineers were conscious that the use of timber, compared to masonry, leads to a lower seismic mass and consequently a decrease of earthquake forces.

Borbone Case Study

Relying on the principles and technical indications of the Borbone Construction Practice Code and Neapolitan engineers, a Borbone case study has been designed that is characterised by a regular development in plan and elevation and two stories, with a maximum height of about 6.50 m.

The vertical load bearing system consists of Borbone walls where the masonry with a depth of about 60 cm is reinforced with an internal timber frame with a maximum depth of around 12 cm, see Fig. 1. The timber frame, without saint Andrew crosses, is stiffened in plane by means of the masonry infill. The masonry is arranged in roughly squared stone blocks mixed with bricks and rubble stones. The horizontal elements of the timber frame are connected by means of half lap joints to the posts running from the sole plate to the wooden ring-beam. A similar layout is used to connect the tie-beam to the ring-beam, relative respectively to the floor or roof level.

[3]1 *palmo reggino* corresponds to 26.5 cm.

[4]«...from their top...».

[5]«...more than the low buildings, the high ones are prone to damage...».

[6]«...put all inhabitants into houses with ground floor only...».

[7]A structural system recommended for public buildings.

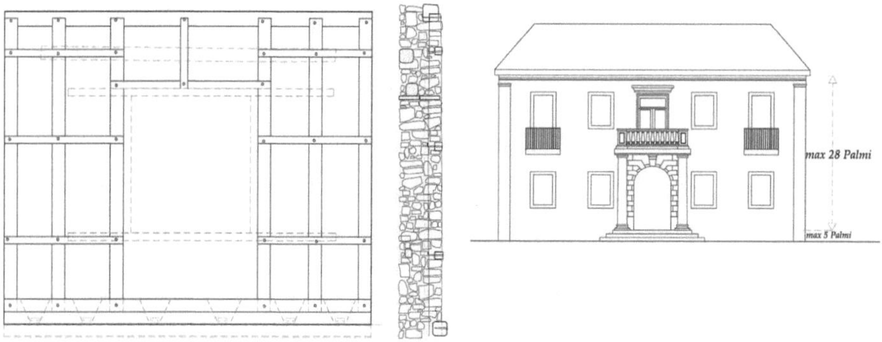

Fig. 1 Borbone shear wall front and section (*on the left*), main facade of the case study (*on the right*)

The Borbone flooring consists of timber beams with a spacing of about 80 cm and wooden boards which are connected to the beams through metallic nails. The roof structure is made of queen post trusses and purlins without any additional board layer. The single trusses are stiffened by means of wooden saint Andrew crosses, a common solution in buildings executed in the South of Calabria after the earthquake of 1783.

The used wood species for all timber elements is chestnut (*Castanasativa Mill.*), according to Article 10 of the Borbone Construction Practice Code.

Quasi-Static Cyclic Tests on Borbone Shear Walls

Re-built shear wall specimens mirroring as exactly as possible a wall of the Palazzo Vescovile in Mileto (Italy), erected according to the Borbone *Istruzioni* after the 1783 earthquake, were tested under quasi-static cyclic in-plane loading [4]. With such an experimental campaign, lateral properties and damage patterns of Borbone shear walls under horizontal loading can be assessed and, even if limited to a single panel, earthquake behaviour patterns can be evaluated. The tested models, in a 1:1 scale, were characterised by an accurate reproduction of the geometrical shape, materials and loads. The tests on the full-scale models were carried out according to the European standard UNIEN 12512:2006 up to a displacement of 80 mm.

The Mileto wall with infill masonry clearly exhibited a non-linear behaviour with high values of energy dissipation (a maximum value of 1579 kN mm related to 80 mm in pulling) and ductility, see Fig. 2 and Table 1. The observed damage pattern consisted of cracks in masonry and falling-out of stones. The tests were stopped at a displacement of 80 mm before a final serious collapse could be observed.

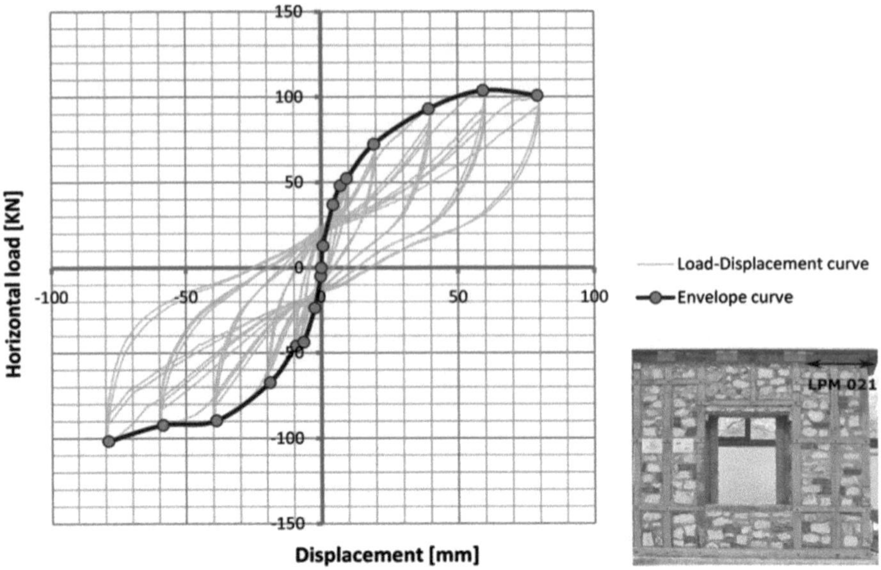

Fig. 2 Hysteretic curve in terms of horizontal load versus horizontal displacement at wall top of tested Borbone shear wall

Non-linear Dynamic Modelling

The results of the quasi-static cyclic tests on shear walls were used to develop a non-linear dynamic lumped mass model of the Borbone case study building presented Fig. 1 in on the right and which was subjected to various earthquake accelerograms. By increasing the single earthquake's peak ground acceleration (PGA) values, the seismic performance of the investigated historic structure could be evaluated.

In this paper, the proposed assessment method by Ceccotti and Sandhaas [3] has been used to assess the earthquake behaviour of the Borbone case study building.

General Approach

The basic idea of the used assessment method is to reduce the complexity of the numerical model as much as possible in order to allow for (i) controllable and transparent models and (ii) the possibility to model entire buildings and not only sub-assemblies. Based on the testing outcomes, a shear wall model was developed, its behaviour calibrated on the test results and then up scaled to be re-used in full-scale buildings. In this paper, the case study building shown in Fig. 1 on the right has been reduced to a 2D model.

Table 1 Results of quasi-static cyclic tests on Borbone shear wall

		L_{max}[a] (kN)	D_{max}[a] (mm)	Lu[b] (kN)	Du[b] (mm)	Uplift (mm)	Uplift (mm)	Ly[c] (kN)	Dy[c] (mm)	α[d] (kN/mm)	μ[e] (Du/Dy)
Envelope curve 1st cycle	P	103.6	59.2	100.7	79.1	−20.6					
	N	−101.6	−79.0	−101.6	−79.0		−32.7				
Envelope curve 2nd cycle	P	93.9	79.3	93.9	79.3	−30.2					
	N	−91.1	−79.2	−91.1	−79.2		−32.3				
Envelope curve 3rd cycle	P	88.6	59.6	88.6	59.6	−20.9		66.1	10.5	6.3	7.6
	N	−86.8	−79.7	−86.8	−79.7		−32.2	60.5	9.11	6.6	8.6

[a]Maximum load and displacement
[b]Ultimate limit load and displacement
[c]Yielding load and displacement defined according to UNIEN 12512
[d]Initial stiffness defined as the angle of the curve between 10 and 40 % of the maximum load
[e]Ductility

Calibration of Shear Walls

Prerequisite to any successful building model is a correct representation of the shear wall behaviour. Such behaviour has to be evaluated experimentally which has been done in the case of the Borbone wall system via quasi-static cyclic loading. The tested Borbone wall deformed primarily in shear, see Table 1. Therefore, all deformation of the shear wall model has been assigned to horizontal behaviour. The uplift, which was smaller, has been neglected. Therefore, the simple model shown in Fig. 3 for historical timber frames [3] can be used here. The rotational springs are calibrated on the cyclic test results (Fig. 2); only these springs represent the wall behaviour under horizontal cyclic in-plane loading. The timber frame is modelled as being rigid in order to be sure that only the springs are working. The masses are added as lumped masses on the upper nodes of the walls. Due to this model approach, the shear walls are assumed to deform in shear only. All energy-dissipating hysteretic behaviour including also additional contributions from friction is hence represented by the rotational springs.

The only numerical elements that need input of mechanical properties are therefore the springs, as all beam elements are considered being infinitely rigid. Correct and reliable spring models are thus a prerequisite to reliable numerical modelling.

Many different spring models are available that represent the typical pinching hysteresis curves of systems with timber elements (e.g. [12–15]). All are based on linear piecewise definitions of the hysteretic curves. In this paper, the Florence pinching hysteresis model as shown in Fig. 4 is used which has been implemented in the finite element software DRAIN 2DX [16]. As the springs are rotational, the experimental load-slip graph from Fig. 2 has to be transformed into a moment-rotation graph. All necessary stiffness (K1–K6), displacement (U1 and U2) and force (F0) parameters, in total 9 parameters, are then derived from Fig. 2 and are iteratively calibrated by loading the shear wall model with the loading protocol from the quasi-static cyclic test until the global load-slip graphs of test and model show a good agreement.

The calibrated shear wall is shown in Fig. 5. The linearised nature of the spring model and the good agreement between test and model in terms of strength and envelope curve can be seen. The difference of the amount of dissipated energy between model and test is 2 %.

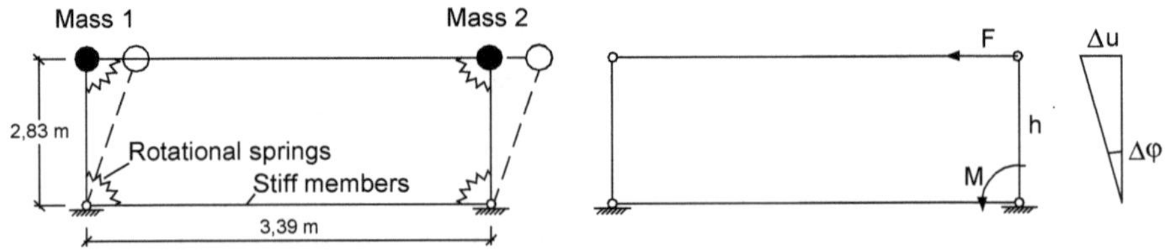

Fig. 3 Simplified numerical model of shear wall

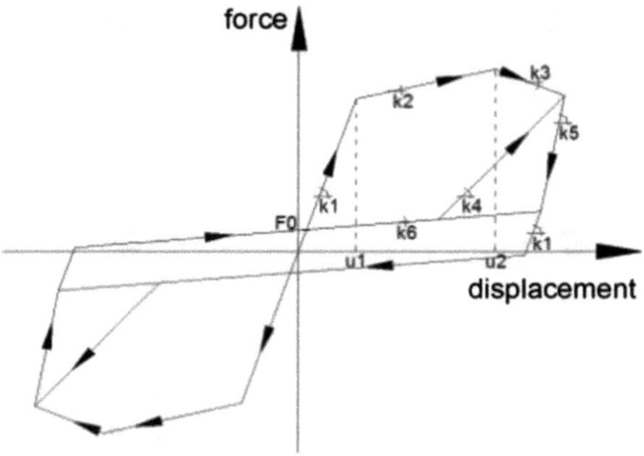

Fig. 4 Florence pinching hysteresis model with six inclinations K1–K6 [15]

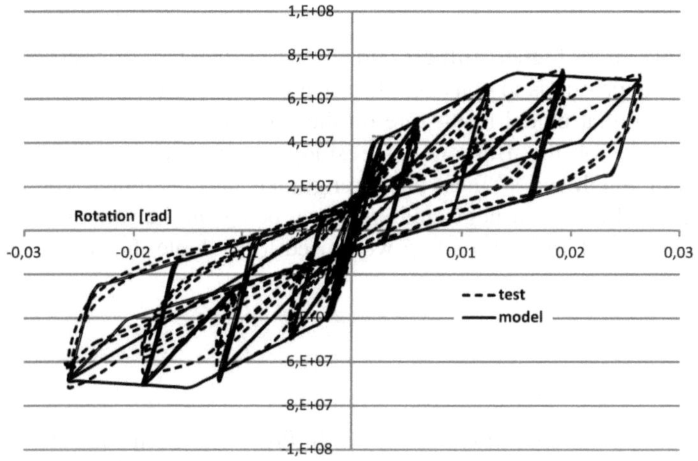

Fig. 5 Overlap of cyclic tests on Borbone wall and numerical shear wall model

Borbone Case Study Building

The transfer of the shear wall model into a 2D building model as shown in Fig. 6 has to be done carefully. Not only the shear wall length, but also the vertical load differs between tests and building model. The extrapolation of test results to longer walls under higher vertical loads is done by modification of the stiffness properties of the spring model. Generally, longer walls have higher stiffness and strength values. Also higher vertical loads generally lead to higher stiffness and strength values. If behaviour and properties of shear walls are linear-proportional to the wall length as it is generally the case when shear deformation is governing, then the

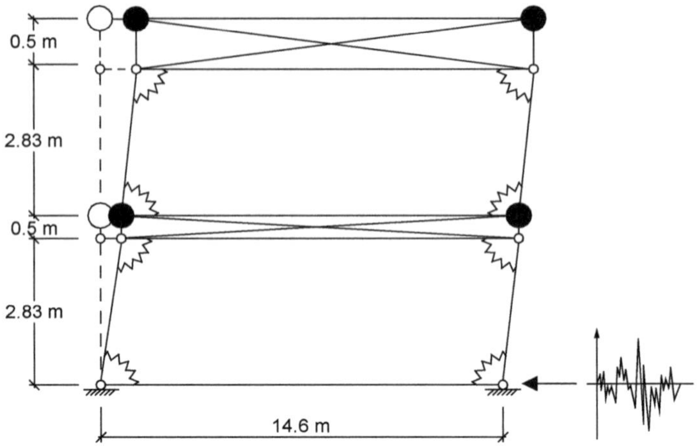

Fig. 6 Simplified numerical model of Borbone model building

adaptation of the values K1–K6 is straightforward (i.e. a wall with two times the length has two times higher stiffness and strength values). As for vertical loading, a conservative approach is to consider shear wall tests with low or no vertical load in order to ensure that model results are on the safe side.

For the sake of the Borbone case study, a linear-proportional relationship between wall length and stiffness and strength has been assumed. The influence of the vertical load has been neglected and thus, being on the safe side, tests results with low additional vertical loading were used to calibrate the shear walls.

Results

The finalised 2D building model as shown in Fig. 6 has been subjected to different accelerograms whose PGA values have been increased until the near-collapse criterion of the building has been reached. This near-collapse criterion has been derived from the cyclic tests and is a maximum horizontal interstorey displacement value of 80 mm which corresponds to 2.71 % drift.

Hence, as soon as a maximum interstorey drift of 80 mm is reached, the calculations are interrupted and the actual PGA is considered as being the PGA at near collapse. Figure 7 shows the modelling results when various earthquakes are applied.

It can be seen that the ultimate PGA and thus the seismic vulnerability of the Borbone case study building against the single earthquakes is very different. Whereas the case study building would have collapsed already at a PGA of 0.25g during the Northridge earthquake, it would have withstood a PGA of 1.52g during the Friuli earthquake at Feltre. Also between the Italian earthquakes, a large bandwidth of sustainable ultimate PGA values can be observed. During the

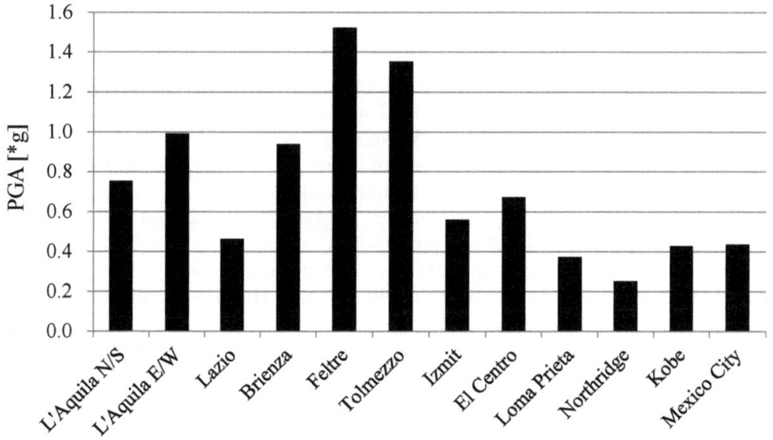

Fig. 7 Maximum sustainable PGA of Borbone case study

Lazio earthquake, the building would have failed at a PGA of 0.46*g*. It is worth to observe that, if the Northridge earthquake is excluded, the maximum sustainable PGA values satisfy the minimum values required by the Italian *seismic hazard zonation map* for the Calabria region, in which the maximum PGA, whose probability of 10 % exceedance in 50 years at 84th percentile referred to rock or stiff soils, is 0.3*g*.

Conclusions

In order to maintain, repair or retrofit Borbone buildings, it is indispensable to gain quantitative information on stiffness, load bearing capacities, energy dissipation capacity, failure modes and sustainable PGA values of these building typologies. The presented paper assessed these parameters by means of a hybrid approach using combined experimental and numerical methodologies.

Quasi-static cyclic tests on Borbone shear walls showed the general suitability of this construction typology in seismic regions due to the walls' non-linear behaviour with high values of energy dissipation and ductility.

The subsequent Borbone case study building, analysed by means of non-linear dynamic modelling in the time domain, reached high values of the PGA at near-collapse state for many of the considered earthquakes confirming the good performance shown in the quasi-static cyclic tests.

The numerical results were derived from simple 2D models and were based on the assumption of a linear-proportional relationship between wall length and wall stiffness and a single Borbone shear wall type only, but nevertheless give valuable insight into the general earthquake behaviour and sustainable PGA values of Borbone constructions.

References

1. Ruggieri N (2015) L'ingegneria antisismica nel Regno di Napoli. AracneEditrice, Roma
2. Ruggieri N (2005) *La casa antisismica*. In: Tampone G (ed) Proceedings of the international conference on the conservation of historic wooden structures. Florence, pp 141–146
3. Ceccotti A, Sandhaas C (2015) A proposal for a procedure to evaluate the seismic vulnerability of historic timber frame buildings. In: Ruggieri N, Tampone G, Zinno R (eds) Historical earthquake resistant timber frames in the mediterranean area. Springer International Publishing, Switzerland. pp 105–118
4. Ruggieri N, Tampone G, Zinno R (2015) In-plane vs out-of-plane "behaviour" of an italian timber framed system: the Borbone constructive system. Historical analysis and experimental evaluation. Int J Cult Architectural Heritage 9:1–16
5. Galassi S, Ruggieri N, Tempesta G (2015) Seismic performance evaluation of timber-framed masonry walls. Experimental tests and numerical modelling. In: Ruggieri N, Tampone G, Zinno R (eds) Historic earthquake resistant timber frames in the mediterranean area. Springer International Publishing, Switzerland, pp 95–103
6. Kouris SLA, Kappos AJ (2012) Detailed and simplified non-linear models for timber framed masonry structures. J Cult Heritage 13:47–58
7. Ramos LF, Lourenço PB (2005) Seismic analysis of a heritage building compound in the old town of Lisbon. In: Proceedings of the international conference on the 250th anniversary of the 1755 Lisbon Earthquake. Lisbon, Portugal, pp 362–368
8. Sarti C (1783) Saggio di congetture su i terremoti. Francesco Bonsignori, Lucca
9. Poleni G (1748) Memorie istoriche della Gran Cupola del Tempio Vaticano e dé danni di essa, e dé ristoramenti loro, divise in libri cinque. Stamperia del Seminario, Padova
10. Mauri-Mori G (1909) Riedificazione di Reggio Calabria. Nuova Antologia, Roma
11. Vivenzio G (1783) Istoria e teoria de' tremuoti in generale ed in particolare di quelli della Calabria, e di Messina del MDCCLXXXIII. Stamperia Regale, Napoli
12. Stewart WG (1987) The seismic design of plywood sheathed shear walls. Dissertation, University of Canterbury, Christchurch, New Zealand
13. Folz B, Filiatrault A (2001) Cyclic analysis of wood shear walls. J Struct Eng 121:1013–1022
14. Pang WC, Rosowsky DV, Pei S, van de Lindt JW (2007) Evolutionary parameter hysteretic model for wood shear walls. J Struct Eng 121:1118–1129
15. Ceccotti A, Vignoli A (1989) A hysteretic behavioural model for semi-rigid joints. Eur Earthq Eng 3:3–9
16. Prakash V, Powell GH (1993) DRAIN-2DX: base program design documentation. Department of Civil Engineering, University of California, Berkeley, USA

The Masonry Timber Framed Load Bearing Structure of "Baraccato" System: A Numerical Model

Ginevra Salerno, Francesca Geremia, Enrico Pagano, Michele Zampilli, Nicola Ruggieri and Stefania Stellacci

Abstract In the modern scientific culture, the seismic vulnerability of the architectural heritage is estimated numerically and need mathematical models of the mechanical behaviour of masonry buildings. The constructive system of "baraccate" houses, whose spread andarchitectural and engineering values have been rediscovered recently, still needs mathematical models. In this piece of paper this first effort has been done for a wall from *Palazzo Vescovile* in Mileto. The comparison between the numerical results and some experimental tests encourages the research efforts to be continued.

Keywords 1783 earthquake · Masonry reinforced with timber frames · Stiffness, strength, ductility · Mathematical models

G. Salerno (✉) · F. Geremia · E. Pagano · M. Zampilli
Dipartimento di Architettura Università Roma Tre,
Via della Madonna dei Monti 40, 00184 Rome, Italy
e-mail: ginevra.salerno@uniroma3.it

F. Geremia
e-mail: francesca.geremia@uniroma3.it

E. Pagano
e-mail: enrico.pagano.89@gmail.com

M. Zampilli
e-mail: michele.zampilli@uniroma3.it

N. Ruggieri
DiBEST, Università della Calabria, Via Pietro Bucci,
87036 Arcavacata di Rende, Italy
e-mail: nicola.ruggieri@unical.it

S. Stellacci
ISCTE—Instituto Universitário de Lisboa, Av. das Forças Armadas,
1649-026 Lisbon, Portugal
e-mail: stestellacci@gmail.com

H. Cruz et al. (eds.), *Historical Earthquake-Resistant Timber Framing in the Mediterranean Area*, Lecture Notes in Civil Engineering,
DOI 10.1007/978-3-319-39492-3_17

Introduction

The constructive system of "baraccate" houses, invented after the earthquake of 1783 in *Ultra Calabria*, was used in the reconstruction of several towns and villages, partially or completely destroyed by the earthquake, immediately after the earthquake and in the aftermath until the advent of reinforced concrete constructions. So it is widespread within that territory, even though a census of the whole "baraccato" architectural heritage is still lacking. Given the large regional seismicity, seismic vulnerability of this architectural heritage needs to be evaluated.

The building system is composed of load-bearing walls, resulting from the interaction of a wooden frame with a load-bearing masonry made of irregular stone blocks or bricks, linked by no-cement mortar and arranged according to a pseudo-squared texture. Within the literature [1, 2] there exist some experimental texts aimed to investigate the mechanic behaviour of this wall under cyclic loading, and evaluating its stiffness, strength and ductility. It is hardly worth noting that, given the variability of both masonry and the wooden frame, the presence of openings, the geometric dimensions etc., the test carried out on a single wall has practically the only function of calibrating some general mathematical models, in order to be able to predict the mechanical behaviour of a generic "baraccata" construction.

Mathematical models for masonry materials are massively present in the scientific literature for walls characterized by periodic arrangement: these are primarily continuous models with constitutive functions obtained through homogenization methods, in order to incorporate information on the constituent materials, the size and the geometric arrangement of the units [3–5]. Continuous models for irregular walls are practically still lacking.

Conversely, when the wall has an irregular texture, the use of discrete systems prevails, where the wall is modelled as a collection of rigid or affine-deformable bodies, interacting through interfaces—the mortar joints—where the whole non-linear mechanical behaviour is concentrated, and characterized by poor tensile strength, not irrelevant compression strength and shear strength governed by Coulomb's friction law [4]. Joints with different constitutive behaviour characterize the interaction between the masonry and wooden frame. Particular attention has been paid for modelling the behaviour of the half wood joints.

Methodological Remarks

In masonry structures, and in particular in those characterized by a non periodic texture, the mechanical behaviour of the assembly depends strongly on some construction details at a very small scale: the interlocking of the units, the units' aspect ratio and their characteristic length, the constitutive behaviour of the constituents, the thickness of the mortar joints etc.

When to this complexity we add also the interaction with a wooden frame, with half wood joints, then the choice of the modelling scale is not so granted. In our case, we opted for a particularly fine scale, which distinguishes the wooden frame from the masonry material, and, within this latter, aims to model each single unit and any joint between any couple of units. In addition to this, particular attention has been devoted to the modelling of half wood joints and the interaction between the masonry and wood.

Following this line of reasoning, each of the constituents of the wall has been modelled from a mechanical point of view: obviously, for providing accurate results, this fine modelling would need some tuned experimental tests on small elements (douplets, triplets and so on) in order to define the constitutive properties of each element. In the absence of these experimental data, partly from the literature (i.e., experimental tests on similar materials), partly by some inverse method from experimental tests on the whole wall, one can calibrate properly not only the qualitative aspects of the constitutive models but also the value of the numerical parameters.

The Numerical Models and the Experimental Tests

The mechanical investigation focused on a wall "extrapolated" from *Palazzo Vescovile* in Mileto, emblematic example of still existing "baraccata" house, already reproduced in a 'imperfectly' identical copy, in scale 1:1, at the laboratory of the CNR-Ivalsa of Trento (see Fig. 1). Full details of the tests are available in [1, 2].

Experimental tests were carried out on two replicas of wooden frame-wall with or without masonry filling. These two copies were built with materials whose mechanical properties were obtained from laboratory tests on specimens taken from the *Palazzo Vescovile*. The wall tested is 295 cm high, a 339 cm wide and about 40 cm thick. The uprights of the frame, 12 × 12 cm of cross section, are linked to the horizontal members, 7 × 7 cm of cross section, through half wood joints reinforced by the presence of a nail with pyramidal cross section of equivalent diameter of 10 mm. The uprights are based onto a wooden beam fixed to a steel beam, reciprocally jointed by means of metal plates.

The masonry filler was made of variously sized limestones from Trentino Region, arranged according to a 'pseudo-squared' texture and with mortar joints with quartz-granite aggregates and hydraulic binder. The wooden frame was made with chestnut from Calabria Region type *Castana sativa Mill.*

A uniformly distributed vertical load amounted to 18.7 kN/m was applied to the wall tested and represents the weight of a wooden truss cover. Horizontally, the loading was impressed via hydraulic actuators that cause positive or negative shifts at the top of the upright at a constant speed of 0.2 mm/s, up to a maximum displacement of 80 mm. The cyclic test was performed following the procedures specified by standard UNI EN 12512-2006.

Fig. 1 The imperfectly identical copy of the wall from the *Palazzo Vescovile* in Mileto: test at CNR-IVALSA

Many are the challenges of modelling that these experiments have placed: all were accepted and not all were won. The first challenge was the modelling of the wood frame. From the experimental results it derives that the rods remain in the elastic phase within the loading range considered. Therefore, the nonlinear behaviour found in the capacity curve, obtained by the envelope of the cyclic tests, has to be attributed to the behaviour of the joints, to which typically all the ductility of wooden frames is due. In Fig. 2 the half wood joint is illustrated.

Our choice was to model it as a multiple constraint, composed of a rigid pin constraint plus a non linear rotational spring. In Fig. 3 on the left side a simple mechanical scheme to assess the rotational stiffness of the spring is shown.

As for the nonlinear part of the constitutive behaviour of the rotational spring, this was estimated by inverse method starting from the frame capacity curve (see Fig. 4). Therefore in Fig. 4, a summary of our modelling effort is presented: on the left side the deformed shape of the frame at a certain load level is depicted, while on the right side the capacity curve obtained by the experimental results is compared with the results of numerical analysis.

Regarding the model of the wall with masonry infill, modelling of the wooden frame remains partially the same as before, while the effort focused on modelling of masonry infill. The masonry material has been modeled as a discrete system [6], in which the units and the mortar joints are geometrically distinguished and differently modeled. Each unit has been discretized by linear elastic finite elements (as shown

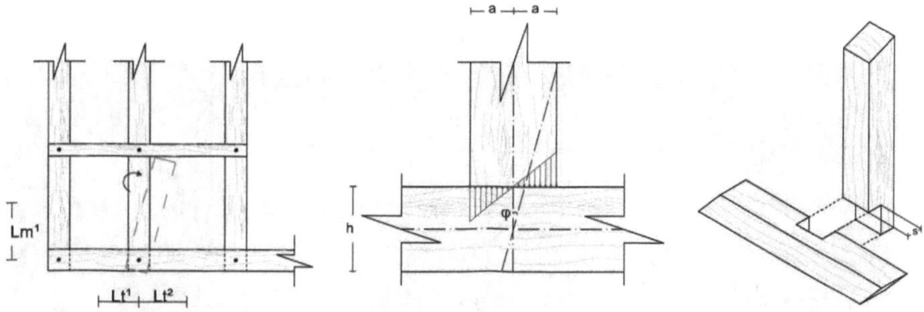

Fig. 2 The half wood joint

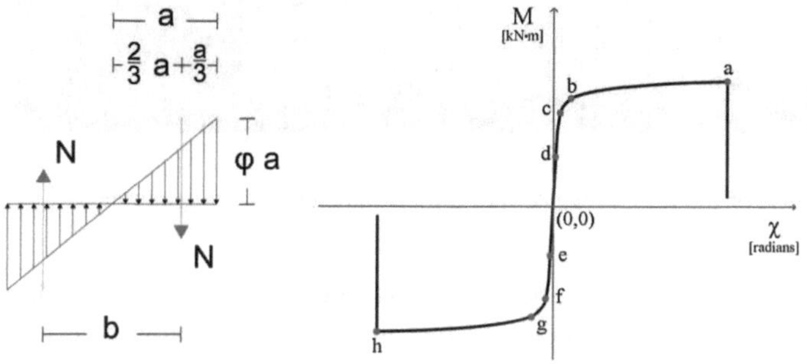

Point	Rotation	Moment
a	7,55E-03	0,550
b	4,76E-03	0,500
c	1,98E-03	0,416
d	7,87E-04	0,256
e	-7,87E-04	-0,256
f	-1,98E-03	-0,416
g	-4,76E-03	-0,500
h	-7,55E-03	-0,550

Fig. 3 Scheme for modelling the stiffness rotational spring (*left side*); the non linear behaviour of the rotational spring (*right side*)

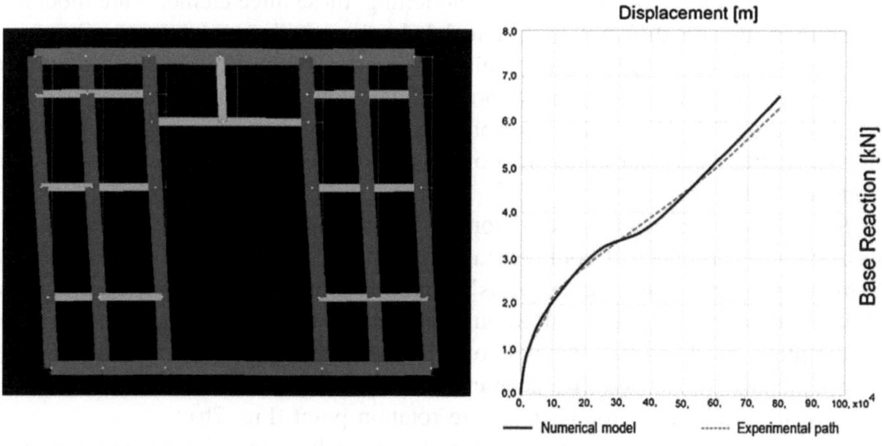

Fig. 4 The mechanical behaviour of the wood frame: deformed shape (*left side*); equilibrium path of the numerical analysis versus experimental capacity curve (*right side*)

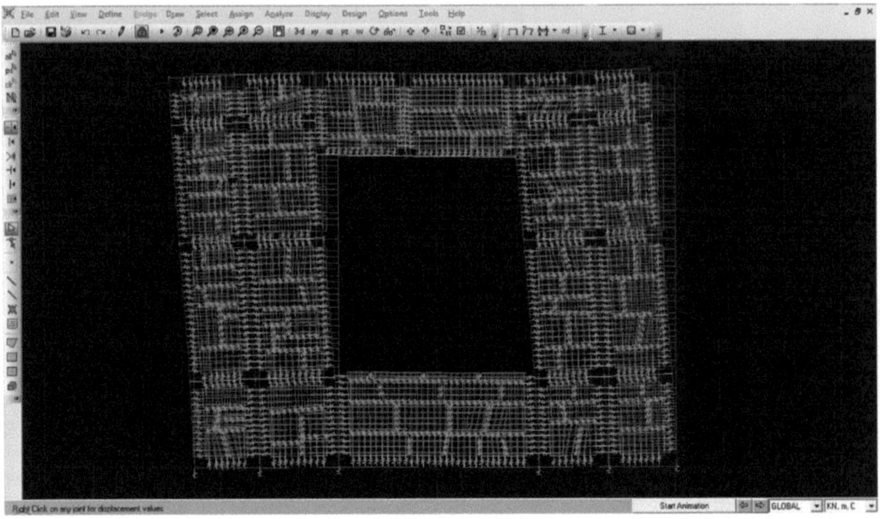

Fig. 5 FEM discretization of the wall with masonry infill in a deformed configuration

in Fig. 5) and the whole non linear behaviour has been concentrated in three essential elements: the mortar joints between the units, the mortar joints between unit and wooden frame, the ground attack section of the posts to the foundation beams.

To make a qualitative hypothesis about the mechanical behaviour of three structural elements is possible; more difficult is to measure the constitutive parameters. From the point of view of modelling, these three elements are modelled as springs, having different mechanical behaviour with respect to tensile, compressive and shear actions. In particular, for mortar joints between stone units, the behaviour under shear action is modeled through the Mohr-Coulomb law. In addition, their behaviour under traction or compression is modeled quite differently, that is, fragile behaviour under traction and moderate ductility under compression, sketched in Fig. 6.

Similar behavior was expected for the wood-stone joints, these latters characterized by lower values of both resistance and stiffness with respect to units' joints. Same reasoning is made for the posts' ground sections, whose tensile/compression strength affects strongly the wall tipping ultimate load. Actually, there are two main deformation mechanisms: the first one is the global tipping of the entire wall (Fig. 7a), and the second one characterized by independent rotations of the two piers of the wall, each around its centre rotation point (Fig. 7b).

In the experimental test the two deformation mechanisms take place one in time, i.e. before the almost elastic global rotation occurs and then, as a result of the cracking of wood-stone joints (see Fig. 8), the second deformation mechanism (see Fig. 7b) occurs.

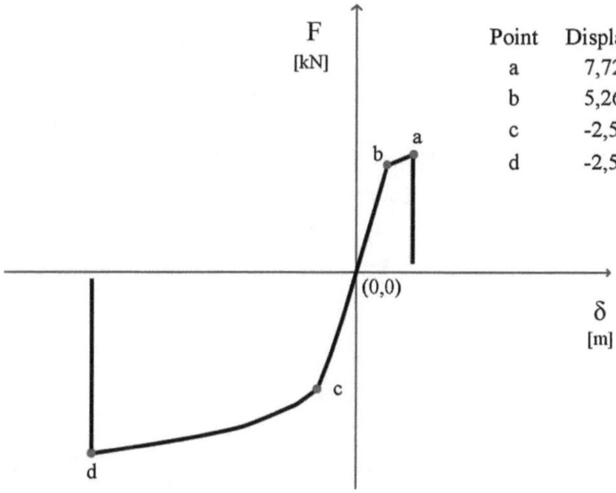

Point	Displacement	Force
a	7,72E-04	7,17
b	5,26E-04	6,57
c	-2,53E-04	-5,57
d	-2,56E-03	-10,24

Fig. 6 Traction and compression behavior scheme of the units' mortar joint

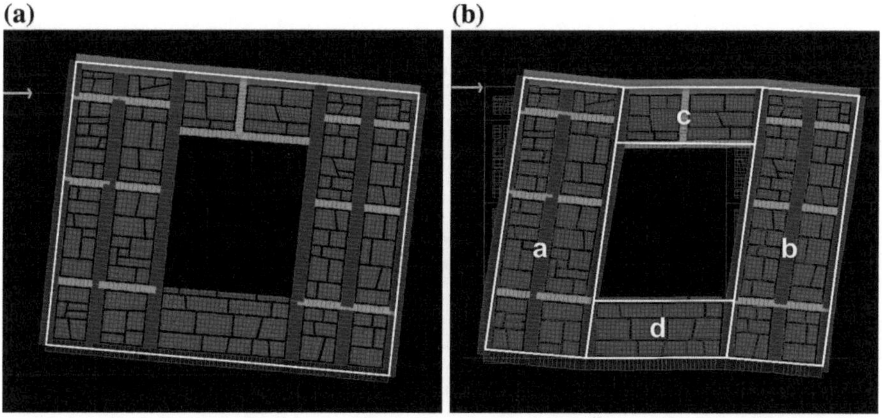

Fig. 7 **a** Global tipping of the wall (numerical simulation); **b** second deformation mechanism

As regards the capability of the numerical algorithm to fully reproduce the experimental test results, a comparison is made between numerical and experimental equilibrium paths. This is shown in Fig. 9, where the experimental path obtained as envelope of the cyclic tests published in [1, 2] is compared with the result of the numerical analysis.

As you can see, even relatively to rough calibration performed on the relevant parameters (especially, tensile strength of the post, strength of stone-wood joints), the numerical equilibrium path captures well the initial stiffness of the wall and also part of the post-elastic equilibrium. However, it fails to determine the ultimate load, and this issue has yet to be resolved.

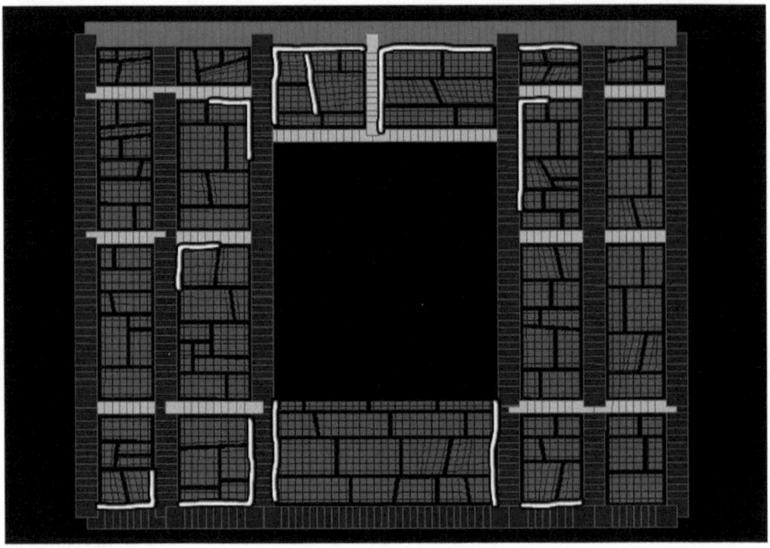

Fig. 8 Damage map

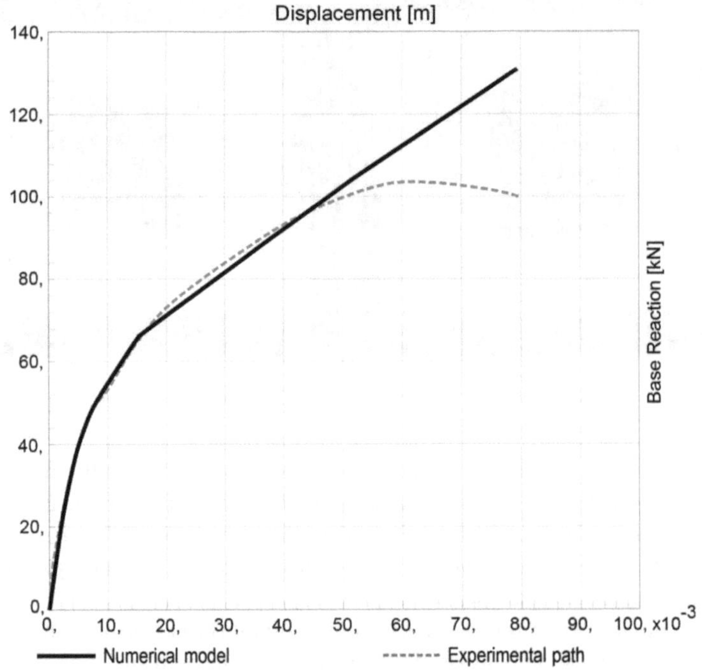

Fig. 9 Numerical path (model in Sap2000) versus experimental path (test CNR-Ivalsa)

Conclusion

The paper contains only a piece of work, which is a first effort to model a "baraccato" wall. In this first effort, which we submit to the attention of other researchers, there are certainly several errors, issues to be investigate, but above all there is the awareness that the proposed approach should be accompanied by an experimental investigation on the small scale samples, which are finely represented in numerical model. Having said that, we think, however, that these numerical analyses aim at the determination of some macro parameters, i.e., mechanical stiffness, strength, ductility of each wall of a "baraccato" building, with the final purpose to carry out seismic vulnerability analysis.

References

1. Ruggieri N, Zinno N (2015) Behaviour of the Borbone constructive system under cyclic loading. In: Ruggieri N, Tampone G, Zinno R (eds) Preliminary Report in historic earthquake-resistant timber frames in the Mediterranean Area, Springer International Publishing, Switzerland
2. Ruggieri N, Tampone G, Zinno N (2015) In-plane vs Out-of-plane "Behaviour" of an Italian Timber Framed System: the Borbone Constructive System. Historical Analysis and Experimental Evaluation. Int J Cult Architectural Heritage 9(6):696–711
3. Salerno G, De Felice G (2009) Continuum modeling of periodic brickwork. Int J Solids Struct 46(5):1251–1267
4. Galassi S, Ruggieri N, Tempesta G, Zinno R (2014) Stability and stiffness contribution of the masonry in the Borbone anti-seismic system. In: Proceedings of the 9th International Masonry Conference in Guimarães
5. Gubana A, Frangipane A, Tomasi G (2012) A simplified procedure for the evaluation of traditional timber frame house seismic response. In: Proceedings of the World Conference on Timber Engineering, vol 2, pp 446–454
6. Chetouane B, Dubois F, Vinches M, Bohatier C (2005) NSCD discrete element method for modelling masonry structures. Int J Numer Meth Eng 64(1):65–94

Load-Bearing Capacity of Traditional Dovetail Carpentry Joints with and Without Dowels: Comparison of Experimental and Analytical Results

Karel Šobra, Coralie Avez, Yasemin D. Aktaş, Richard de Rijk, Isabela Burawska and Jorge M. Branco

Abstract Post-disaster field studies widely suggest that historical timber structures are seismically resistant, and a growing number of experimental studies support this observation. The joints between structural members, which are the major energy dissipation mechanism within the structure, play a crucial role in the overall robustness and the way that a structure handles the seismic demand. Joints mostly fail when the timber members are still in the elastic range, therefore a thorough understanding of their behaviour under various loading schemes is of utmost importance to gain deeper insight about the overall structural performance of timber structures. This paper summarizes the findings from a series of testing carried out on dovetail joints, which is one of the most common traditional carpentry joints, during the 5th COST FP 1101 Training School, held in University of Minho, Portugal. Within this framework, a dovetail joint (with and without dowel) was tested under compression and tension. The experimentally obtained load-bearing capacity of the joints was then compared to the capacity values calculated using

K. Šobra (✉)
Czech Technical University in Prague, Thákurova 7, 166 29, Prague 6, Czech Republic
e-mail: karel.sobra@fsv.cvut.cz

C. Avez
University of Mons, Rue du Joncquois 53, 7000 Mons, Belgium
e-mail: coralie.avez@umons.ac.be

Y.D. Aktaş
University College London, Gower Street, WC1E 6BT London, UK
e-mail: y.aktas@ucl.ac.uk

R. de Rijk
Eindhoven University of Technology, 5612 AZ Eindhoven, Netherlands
e-mail: r.j.c.w.d.rijk@student.tue.nl

I. Burawska
Warsaw University of Life Sciences, Nowoursynowska 166, 02-787 Warsaw, Poland
e-mail: izaburawska@o2.pl

J.M. Branco
ISISE, University of Minho, Campus de Azurém, 4800-058 Guimarães, Portugal
e-mail: jbranco@civil.uminho.pt

H. Cruz et al. (eds.), *Historical Earthquake-Resistant Timber Framing in the Mediterranean Area*, Lecture Notes in Civil Engineering,
DOI 10.1007/978-3-319-39492-3_18

analytical models, and the failure modes were further discussed. The results showed that the experimentally obtained capacity values can be successfully reproduced by analytical models for dovetail joints without dowel. On the other hand, the capacity of a dovetail joint with dowel under compression or tension is always underestimated by analytical models.

Keywords Dovetail joints · Testing · Analytical models · Capacity

Introduction

Around the world, there is a variety of traditional/vernacular timber architecture that is widely considered as seismically resistant based on post-disaster field reconnaissance studies throughout centuries. Although there are cases where poor performance of timber buildings has been reported, mainly due to (1) aging, material degradation and bio-deterioration in timber, (2) lack of maintenance, (3) structural alterations that follow changes in the function of a building and (4) cumulative structural damage due to successive earthquakes as well as (5) disproportionately heavy roofs and (6) failure in the masonry sections of masonry-timber hybrid buildings which triggers an overall structural collapse, the performance of timber structure has mostly been found sufficient to bear seismic demand in areas where they are built. Thanks to the significant amount of static and dynamic experimental work in the last decades, both in the laboratory and on-site, this observation is no longer anecdotal [e.g. 1–3]. The testing efforts as well as in situ observations confirmed that timber structures' seismic performance is governed to a large extent by the joints. Joints are the weakest points within the structure, and they mostly fail when the timber members are still in the elastic range. Furthermore, they are the major energy dissipation mechanism within the structure; therefore they play a crucial role in overall robustness and the way that a structure handles the seismic demand.

Carpentry joints occupy an important place among most widely used connections in traditional timber buildings. Most studies regarding semi-rigid carpentry joints use testing and modelling tools to determine performance under axial or flexural loading and how the behaviour can be improved via different reinforcement techniques. Most studies regarding dovetail joints focus on rounded dovetail joints [e.g. 4–6]. A few remaining studies on lap dovetail connections include [7], where the authors carried out cyclic tests on log walls built by this type of joints to assess the in-plane response. In [8], authors test a frame with lap dovetail joints (inverted-trapezoid sectioned dovetail mortise-tenon joints, as authors name it) and discuss failure modes and obtained force-displacement relationships. The impact of geometrical features on mechanical behaviour was investigated using finite element models [e.g. 9] or experimentally [e.g. 10].

This paper summarizes the findings from testing of lap dovetail joints (with and without dowel) under compression and tension. It then discusses the observed

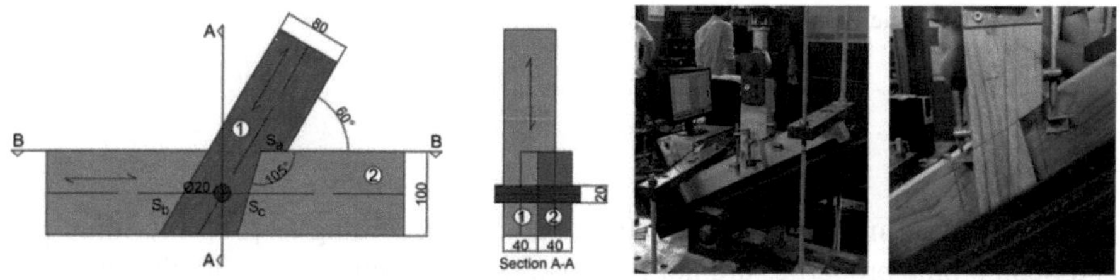

Fig. 1 Dovetail joints with and without dowel

Table 1 Test results (one sample per configuration)

	Compression (kN)	Tension (kN)
Dovetail without dowel	17.51	4.74
Dovetail with dowel	51.64	13.9

failure mechanisms and compares the obtained capacity values to those calculated using models proposed in the relevant codes.

Tests

The tests described here were carried out on dovetail connections, which is among the most common traditional carpentry joints, during the 5th COST FP 1101 Training School that was held in University of Minho in Guimarães, Portugal, between the 11th and 14th of May 2015. The joints with and without dowel were tested under compression and tension (Fig. 1).

Dovetails used in this study for testing purposes were made of Scotch Pine (*Pinus sylvestris*) and Maritime Pine (*Pinus pinaster*) and the dowels were made of Massaranduba (*Manilkara spp.*, tropical wood, assumed to be D60). The capacity values obtained at the end of tests under tension and compression are given in Table 1.

Calculations

Although there is no guideline in the Eurocode 5 [12] on how to design carpentry joints, calculations made here follow the Eurocode 5 (and Johansen's equations) philosophy: to every possible failure mode of the dovetail joint matches a failing load, and the minimum of all failing loads corresponds to the capacity of the connection. Material properties used in calculations for Scotch Pine and Maritime Pine were taken as mean values for *Lodgepole Pine*, which is the pine species with

Table 2 Scotch pine or maritime properties [11]

ρ_{mean} (kg/m^3)	$f_{v,m}$ (MPa)	$f_{c,90,m}$ (MPa)	$f_{t,90,m}$ (MPa)
410	6.1	4.2	2

the lowest mean mechanical properties (Table 2) according to [11]. Maritime pine and Scotch pine properties are not fully available, therefore choosing the species of the same family exhibiting the lowest mean mechanical properties leads to conservative but still realistic results. Mean properties are preferred in calculations *because design values and characteristic values are too conservative and probably do not reflect real properties of the wood samples tested here.*

Possible Failure Modes

(a) **Failure of the dowel** (in tension and compression; Fig. 2)

The capacity of the dowel can be estimated using Johansen's equations (in single shear, timber-to-timber connection, EC5 equations given in Sect. 8.6), provided that the value of characteristic yield moment ($M_{y,r,k}$) is adapted to timber dowel. This method was proposed and successfully used by authors of [13], who adapted Johansen's equations to account for design of all-wood connections (namely, plywood flitch plates and oak peg). Since a plastic hinge cannot develop with wood pegs, the adapted yield moment was defined as the moment for which rapid loss of load resistance was observed (since with wood dowels a ductile behaviour is not likely to occur). Since pegs were not tested in the present research, this adapted yield moment $M_{y,r,k}$ is evaluated through theory of material strength under plastic behaviour, but using a low value (i.e. 70 MPa) for "yield" strength (to take into account the rather brittle behaviour of bended wood dowels), as defined in (1).

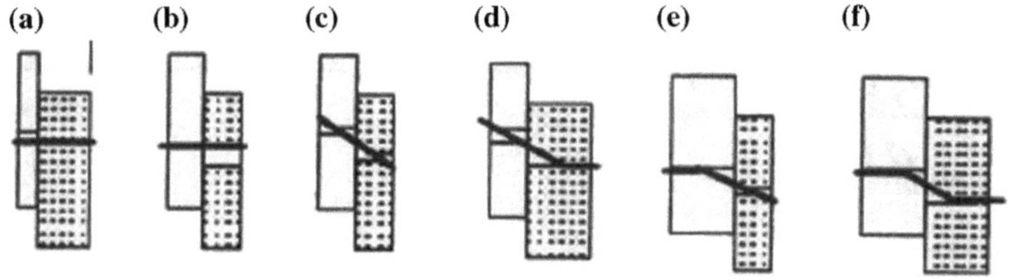

Fig. 2 Failure mode 1—fastener in single shear—according to Johansen [EC5, Sect. 8]

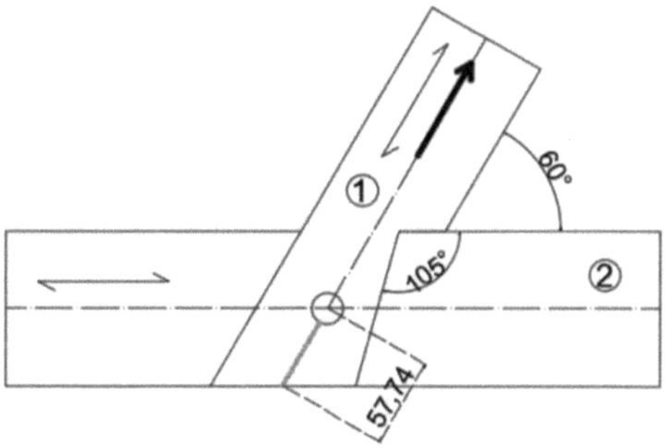

Fig. 3 Failure mode 2—splitting

For cylindrical rods, the definition of the yielding moment is:

$$M_{y,r,k} = f_{y,k} * \frac{D^3}{6} \quad (1)$$

where $f_{y,k}$ is the yield strength, assumed to be $f_{m,k}$, i.e. 70 MPa. This estimation is conservative since we use the characteristic value for bending strength instead of the mean value; and D is the diameter of the peg, i.e. 20 mm.

For dovetails studied here, the failure mode identified using those equations (8.6 in EC5) is rotation of the dowel inside the timber block (i.e. failure mode (c) in EC5 8.6, see Fig. 2, with a calculated capacity equal to 8.53 kN), followed by "yielding" (i.e. dowel fracture) in one point of the timber dowel (i.e. failure mode (e), with a characteristic load-carrying capacity equal to 8.97 kN).

(b) **Splitting near the dowel** (in tension and compression; Fig. 3)

Splitting failure is prevented by following prescriptions regarding minimum edge/end distances from the dowel. Values given in the Eurocode 5 are for steel dowels (and are therefore too conservative for wood dowels, as highlighted and discussed in [13]), so in this paper rather minimum edge distances determined in [13] will be referred to: distance between the dowel and the end of the skew element must be greater or equal to 2.5 times the dowel diameter. This minimum end distance was respected in samples tested here, and by doing so splitting failure was prevented.

(c) **Extraction of a shear plug around the dowel** (in tension; Fig. 4)

Dovetail joints in tension may also fail by extraction of a plug around the dowel, failure due to excess of shear stresses in the skew element. The corresponding failure load can be estimated by multiplying the wood shear strength by the surface available to transfer shear.

(d) **Excessive compression perpendicular to the grain on surface S_a** (in compression; Fig. 5)

In dovetail joints in compression, the load is transferred by compression along S_a and compression and friction along S_b. This failure mode load is evaluated through

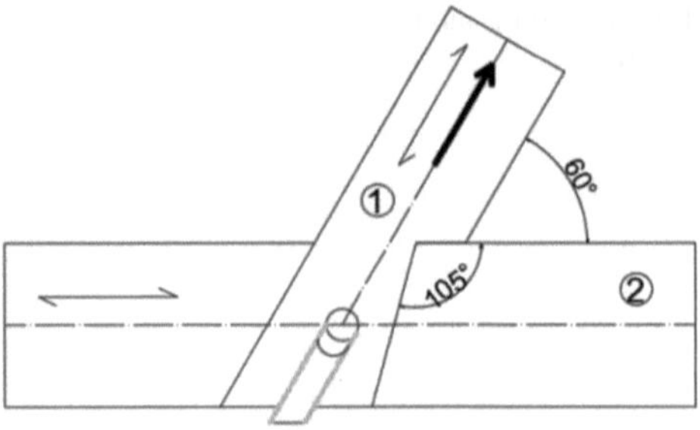

Fig. 4 Failure mode 3—shear plug

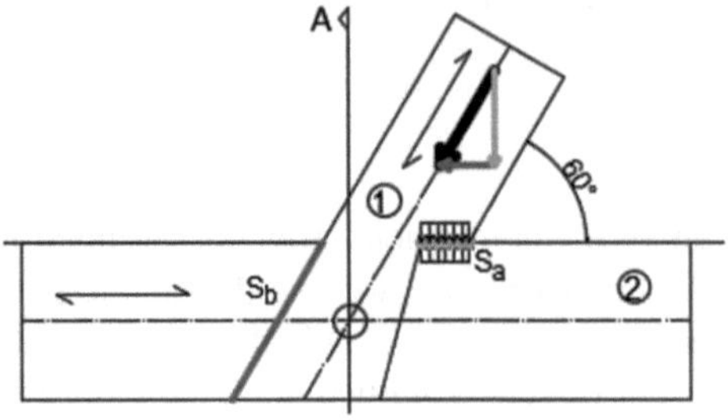

Fig. 5 Failure mode 4—compression along S_a

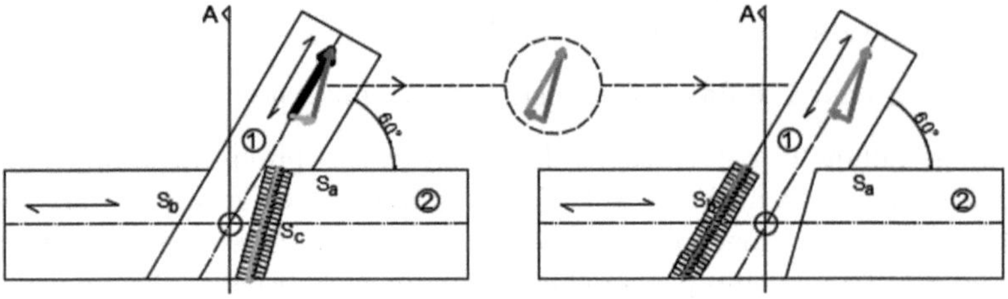

Fig. 6 Failure mode 5—compression on both sides of the dovetail

maximum permissible compression at an angle to the grain along S_a. It should though be noted that usually failure is due to incidental cracks (caused by tension perpendicular to the grain) and not to excessive compression.

(e) **Excessive compression at an angle to the grain on both sides S_b and S_c of the dovetails** (in tension; Fig. 6)

Tension of the dovetail leads to compression at an angle to the grain on surfaces S_b and S_c. To evaluate the corresponding capacity, the tensile load is divided into forces inducing (i) compression perpendicular to the grain on S_c and (ii) compression and friction along S_b. However, the load-bearing capacity is difficult to assess since this process of decomposition is iterative, and the wood crushing is rarely the

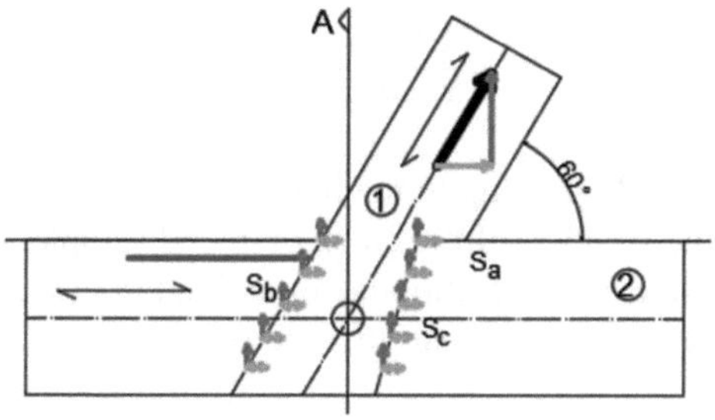

Fig. 7 Failure mode 6—interaction tension perpendicular and shear—method 1

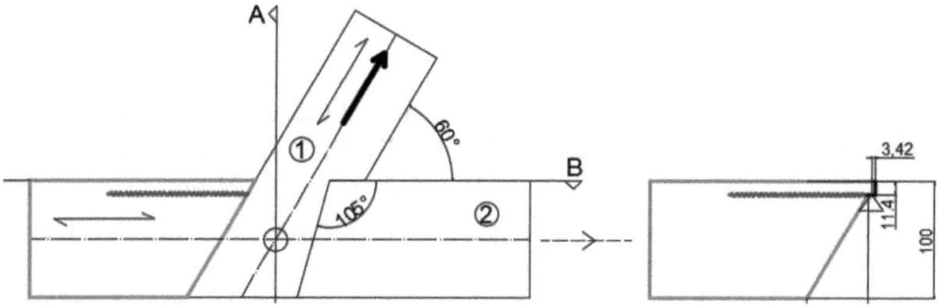

Fig. 8 Failure mode 6—interaction tension perpendicular and shear—method 2

cause of a joint failure because of its ductility. Usually another failure mode (e.g. cracks due to shear parallel to the grain or tension perpendicular to the grain, see failure mode 6) occurs before wood completely crushes in compression (since the extra reserve of capacity is high).

(f) **Failure by tension perpendicular to the grain and shear**
(in tension; Figs. 7 and 8)

Interaction between tension perpendicular to the grain and shear parallel to the grain is highly dangerous for wood joints since it can induce cracks in the connection. The verification of the connection against this failure is made using two different calculations: (i) Calculations focusing on tensile stresses perpendicular to the grain and shear stresses separately (Fig. 7) and (ii) calculations inspired by Eurocode 5 guidelines for notched members, which take into account both tension perpendicular to the grain and shear in a single formula (EC5, Eq. 6.60). Here, we transpose this philosophy of design to the verification of the horizontal element, as illustrated on Fig. 8. The skew element is regarded as a support for the horizontal element.

Since the capacity calculated using method (ii) is lower (and therefore conservative), it will be regarded as the load inducing failure by tension perpendicular to the grain and shear (i.e. mode 6). The failure load is thus defined through expression (2):

$$\tau = \frac{1.5 * V}{b * h_{ef}} \leq k_v * f_{v,m} \quad (2)$$

where V is the shear load applying to the horizontal element (which is linked to the failure load), h_{ef} is the position of the crack on failed sample, $f_{v,m}$ is the mean value for shear strength, and k_v is a reduction factor defined in the Eurocode 5, Eq. (6.62). By calculating V, the failure load can then be obtained.

Comparison with Experimental Results

In Compression

For dovetails without dowel, the only possible failure mode is excessive compression on S_a, with a capacity of 17.75 kN (Table 3). Experimental results agree with calculations since the failure was observed at 17.51 kN due to compression—more precisely to cracks of tension perpendicular to the grain induced by the compression deformation, see Fig. 9a.

For dovetails with dowel, the joint should first fail at the dowel. According to calculations, the first dowel failure mode should be the rotation of the dowel, followed by yielding of the dowel. Those dowel failures should occur at around 8.5 and 8.97 kN, respectively. The tested sample confirms failure modes since it shows both dowels "yielding" (Fig. 9c) and elongation of the holes in which the dowel is inserted (due either to dowel yielding or rotation), and deformation due to excessive compression on the horizontal element (Fig. 9b). We also see a crack due to tension perpendicular to the grain along the compressed surface S_a on Fig. 9b. However, the load reached experimentally (51.64 kN) is much higher than the calculated value (8.53 kN) (Table 3).

Table 3 Possible failure modes and corresponding calculated capacities (kN) for dovetails in compression

Failure modes	With dowel	Without dowel
Failure mode 1: dowel failure (by rotation)	**8.53**	–
Failure mode 4: excessive compression perpendicular to the grain on surface S_a	17.75	**17.75**

Bold values are considered as limit failure capacity. Those values are minimal calculated values for the joint

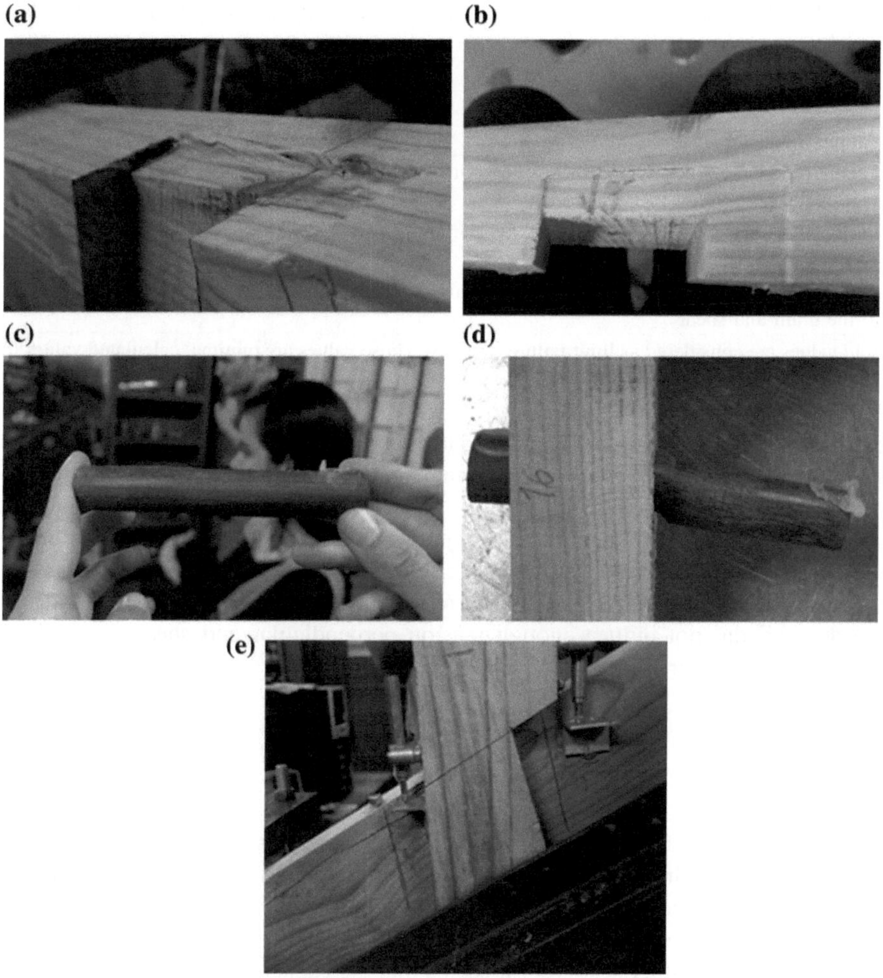

Fig. 9 Failures observed in the tested specimens. **a** Excessive compression on S_a (+tensile cracks), **b** excessive compression on S_a, **c** dowel yielding, **d** failure of the dowel, **e** dovetail in tension: Tension + shear crack

In Tension

According to calculations, both dovetails with and without dowel should fail by cracking of the horizontal element due to combination of tension perpendicular to the grain and shear (Table 4). Regarding dovetails without dowel, experimental results are consistent with the analytically obtained values since the capacity is 4.74 kN and the failed specimen exhibits a crack in the horizontal element (Fig. 9e).

However, for dovetails with dowel, the failure occurred by shear of the dowel at the interface between the two timber elements (see Fig. 9d), and the obtained

Table 4 Possible failure modes and corresponding calculated capacities [kN] for dovetails in tension

Failure modes	With dowel	Without dowel
Failure mode 1: dowel failure (by rotation)	8.53	–
Failure mode 3: extraction of a shear plug around the dowel	28.18	–
Failure mode 5: excessive compression at an angle to the grain on both sides of the dovetails (i.e. on surfaces S_b and S_c)	53.90	53.90
Failure mode 6: failure by tension perpendicular to the grain and shear	**3.21**	**3.21**

Bold values are considered as limit failure capacity. Those values are minimal calculated values for thejoint

capacity is 13.9 kN. This is because the load is transferred not only by compression on both sides of the dovetail (inducing tension perpendicular and shear responsible for cracking failure) but also through the dowel. Therefore, the dowel alleviates the part of the horizontal element that is subject to tensile and shear stresses and delays the failure (hence the load-bearing capacity is higher). It seems that the stiffness of the dowel is so high that it transferred most loads; hence, compression on sides of the dovetail did not induce enough tension perpendicular and shear on the horizontal element to crack it.

Discussion

As seen from the capacity values obtained at the end of tests, the dowel seems to triple the load-bearing capacity, both for compression and tension. Also, as seen in Fig. 10 below, the capacity of a dovetail joint without dowel can be quite accurately predicted, while the values obtained for a dovetail joint with dowel for different failure modes are rather scattered. It seems that the capacity of a dovetail joint with

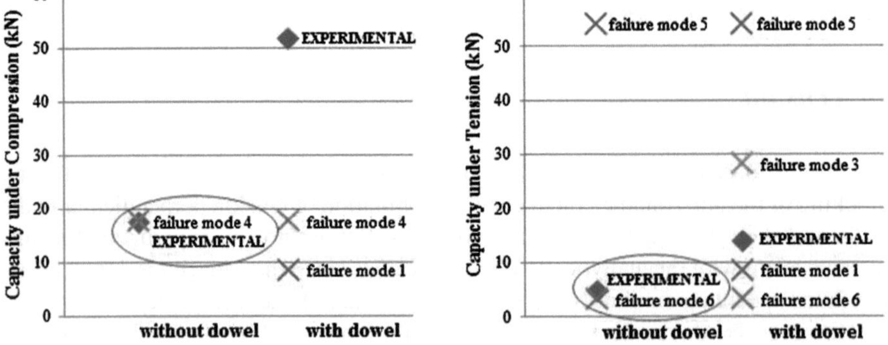

Fig. 10 Experimentally and analytically obtained capacity results

Table 5 Comparison between experimental and analytical results

	With dowel		Without dowel	
	Tests (kN)	Calculations (kN)	Tests (kN)	Calculations (kN)
In compression	51.64	8.53	17.51	17.75
In tension	13.9	3.21	4.74	3.21

dowel under compression or tension (i.e. the minimum value among capacities calculated for each failure mode) is always underestimated by analytical models.

An important point to bear in mind while evaluating this result is that as dovetail joints with dowels are statically indeterminate structure, loads absorbed by each part of the joint depend on the relative stiffness of the constitutive parts of the joint. For instance, dovetail with dowel in compression transfers the load from the skew element to the horizontal element through the dowel and surface S_a (spring analogy: parallel circuit, see Fig. 5). The fraction of the load transferred by the dowel depends on the stiffness of this dowel compared to the stiffness of the surface S_a. The relative stiffness, and therefore the fraction of the load absorbed by each part of the joint is difficult to assess analytically. Failure loads determined in this paper do not take into account this phenomenon, and thus underestimate the load-carrying capacity of dovetails with dowel.

Conclusions

Calculations based on the EC5 philosophy can predict accurately both capacity and failure mode of dovetails in tension or compression without a dowel, as recapped in Table 5.

However, behaviour of dovetails with dowel is more difficult to assess with a simple equation since the stiffness of the dowel and the surfaces under compression is of crucial importance to correctly identify the failure mode and estimate the corresponding ultimate load. Moreover, wood properties and yielding moment to be used at Johansen's equations need be estimated through approximated equations, which induce additional errors in the calculations. Hence, further investigations that take into account the relative stiffness of the dowel and the areas under compression are needed in order to make better estimates. Also, the dowel yielding moment should be determined experimentally.

The engineers who are commissioned for the condition assessment and retrofitting of traditional timber frame structures with carpentry joints should pay utmost attention to the discrepancy between the results obtained from testing and theoretical models. The carpentry joints are not standard, and this study clearly shows that the proposed models should be further fine-tuned to reflect their load bearing capacity and behaviour under certain actions.

Acknowledgments The authors are grateful to COST Action for funding the FP1101 and RILEM 245 Training School at University of Minho in Guimarães, Portugal, thus, making this study possible.

References

1. Aktaş YD, Akyüz U, Türer A, Erdil B, Şahin Güçhan N (2014) Seismic resistance evaluation of traditional Ottoman timber-frame Hımış houses: frame loadings and material tests. Earthq Spectra 30:1711–1732
2. Poletti E, Vasconcelos G (2015) Seismic behaviour of traditional timber frame walls: experimental results on unreinforced walls. Bull Earthq Eng 13:885–916
3. Vieux-Champagne F, Sieffert Y, Grange S, Polastri A, Ceccotti A, Daudeville L (2014) Experimental analysis of seismic resistance of timber-framed structures with stones and earth infill. Eng Struct 69:102–115
4. Tannert T, Lam F, Vallée T (2011) Structural performance of rounded dovetail connections: experimental and numerical investigations. Eur J Wood Wood Prod 69:471–482
5. Tannert T, Haukaas T (2013) Probabilistic models for structural performance of rounded dovetail joints. J Struct Eng 139:1478–1488
6. Xu B-H, Bouchaïr A, Taazount M, Racher P (2013) Numerical 3D finite element modelling and experimental tests of rounded dovetail connection. Eur J Environ Civ Eng 17:564–578
7. Branco JM, Araujo JP (2012) Structural behaviour of log timber walls under lateral in-plane loads. Eng Struct 40:371–382
8. Li X, Zhao J, Ma G, Chen W (2015) Experimental study on the seismic performance of a double-span traditional timber frame. Eng Struct 98:141–150
9. Jeong GY, Park M-J, Hwang KH, Park J-S (2012) Effects of geometry on mechanical behaviour of dovetail connection. World Conference on Timber Engineering, Auckland
10. Kunecký J, Arciszewska-Kędzior A, Sebera V, Hasníková H (2015) Mechanical performance of dovetail joint related to the global stiffness of timber roof structures. Mater Struct. Available at http://link.springer.com/article/10.1617/s11527-015-0651-1
11. USDA (2010) Wood handbook: wood as an engineering material. United States Department of Agriculture—Forest Products Laboratory, Wisconsin
12. BSI (2004) Eurocode 5: BS EN1995-1:2004—Design of timber structures. CEN, Brussels
13. Chang WS, Thomson A, Harris R, Walker P, Shanks J (2011) Development of all-wood connections with plywood flitch plate and oak pegs. Adv Struct Eng 14(2):123–132

Part III
Codes Analysis and Safety Assessment

Reliability of Timber Framed Constructions in Seismic Prone Areas in the Practice Codes Issued from 18th to 19th Century

Nicola Ruggieri

Abstract The analysis of a few significant events of the modern and contemporary history, carried out examining not only the characteristics of the catastrophes but the phases following the occurrence, concerns the earthquakes of Guatemala (1541), Lisbon (1755), Calabria (1783), Lefkas (1826), Casamicciola (1883) and others. Apart from the concern for the humanitarian assistance to injured people, the tax exemption for a certain period, the indications of sanitarian policy in an emergency situation etc., that is a common factor of reports, rules or codes, it is to be noticed that only in a very restricted number of cases technical indications or rules are given for the construction/reconstruction in order to prevent or at least reduce the risk and the amount of damage caused on the buildings by seismic activity. The survey and analysis of the construction practice "codes" presently known, generally issued in seismic prone areas shortly after important earthquakes from the 18th century on, or the reconstruction really put into practice allow to perceive the evolution of the research concerning the origin of the seismic phenomena, the filing, classification, evaluation, interpretation of the damage as well as the slow evolution of the strategies of damage prevention and failure repair. The use and the role, documented since prehistoric times, of the reinforcement of the masonry bearing walls with timber elements or frames is documented in all the seismic zones of the Mediterranean basin, in the widest meaning of the term, as well as all over the world without any apparent mutual influence; then the recommendations, common to some national codes, to make new constructions adopting this building system, is to be regarded as a reference to the traditional practice that had proved to be efficacious at least in preserving the constructions from total collapse and reducing the loss of human lives. The concern for the conservation of the historic buildings, conflicting with the supposed necessity of demolishing the buildings that had been severely damaged, and the rules for the correct repair intervention are only dealt with in the contemporary codes.

Keywords Heartquake · Earthquake-resistant · Timber · Frames

N. Ruggieri (✉)
DIBEST, Università Della Calabria, Cosenza, Italy
e-mail: nicola.ruggieri@unical.it

H. Cruz et al. (eds.), *Historical Earthquake-Resistant Timber Framing in the Mediterranean Area*, Lecture Notes in Civil Engineering,
DOI 10.1007/978-3-319-39492-3_19

The First Government Indications

Provisions, although perhaps it is legend, for mitigation of earthquake effects would been ordered in Central America after the earthquake that struck Guatemala in 1541, where: «*...le gouvernement espagnol avait édicté le peine de morte contre qui aurait été assez téméraire pour élever des maisons étages...*»[1] [1].

A terrible earthquake struck Catania, South Italy, in 1693. The Duke of Camastra, appointed general vicar for the *Val di Noto* and *Val Demone*, immediately intervened with measures concerning the removal of the ruins, victualling for the population and other provisions. Furthermore, in 1694 he promulgated *Consiglio ed Istruzioni*[2] containing rules and instructions for the reconstruction of the city of Catania. That document seems not to recommend the use of timber frames, neither in general a punctual constructive method aimed at improving the building response in case of earthquake, merely recommending «...una più soda sicurezza per evitare le rovine che potessero cagionare li terremoti...».[3] However, the regulations includes some interesting and innovative instructions of urban and modern civil protection providing the minimum dimensions of the viability[4] for the new city of Catania «...avendosi sperimentato nel successo terremoto che il macello delle persone fu effettuato nella loro strettezza, che cadendovi dall'una e dall'altra parte la sommità dei palazzi racchiudevano il passo ai passeggeri e l'aprivano al sentiero della morte...» [5] [2].

Chesneau reported about some technical provisions ordered by the governor of Algiers after the 1716 earthquake, «...le maisons d'Alger ... s'appuyant autant que possible les unes contre les autres; et les planchers des étages supérieurs reposant sur des poutrelles de cèdre dépassant de plusieurs pieds les murailles de façon à ne pas s'effondrer...» [6] [3], therefore wooden beams, that crossing deep into the masonry creating a constraint at the top and improving the panel behaviour if it is

---

[1] «...the Spanish government decreed the death penalty against who was rash to raise the building storeys ....».

[2] *Consiglio ed istruzioni fatte dal Vicario Generale Duca, che fu di Camastra, col voto dell'Ill.mo Senato, e corpo Ecclesiastico, per la nuova riedificazione della città di Catania 18 Aprile 1694* [2].

[3] «... a firmer safety to prevent the ruins that could be caused by earthquakes ...».

[4] Variable from 8 *canne* (around 16 m) to a minimum of 4 *canne* (around 8 m).

[5] «...having experienced in the occurred earthquake the people massacre happened due to their narrowness, and that the falling of the binding's tops on them from both sides obstructed the passage to the people and opened the death path for them...».

[6] «...Algiers buildings ... support as much as possible each other; and the floors of the upper storey bear on cedar beams that cross several *feet* beyond the wall aimed at not collapsing...».

stressed by actions perpendicular to the plane.[7] It is interesting the indication aimed at reducing vulnerability in the arches, where between the impost and the capital of the column it is recommended to interpose small logs [3]. Thanks to this technical device, for low accelerations, the constraint no more stiff can "absorb" the induced movements preventing the pier rotation and consequently the collapse of the arch.

Another indication in the regulations is to insert in the masonry, courses constituted by horizontal wooden elements [3]. Hence, it is reached the objective of realising coplanar laying and consequently facilitating the installation of the following layers of masonry. Furthermore, above all, it induces a preferential modality of failure for actions perpendicular to the plane, that is, the generation of a cylindrical hinge. Montessus de Ballore referring to Chesneau, affirmed that the execution of earthquake-resistant buildings in Algiers would represent «*...le premier exemple authentique connu de règles précises d'édilité...de ce règlement permet de lui supposer une grande sévérité ...été longtemps respecté...*» [8] [1].

The Eighteenth Century and the Birth of the Seismic Engineering

The approach to the earthquake clearly changed in the Eighteenth century,[9] no more seen as divine punishment but as a natural event whose effects can be mitigated by implementing some constructive devices. The event that emphasised that change towards a different and more scientific orientation was represented by the days following the Lisbon earthquake of November 1st, 1755. The earthquake in fact,

---

[7]A constructive device foreran by Leonardo «…ogni trave vole passare i sua muri e essere ferma di la da essi muri con sufficiente catene, perchè spesso si vede per tremoti le travi uscire de' muri e rovinare poi i muri e solai, dove, se sono incatenate, teranno i muri insieme fermi, e muri fermano i solai…» («…every beam has to pass the walls and to be firm with sufficient ties, because the beams come out from the walls and then ruined walls and floors are often seen due to earthquakes, whereas, if they are tied, they will firmly hold the walls, and the walls will fasten the floors…»).

[8]«…the first authentic known example of precise building regulations … of that regulations we suppose a great severity… for a long time respected…»

[9]A "scientific" approach to the earthquake begun since Greek and Roman Age with Aristotle and Pliny and the "seismological" theories about the subterranean vapours (*pneuma*) as earthquake triggering cause. An explanation of the earthquake origin that remained almost totally unchanged up to the Eighteenth century. Furthermore, in order to reduce the buildings seismic vulnerability Pliny, in *Naturalis Historia*, provided constructive devices i.e. wells execution with the scope of realizing "vent" for the subterranean vapours and the well known description of the Artemide temple in Efeso. It was erected on sheep's woolly coats and coals aimed at attempting to decouple the superstructure from the soil, namely the first case of seismic base isolation.

with disastrous consequences in many countries,[10] caused a deep shock with the beginning of a heated debate[11] across the whole Europe on topics such as seismology[12] and seismic engineering.

A widespread empiricism ruled the Eighteenth century research supported by several experimental tests on building materials and in particular on timber frames which involved a refining, even if slow, in knowledge around the mechanical and the solution to the "beam resistance problem".[13] Relatively to wooden structures studies, interesting are those of Du Hamel and De Buffon [5] and, with a predominant research in topics concerning the rising statics applied to buildings, the investigations of Eulero and Musschenbroek with an experimentation made valid by a large number of tools inventions for testing. Although these authors in a limited way investigated around topics concerning prevention methods and buildings behaviour, leading the analysis of vulnerability to earthquakes, in agreement with other 18th century scientists, to factors generally related to the Vitruvio *firmitas* and therefore proper constructive rules. A resistance shown by the building against dynamic actions that could not prescind from the use of timber, material indicated by some writers of treatises[14] as an essential element in the connection of the masonry walls constituting the building. Timber frames, arranged with the aim of achieving a "cage" and constraining the constructive elements of a building, characterised the governmental system (i.e. *Gaiola*) developed after the 1755 Lisbon earthquake with the main purpose of creating buildings resistant to earthquake. The *Gaiola* consisted in a wooden framing braced by Saint Andrew's crosses and masonry infill (Fig. 1). Such a system is delimited toward the external of the building and the adjoining constructions by a masonry with thickness up to

[10]The aftershocks caused widespread damage elsewhere in Portugal and were felt as far away as Venice and southern France and also reached Morocco and northern Africa.

[11]From Kant, mainly interested in the earthquake origin scientific aspect, to Rosseau that piercingly observed «...que la nature n'avoit point rassemblé là vingt mille maisons de six à sept ètages & que si les habitans de cette grande ville eussent été dispersés plus également & plus légérement logés, le dégât eût été beaucoup moindre & peut-être nul...» [4].

[12]In particular speculations relative to the earthquake origin, among the several theories, that relied upon Aristotle's *pneuma* and the most at that time accredited hypothesis of an "electrical" origin founding on the Franklin and Bertholon widespread theories.

[13]On that purpose see, among the numerous publications on the topic, the accurate and synthetic analysis from Galilei to Coulumb contained in Heyman, J., 1997, *Coulomb's Memoir on Statics, An Essay in the History of Civil Engineering, Imperial College Press, London.*

[14]At the beginning of the 18th century, in order to improve the building seismic response, Christian Wolf suggested the connection among the various masonry panels. The German scholar, the author of an encyclopaedic treatise particularly wide spread in South Italy, dealt with «... *Craticula ad firmitatem fundamenti quomodo paranda...*» in which described a timber frame for stone foundations with the scope of «...*In Terrae motibus hac ratique partiam dissociatio impeditur...*» [6].

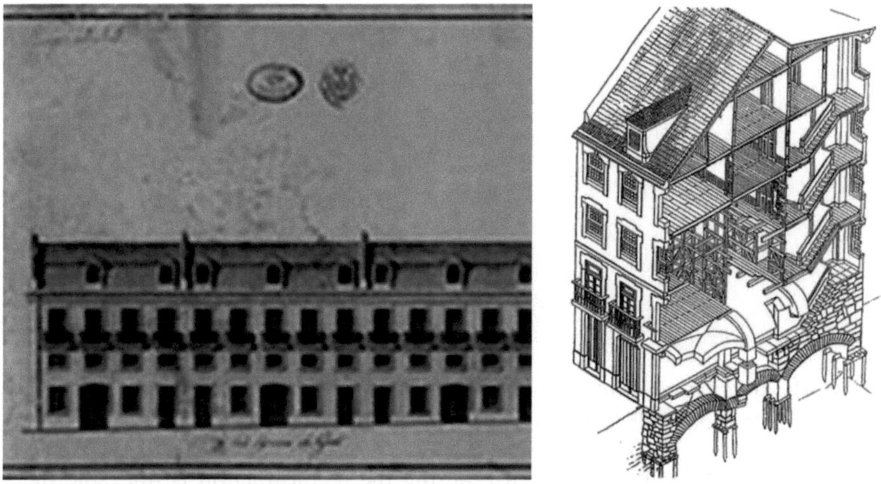

Fig. 1 A Pombalino building façade (cartulario arquivo municipal de lisboa) and axonometry (from Mascarenhas [7])

about 50 cm, this measure is essential in order to compartmentalise the building and limit the development of fires.[15]

The presence of depressed vaults, therefore with a high horizontal component,[16] at the ground floor, provides a further assurance in isolating the ground floor of the building towards fire. On the contrary the inter-storey floors show beams with reduced span, not justified, at least for what is to our knowledge, by the presence of significant dead and live loads; that might be interpreted in the research will for an improvement of the floor stiffness in order to obtain a distribution of the seismic action also to the shear walls.

The same aim is pursued through the application of a double boarding localised at the lower edge and upper edge of the main wooden floor framework. Such an organization, by means of effective connections, besides limiting the beam buckling, entails a beneficial stiffness increasing of the floor and a consequent significant contribution of the seismic action parallel walls. The high number of the beams that compose the floor entails a restraint action towards the masonry panels to which are constrained, constituting an additional element, besides the wooden framing presence in the vertical structure, suitable to avoid possible overturning due to orthogonal actions perpendicular to the wall.

Despite the presence of several graphic works that describe edifices façades (Fig. 1) erected afterwards the 1755 earthquake are still kept in the *Cartulário Pombalino* in Lisbon, however neither indications about the constructive technique,

[15]Lisbon was characterized by several fires consequent to the earthquake such as in the Baixa area; from St. Paul's quarter to St. Roch, and from Carmo and Trindade to the Rossio square area were burnt.

[16]However it is obtained a beneficial reduction of the inter-storey height.

Fig. 2 The *Istruzioni sul metodo da tenersi nella Riedificazione dé Paesi diruti della Calabria* and an example of Borbone building

nor relative to the regulations in use during the reconstruction is traceable in the historical archive. Hence we probably witness, rather than a code drafting, the carrying out of a constructive practice developed in the *Casa do Risco das Obras Públicas* coordinated by General De Maia[17] and desired by the enlightened De Pombal, prime minister of the kingdom, in which one of the first examples of earthquake resistant structure was invented and actuated.

In the South Italy, the regnant of that era Ferdinando IV of Borbone, was the author, after the catastrophic earthquake that struck the Calabria Ulteriore in 1783, of a series of political initiatives aimed at alleviating the terrible tremor consequences. Furthermore, on disposal of Borbone, the Neapolitan engineers ordered the reconstruction and the damaged building repairs relying on *Istruzioni e Suggerimenti (1784?)* suggested by La Vega and Winspeare, works superintendents on the whole Calabrian territory. Such a seismic code includes numerous points regarding the hygiene, urban planning and the distribution of land for new buildings, civil protection measures but above all, in order to decrease the buildings earthquake vulnerability, it provides recommendations following the most innovative ideas in the field of the seismic engineering (limitation of height, decreased weight, box-like behaviour, etc.). Hence, constructive modalities that had the aware scope of reducing the building seismic vulnerability [6]. The article 10, in which is described the masonry fabric reinforced by timber frames,[18] represents the most interesting recommendation included in the *Istruzioni sul metodo da tenersi nella riedificazione dè paesi diruti della Calabria* (Fig. 2).

The constructive typology is composed «... con ossatura di grossi travi di Castagno o Quercia, secondo la natura dei boschi vicini. Questi si situeranno negli angoli ed a tali ragionevoli distanze, e saranno in tale maniera legati con altri travi

[17]A design group composed by military engineers, among the others, Eugénio dos Santos e Carvalho, Carlos Mardel, José Monteiro de Carvalho, António Carlos Andreas.

[18]Hence this involves significant behaviour differences, if compared to the Pombaline system, under static loads and seismic action [8].

trasversali … detta ossatura di legnami abbracciata di fabrica, in maniera che non resti esposta la menoma parte alle varie impressioni dell'aria ed ad un tale effetto non restino le mura men grossi di palmi due e mezzo…»,[19] with a founded concern that the timber can be, if exposed directly to the environment with continuous hygro-thermic variations and water in a liquid state, subject to decay due to biotic agents. The members will be placed also in the corners, providing particular care to the quoin execution, now consolidated constructive rule. The point 10 represents the most significant provision, fundamental tool to achieve the building components cooperation, ensuring a "box behaviour" and improving the seismic response of the building.[20] The article 10 continues recommending that «…La fabrica sarà di mattoni o di pietre piccole…»,[21] constant indication contained in several treatises used in the Neapolitan academies, probably motivated by the will, through the reduced size, of decreasing the masonry mass and expressly avoiding «…di servirsi di qualunque pietra viva liscia nella superficie[22]… legate con calce spenta all'uso Napoli, ed impastata con arena di torrente, o terra aspra al tatto, non cretosa…»,[23] prohibiting the round stones use that do not ensure a sufficient adhesion namely a monolithic behaviour of the panel during the earthquake. Furthermore, a particular care is given to the mortar that must be characterized by an excellent quality indicating in a detailed manner the component elements [6].

1826, The Lefkas Code

The Lefkas island was struck by a severe earthquake, in 1825. The high English commissioner,[24] general Sir Frederick Adam, basing on framing constructions existing previously the earthquake and executed by the Venetians [9], ordered that the reconstruction had to be founded upon a regulation that was specifically drawn up shortly after the tremor [10].

Such a code included a constructive typology that, even though characterised by wooden framings, has a diverse structural organization compared to those developed in Italy and in Lisbon, with a consequent different answer, in particular to the

[19]«…with skeleton of big chestnut or oak members, according to the nature of the near woods. Those will be located in the corners and to such reasonable distance, and in such way will be linked to other transversal members … the above-mentioned skeleton embraced by masonry, so that all even the smallest part not remain exposed to the environment and for such effect not will be the walls less thick of two and half *palmi* …».

[20]In Italy, during the 16th century, Pirro Ligorio, engrafting in a widespread Renaissance tradition identifiable in Alberti and Palladio, was a precursor of the connection among the walls as a device to prevent earthquake damages.

[21]«…The wall fabric will be constituted by bricks or small stones…».

[22]The roughness increases and therefore the frictions.

[23]«…make use of any sharp stone and with smooth surface … bounded by slaked lime as the Naples usage, and mixed with torrent sand, or irregular earth to the touch, not clayey…».

[24]The Ionian islands were subjected, after the 1815 Paris Treatise, under British protection.

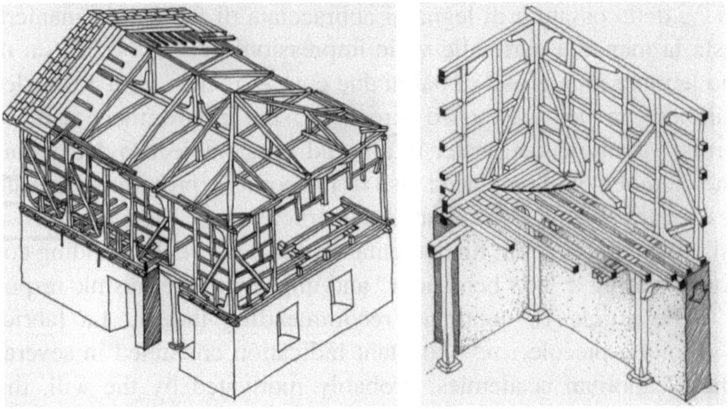

Fig. 3 Axonometry of the Lefkas constructive system (from Vintzeleou et al. [11])

dynamic actions. In fact, the Greek hearthquake-resistant system is constituted by a timber frame at the ground floor, detached from the masonry and devoid of saint Andrew's crosses (Fig. 3), whose independent frame, if sufficiently in plane stiffen, represents a safety structure that preserve people life from the collapse of the floors and in case of wall overturning [11]. At the upstairs the typology of Lefkada is characterised by frames with masonry infill and the presence of a unique diagonal bracing. The arrangement of the nodes is peculiar for both floors of the building, where the concurrent members benefit from the stiffening presence constituted by curved wooden elements.[25]

The same technical device is present in the roof localised perpendicularly to the timber trusses, i.e. an aid in improving the longitudinal stiffness of the structural system. At the roof level, near the hipped roof extremity, the structure is characterised by an inclined wooden element that, somehow, reduce the horizontal in plane deformability.

The "Seismic" Code Development from the 19th to 20th Century

In Italy the Nineteenth century was characterised by numerous high magnitude seismic events, among these the main ones were recorded in the *Valnerina* belonging to the *Stato Ponteficio* and in the Ischia island in the South Italy.

Ordinances aimed at regulating the reconstruction of the sites destroyed by the earthquake followed to those terrible events, as in the case of *Rescritto del papa*

[25]Such a technical artifice relates back to the transversal junction element of the boat planking. From this it can be assumed, considered that Venetia occupation of the islands has deeply influenced Lefkas culture, a derivation from the Venetian shipwright constructive knowledge.

Pio IX relativo al regolamento della nuova legge edilizia statutaria della città di Norcia, a seismic code formulated thereafter the 22nd August 1859 earthquake. The new regulation, structured in three parts, *Disposizioni generali, Regolamento edilizio, Disposizioni transitorie*, a total of 33 paragraphs, relied on the technical indications provided by the Architect Luigi Poletti and by Father Secchi,[26] sent by the government on the sites struck by the quake.

The first great catastrophe of the united young Italian state is represented by the 28th July 1883 earthquake that struck the Ischia island. The prompt promulgation of *Norme di costruzione, che si propongono per i nuovi edifizi pubblici nei comuni dell'isola d' Ischia minacciati dai terremoti,* followed that seismic event. These codes enacted in the Nineteenth century, the Norcia one and the regulations relative to Ischia, present basically similar measures[27], i.e. the height building limitation (around 8 m for the Norcia code and 9.8 m relatively to the Ischia island) the ban on erecting vault and mainly the adoption of «...sistema di costruzione detto baraccato, ad intelajature di legno o di ferro secondo i casi, in modo che esse intelajature, le travature dei solai e le incavature dei solai e le incavallature del tetto riescano solidamente connesse fra loro…»[28] [12], reasserting and following, in a nearly slavish way, the indications contained in the 1783 *Istruzioni.* The 1906 laws of the Italian state will have the identical inspiration source, characterised by few innovation points compared to the Borbone code with the exclusion of recommendations relative to the repairing modalities of the damaged buildings,[29] fruit of

[26]*Relazione della Commissione speciale nominata con dispaccio del Ministero dell'interno del 24 agosto 1859 N.33.098 per provvedere ai danni cagionati in Norcia dal terremoto del 22 di esso mese, e per amministrare ed erogare le somme avutesi in favore dei danneggiati; corredata di allegati comprovanti l'esercizio dell'incarico dal giorno del disastro, al 30 novembre 1860 e rassegnata a sua Eccellenza Reverendissima Monsignore Andrea Pila Ministro dell'interno il 15 dicembre 1860.* Roma 1861.

[27]The Nineteenth century codes differ from the Borbone one for some indications regarding the building plan. In the Ischia island regulations is recommended the «*...forma quadrata o che poco si scosti dalla medesima. Nelle località dove sia constatata una direzione predominante nelle scosse pei terremoti, verrà preferibilmente disposta in tale direzione una delle diagonali…*» (*… square shape or that differ a little from this. In places where a predominant earthquake shake direction is ascertained, to such a direction is preferably disposed one of the building diagonal…*). However it is a principle already contained in the Italian Eighteenth century treatises, i.e. Vivenzio, implemented with Torres and Ruffolo's research of the 19th century.

[28]«...Construction system called *Baraccato* constituted by wooden or iron framings according to the cases, in order that the framings, the floors beams and the notches floors and the roof trusses are firmly connected among them…».

[29]Of particolar interest is the article 31, «Per le riparazioni degli edifici di carattere nazionale, in specie per valore artistico, storico ed archeologico sarà stabilito, caso per caso, il partito da seguire pel suo consolidamento» (To repair buildings of national importance, in particular due to their artistic, historical and archaeological value will be established, case by case, the solution to follow for their strengthening) in which the Italian principle of the modern seismic engineering applied to the historical architectural heritage of "seismic improvement" is *in nuce* contained. Furthermore, among the possible seismic reinforcement and repair interventions is included wooden members installation aimed at reproducing a Borbone timber frame. The article 35 is peculiar, «Gli edifici di cemento armato, che presentino lesioni tali nelle membrature dell'ossatura resistente da renderli

Fig. 4 Timber frame structure executed after the 1905 earthquake relying on the Borbone constructive system

a natural scientific advance. However, it is worth to emphasise that the applications of the *Baraccato* system of the 19th and 20th century were characterised by the presence of metal wares, plates, bolts, stirrups and were differentiate for a greater degree of the nodes stiffness compared to the Borbone system (Fig. 4).

A protagonist role of the timber persisted in the 1909 Italian provisions,[30] relative to appropriate materials in reducing the construction seismic vulnerability. In fact, the Borbone system was indicated in the code as a possible substitute for the r.c. structures, even if with persisting implementation difficulties due to the lack of suitable components. This again demonstrated the recognised effectiveness of the timber framing system during tremors [13].

The difficulties in the definition of the mechanical and performance level of such a system thereafter entailed, at least as regards Italy, the abandonment of the framed masonry as earthquake resistant device with the almost complete disappearance from the construction practice codes (it was defined "legname" in a denigrating way), even if it works under gravitational charges.

The timber frames structures reacquired equal status, compared to other earthquake-resistant constructive typologies, with the drafting of the Eurocodes and in particular, of the Eurocode 5 *Design of timber structures* and above all Eurocode 8 *Design of structures for earthquake resistant,* in which, among other indications,

(Footnote 29 continued)

inutilizzabili, debbono essere demoliti e rifatti … quelle … semplicemente baraccate, che si ritrovino nello stesso caso possono essere riparate …» («The r.c. buildings, that present cracks in the resistant skeleton members which make them ineffective, must be demolished and remade … those … simply *baraccate*, that are found in the same condition can be repaired…»), in that case the reliability of the timber framed system is considered more significant than the r.c. one. As it is supported by timber structures properties, frequently characterized by statically rendundant condition and a structural collaboration. Those features lead to a peculiar behaviour in which the rupture of a single element does not entail the collapse of the global structural system.

[30]*Norme edilizie per le regioni colpite dai terremoti del 28 dicembre o anteriori*, 1909.

the timber framed constructive typologies are described as a structural system with a high ductility.

Conclusion

The first governmental indications of the history were, obviously, often fragmentary, reported by the successive literature and in general without source quote.

The rulers attitude that the State became protector of its subjects got stronger during the Eighteenth century. Lisbon before and the *Regno di Napoli* later, marked a real rupture in the seismic design, with regulations relied on timber frames constructive system, unique technology that will remain unsurpassed for the following 250 years and till the overflow of the reinforced concrete constructions. Three main structural typologies stand out from the above analysis, the Pombaline system, the Borbone one and the Lefkada constructive technique. Although such structural systems are characterised by the common use of wooden frames, present significant differences both in the answer to gravitational charges and to inertial forces seism-induced.

References

1. Montessus De Ballore (1907) La science séismologique les tremblements de terre, Librairie Armand Colin
2. Fichera F (1934) G.B. Vaccarini e l'architettura del Settecento in Sicilia, Roma
3. Amina Abdessemed-Foufa A, Benouar D (2005) Atlas of earthquake-resistant traditional techniques in Algeria: the case of the Casbah of Algiers. Eur Earthq Eng J EEE 2(05):2–29
4. Rosseau JJ (1853) Oeuvres complètes de J.J. Rosseau avec des notes historique, Tome quatrième, chez Alexandre Houssiauz, Librarie, Paris
5. Ruggieri N (2011) Il legno nel '700: aspetti meccanici e d'anatomia, in Bollettino Ingegneri n. 6, pp 3–16
6. Ruggieri N (2015) L'ingegneria antisismica nel Regno di Napoli. Aracne Editrice, Roma
7. Mascarenhas J (2005) Sistemas de Construção V—O Edifício de Rendimento da Baixa Pombalina de Lisboa. Processo evolutivo dos edifícios; inovações técnicas; sistemas construtivos. Livros Horizonte, Lisboa
8. Ruggieri N (2014) In-plane cyclic tests on historic timber framed wall. Outcomes Comparison. In: Schober KU (ed) proceedings of COST action FP1004 Experimental Research with Timber, Prague, pp 56–61
9. Makropoulos C, Kouskouna V (1994) The Ionian Islands earthquakes of 1767 and 1769: seismological aspects. In: Albini P, Moroni A (eds) Contribution of historical information to a realistic seismicity and hazard assessment of an area. Review of Historical Seismicity in Europe, CNR, Milano
10. Vintzileou E, Zagkotsis A, Repapis C, Zeris Ch (2007) Seismic behaviour of the historical structural system of the island of Lefkada, Greece. Constr Build Mater 21(2007):225–236
11. Vintzeleou E, Touliatous P, Zeris C, Repanis K, Zagotsis A, Leonardos E, Palieraki V, Adami C (2007) Earthquake Planning and protection organization, Public library of Lefkada

12. Estratto della relazione per le prescrizioni edilizie dell'isola d'Ischia, in atti del Regio Istituto e Incoraggiamento alle Scienze Naturali Economiche e Tecnologiche, Napoli, 1884
13. Ruggieri N, Tampone G, Zinno R (2015) In-plane vs Out-of-plane "Behaviour" of an Italian Timber Framed System: the Borbone Constructive System. Historical analysis and experimental evaluation. Int J Cult Archit Heritage 9:1–16

Assessment and Safety Requirements for Existing Structures

J.M. Catarino, Teresa O. Santos, Ana S. Louro and Helena Cruz

Abstract Extending life of existing structures should be supported by guidelines or technical rules. In the case of historic buildings, besides the technical, economic and social aspects to be taken into account, the historic importance may determine the level of intervention to be performed. In spite of that, the integrity of the building itself and protection to people must be safeguarded and structural safety must reach a minimum level. However, safety requirements are not always established. Some European countries have national regulations and standards regarding assessment and retrofitting of existing structures. There is a need for harmonised European technical rules on this subject which can be broadly accepted. This paper refers the existing design codes and the work under way to extend them to existing structures and discusses their application and limitations regarding historic buildings. Based on LNEC experience, an approach is also proposed concerning the safety requirements applicable to heritage buildings, discussing the suitability of specifying different target reliability levels. Those levels should depend on the heritage value of the structure and the consequences of a possible failure of the structure, thus linked to its acceptable use.

Keywords Eurocodes · Assessment · Retrofitting · Existing structures · Timber · Masonry

J.M. Catarino (✉) · T.O. Santos · A.S. Louro · H. Cruz
Laboratório Nacional de Engenharia Civil (LNEC), Av. do Brasil, 101, 1700-066 Lisbon, Portugal
e-mail: jmcatarino@lnec.pt

T.O. Santos
e-mail: tosantos@lnec.pt

A.S. Louro
e-mail: analouro@lnec.pt

H. Cruz
e-mail: helenacruz@lnec.pt

H. Cruz et al. (eds.), *Historical Earthquake-Resistant Timber Framing in the Mediterranean Area*, Lecture Notes in Civil Engineering,
DOI 10.1007/978-3-319-39492-3_20

Introduction

Maintenance, repair and rehabilitation of existing structures have gained increasing importance in the construction industry in the last years.

The assessment of an existing structure may be required for many reasons: feasibility analysis of change of use, including increased load requirements, concerns about design or construction errors, quality of building materials and serviceability, evaluation of aging/deterioration effects, existence of damage due to extreme loads. Decision about the need of retrofitting and the choice of the retrofitting solution will be based on the assessment, considering benefits, costs and target safety values.

The approach to safety requirements and design of retrofitting of existing structures has much less support in codes or guidelines than the design of new structures. The retrofitting needs, techniques and design must also take into account the subsequent life extension and intended use of existing structures, as well as specific concerns regarding their economic, strategic or heritage values. Assessment and intervention of historical buildings have a particular context related with the longer expected life and the need to respect the original character and fabric of the building.

European and Portuguese Regulations

Some European countries have national regulations and standards regarding assessment and retrofitting of existing structures, but there is a need for harmonised European technical rules on this subject which can be broadly accepted. Current Structural Eurocodes are primarily focused on the design of new structures. Nowadays only Eurocode 8 Part 3 [1] provides rules applicable to all categories of buildings. But not only the seismic assessment and retrofitting of monuments and historical buildings often require different types of provisions and approaches, depending on the nature of the monuments, as other actions besides the seismic should be considered whenever the assessment is needed. To answer to the urgent need to bring together different national approaches, a working group on Assessment and Retrofitting of Existing Structures (WG2) was created by the Technical Committee CEN/TC 250 "Structural Eurocodes".

The main purpose of this WG2 is the development of a European-wide harmonised and acknowledged coherent set of technical specifications or Eurocode parts for the assessment and retrofitting of existing structures. In that way the Joint Research Centre published a policy support document on *New European Technical Rules for the Assessment and Retrofitting of Existing Structures* [2]. Although this document is in principle applicable to all existing structures, particular aspects of historic structures are not specifically addressed.

As first milestone for the preparation of a CEN Technical Document, that report [2] presents scientific and technical proposals intended to serve as a starting point for further work towards a harmonized European view and it contains preliminary proposals for technical provisions and identifies several key issues requiring resolution. It is important to clarify if it will be appropriate to have different target reliability levels for the assessment of existing structures compared to the design of new structures. Regarding the partial-safety factor format, member's states should also take a position about the idea of update partial-safety factors, based on knowledge of the structure and its condition (materials, actions etc.), recognising the uncertainties associated with the updating process. Other question is how to model those uncertainties when undertaking non-linear structural analyses.

European countries also have to make clear if the remaining working life should be taken into account when determining live loads, variable actions etc. and which circumstances may lead to be necessary to undertake a structural assessment. A change in the design requirements arising from a revision of a design code for new structures must determine assessment of existing structures? A structural assessment based solely upon satisfactory past performance of a structure would be appropriate?

So, the future CEN Technical Specification will describe the basis for the assessment and retrofitting of existing structures, aligned with all relevant principles and requirements of EN 1990 [3], and provide guidelines on the determination of specific actions for use in the assessment and retrofitting of existing structures. It is intended that this specification will be applicable to structures constructed using materials within the scope of the structural Eurocodes. It will not include recommendations on when assessments should be undertaken and do not cover assessment and retrofitting of existing structures subject to seismic actions, which is the subject of Eurocode 8 [1].

In Portugal there are several codes and standards covering the design of new construction. However some of them are outdated and there are no national design codes for timber or masonry structures. For existing structures, until April 2014, there were no rules for retrofitting. Then, it was established than until April 2021, for residential buildings more than 30 years old or located in areas of urban renewal, some requirements might not be applied due to the associated costs that could make it unfeasible. Concerning structural behaviour, there is no obligation to improve the structural performance of the building, as long as interventions do not reduce its structural and seismic safety (even with low safety level) [4]. This could lead to unacceptable high risk levels, particularly in earthquake prone areas, if designers and promoters do not evaluate properly the consequences of rehabilitation.

Safety Levels in Design of New Structures

The adoption of a certain safety level or reliability index in design is a complex procedure of balancing different factors. Decision rules are intended to define criteria for obtaining the most suitable solution and the optimization criteria to be used must take into account technical aspects and utility. A simple rule of maximization of estimated gains can be suitable if the rule is applied for the foreseen lifetime of a project with an estimation of costs and benefits not biased by short term advantages for some stakeholders. Such rule implies the definition of generalised cost as the sum of initial cost and the estimated costs associated to the different types of failures [5]. In this context, cost applies to all negative failure consequences, namely economic, social, life threat, etc. In the simplest case, with just one single type of failure, the generalised cost, C_g, can be estimated by:

$$C_g = C_i + P_f C_f \tag{1}$$

where

C_i is the initial cost

P_f is the probability of failure

C_f cost associated to the failure.

The estimation of generalised costs must consider timing of costs, using discount factors or another equivalent procedure intended to report all costs to a specific moment (normally the end of the expected lifetime of the structure). The difficulty is that participants in the construction process have different options in such estimation, as shown in Fig. 1. For example the builder (case 1 in Fig. 1) is led to use high values of discount factors, amplifying the weight of initial costs. The long term owner and public authorities (case 2) give more importance to costs which occur during the lifetime of the project, using lower discount factors. The designer (case 3) has the tendency to over evaluate costs of failures to defend himself.

The consequence of these different views of the problem is shown in Fig. 1: the factor of safety corresponding to the minimum generalised cost is lower for the builder (Fs_1) and higher for the designer (Fs_3). Compatibilizing these different options is an ethical problem of codes, decision makers and designers. The lower the discount factors the higher the contribution of present generation in benefit of future ones [5].

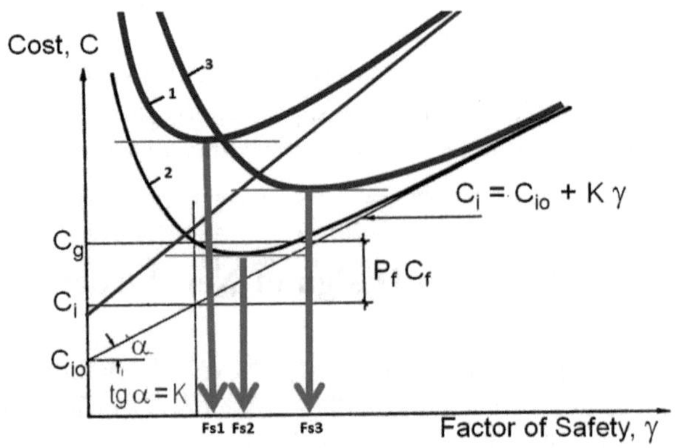

Fig. 1 Influence of variation of factor of safety on initial and generalized costs estimated by different participants in construction process (*1* builder, *2* owner, *3* designer)

Besides such consequences in the estimation of factors of safety, another discussion can be made: it is not economic to have solutions with factors of safety lower than those corresponding to the minimum values of generalised costs ($F_{s,min} = Fs_1$, Fs_2 or Fs_3) as such solutions should correspond to higher costs with lower safety; however, cost-benefit analysis could support the choice of solutions with factors of safety higher than $F_{s,min}$ and higher costs. Also a judgement must be made if $F_{s,min}$ obtained like this is acceptable for the society. Attempts have been made by other authors, using concepts of utility and global felicity, with the same type of difficulties above mentioned.

Safety Assessment and Safety Levels in Retrofitting of Structures

According with SAMCO report [6], structural assessment routines can be classified in five levels, from performance assessment which compares directly measured performance values with defined threshold values (level 1), to probabilistic assessment routines which calculates probability of failure or equivalent reliability index (level 5). Levels 2–4 use partial factors, based on documents and visual inspection and on supplementary investigations.

Uncertainties in structural assessment are modelled using appropriate probability distribution functions for each basic value, considering different types of uncertainties related to human activities, due to inadequate knowledge and statistical uncertainty.

Additional information gathered on an existing structure can be applied to improve any previous ("prior") estimate of the structural reliability. This information may come from inspections, tests and monitoring, proof load testing and past service loading.

Structural design is based on models of structural or material strength. Tests can be performed on components or samples be taken from the existing structure. Proof load tests of the whole structure are only used in special cases. Design code rules for proof load testing are rather arbitrary but probabilistic arguments can be invoked to make sensible use of the information [7]. In order to have practical consequences in assessing structural integrity, proof loads must be reasonably high so that posterior failure probability can be significantly reduced.

Another type of tests with increasing potential for global structural health monitoring is the use of vibration inspection methodology. The characterization of the dynamic response of a structure with ambient or forced vibration tests, in terms of natural frequencies, corresponding mode shapes and modal damping factors, gives a better understanding of the global structural behaviour and reduces uncertainty. The National Laboratory of Civil Engineering (LNEC) of Portugal has a large experience in this domain [8].

Fig. 2 Sta. Isabel Bridge, Coimbra, Portugal. Proof load test made by LNEC for Infraestruturas de Portugal

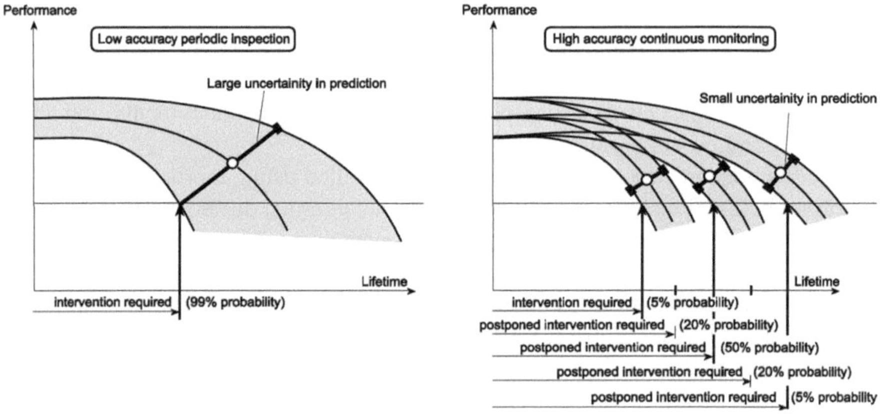

Fig. 3 Influence of the uncertainty of the input data in decisions of intervention [9]

The selection of available methodologies for health monitoring and assessment of structures can be based in cost-benefit models, where all the benefits and costs of the possible strategies are computed with a generalised cost function, choosing the strategy associated with an instrumentation coverage that minimizes global costs [8]. Ambient vibration tests, having very low costs and giving important information, can be an important part of a strategy that intends to optimize global costs, combined with visual inspections, material sampling and testing and other semi or non-destructive and non-invasive testing methods (Fig. 2).

Figure 3 shows how estimated timing of required interventions can be influenced by the reliability of the data gathered and the assessment made. The conclusion is that the level of uncertainty in the data and assessed outcome can influence significantly the decision concerning timing for interventions [9].

In certain circumstances, satisfactory past performance combined with a structural system review can be sufficient to assume safety for loads other than accidental, if inspections do not reveal significant damage or deterioration, no significant changes are anticipated of loads or durability parameters and predicted deterioration ensures sufficient durability.

The *Guide for the Structural Rehabilitation of Heritage Buildings* of the International Council for Building, CIB, [10] introduces the rule for the calculation of design values of material properties using confidence factors (CF) depending on sample size, type of building and materials used. The values of CF are in the range between 1.0 and 1.35.

$$f_d = f_k/CF \tag{2}$$

where

f_d is the design value of the material property

f_k is the characteristic value of the material property.

EN 1990 [3] accepts, in special circumstances, that design values of a material or product property can be obtained by direct determination. Part 3 of Eurocode 8 [1] presents a similar procedure with the above described from CIB publication [10], for the calculation of values of properties of existing materials to be used in the calculation of structural capacity using confidence factors. The range of values of confidence factors is also between 1.0 (comprehensive in situ testing) and 1.35 (limited in situ testing).

The revaluation of structural safety within an assessment procedure may include an updated calculation of the reliability index or failure probability, considering the best estimation of material properties and a review of all possible types of loads. Using again the simplest case with just one single type of failure shown in Fig. 4, it is possible that the probability of failure can be within acceptable limits, even taking into account materials deterioration and the mentioned assessment judgements of loads and material properties.

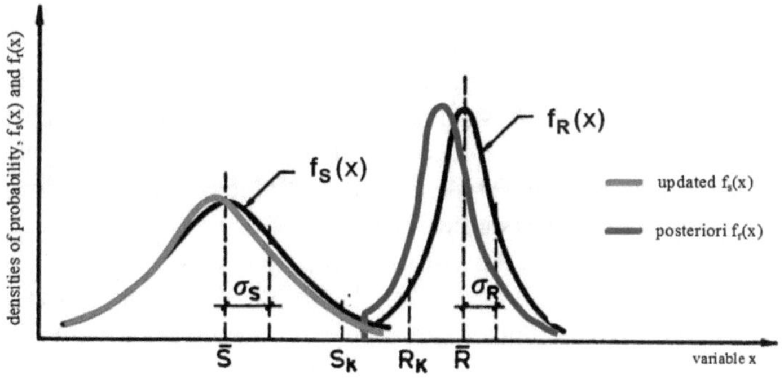

Fig. 4 Computation of probability of failure (initial and updated)

A consensus can be verified in bibliography about the importance of information gathered in structural assessment activities for judgements about reliability and retrofitting calculations. A lot of research is still needed to support harmonised safety criteria, to be used by designers and consultants. The ambition is not necessarily to have codes covering all possible situations but guides or manuals that can help engineers, leaving them the judgement about possible options, taking into account their sensibility to all information gathered within a structural assessment.

Safety in Heritage Buildings

Assessment and intervention of historical buildings pose frequent problems. Original structures were frequently complex and irregular, in many cases suffered many intentional modifications along their life, and in general there is no information on the building construction and subsequent works.

Even if standing in apparently good conditions, existing buildings may have major structural problems and safety deficiencies. This may result from poor design or construction, from materials damage or previous interventions determining load paths that are different from the ones intended and to be expected.

Understanding historical buildings will also require the identification of geometrical features and boundary conditions of the structure, the assessment of mechanical properties of elements, decay agents and residual cross section of timber due to damage.

This obviously require knowledge and skills from the professionals involved in assessment and interventions, on traditional building techniques, particularly as concerning structural connections, common building materials' properties and damage processes. Lack of knowledge may be responsible for misunderstanding and ignoring problems that should be taken care of, or for inadequate interventions, sometimes creating more problems than the ones solved.

These difficulties have led in the past to extreme attitudes concerning intervention in historical buildings (either the "touch nothing" or the "leave nothing" approaches), where intermediate approaches should have been considered.

When modifications become heritage and several layers of history are present, one may even debate to which "original state" the building should be reverted to. Besides, when a specific building is a poor example of the building know-how and techniques used at its time of erection, it may be argued whether the planned intervention should respect the original building or the original building principles (good practice at the time of construction).

No doubt, whatever may be the historical importance of a building it should not pose any risk for life and health of persons. Therefore, if a serious deficit at the load bearing structure is noticed, either some degree of retrofitting has to be done or else access to the building (and possibly its neighborhood) should be restricted. And, although respecting the principle of the minimum intervention, it is also advisable to adopt measures to prevent it from collapse.

Can we do more than recover the original safety level? Some construction details are determinant for the stability and global performance of a building, namely under seismic actions, and their benefit highly compensates for the possible invasion of the original building fabric.

It is clear that both ultimate and service limit states may be different in the case of heritage buildings. So, what is the accepted level of safety for historical buildings?

EN 1990 [3] recommends different target reliability levels (high, normal and low) for new structures depending on the consequences of structural failure. These levels may apply, respectively to: bridges and public buildings, residences and offices, agricultural buildings and greenhouses.

ISO 2394 [11] constitutes a risk and reliability informed basis for decision making concerning design and assessment of structures. For ultimate limit states according to Joint Committee on Structural Safety (JCSS), it proposes different target reliability indexes as a function of both the consequences of failure (small, some, moderate, great) and the relative cost of safety measures (high, moderate, low).

ISO 13822 [12] establishes, for a number of limit states, target reliability indexes depending from the consequences of failure (in the case of ultimate LS), or from their reversibility (in the case of serviceability LS) and the possibility of inspection (in the case of fatigue).

Schueremans and Verstrynge [13] proposes an equation to derive target probability of failure from numerical values given to factors like residual service life, lives exposed to danger, economical factor, warning factor, activity factor, and social criterion factor (that may be taken as preservation value in the case of heritage buildings).

It is not always possible (and certainly not advisable) to strengthen an historical structure to meet requirements set for the intended use. A reasonable alternative is to adjust the use to the existing structure, for example by limiting the number of persons allowed in a room, or by closing down a building under extreme weather conditions.

Table 1 Minimum target reliability levels *versus* uses

	Reliability class			
Construction type	A+	A	B	C
New		☺	no	no
Existing < 50 years			☺	no
> 50 years – current			☺	no
> 50 years – heritage value				☺
	Strategic buildings	Current buildings	Limited use (occasional)	No public access
	Possible uses			

It is reasonable to set different safety requirements (reliability classes) depending on the heritage value of the building and the consequences of a possible failure. This should be linked to its acceptable use, as similar minimum reliability indexes should be sought e.g. for equivalent strategic buildings or for public buildings of some sort (Table 1).

Conclusions

Extending life of structures should be based in reliability assessments and supported by guidelines or technical rules. If a reliability assessment concludes with the need of a structural intervention, the first ambition should be to design retrofitting solutions with equivalent safety level of new structures. Considering gathered information about the performance and materials of an existing structure and the associated reevaluation of design values of material properties, together with a review of all types of actions, it is so possible to design more economically with the reduction of uncertainty provided by structural assessment.

Simple rules of maximization of estimated gains can be applied both in the design of new structures and retrofitting solutions, provided that such analysis covers the foreseen lifetime of the project with non-biased estimation of costs and benefits. Designers should also be encouraged to perform cost-benefit analysis besides checking minimum code safety requirements.

In the case of historic buildings, besides the technical, economic and social aspects to be taken into account, the historic importance may determine the level of intervention to be performed. In spite of that, the integrity of the building itself and protection for people must be safeguarded and structural safety must reach minimum levels.

References

1. EN 1998-3 (2005) Design of structures for earthquake resistance—Assessment and retrofitting of buildings, CEN
2. Joint Research Centre (2015) Report EUR 27128 EN—New European technical rules for the assessment and retrofitting of existing structures. Publications Office of the European Union
3. EN 1990 (2004) Eurocode—Basis of structural design. CEN
4. Decree-Law no. 53/2014 [Exceptional and temporary legal framework for the rehabilitation of existing buildings or fractions]. In: Official Journal (PT), 1st Series, no. 69, 2014-04-08, pp 2337–2340. Available at https://dre.pt/application/file/25344956
5. Ferry Borges J, Castanheta M (1985) Structural Safety. LNEC, Lisbon
6. SAMCO (2006) Guideline for the assessment of existing structures. Final Report—F08a. Available at http://www.samco.org/network/download_area/ass_guide.pdf
7. Melchers RE (1999) Structural reliability analysis and prediction, 2nd ed. Wiley

8. Campos Costa A, Rodrigues J (2001) Structural health assessment of bridges by monitoring their dynamic characteristics. In: Seminário Segurança e Reabilitação das Pontes em Portugal, FEUP Editions, pp 243–260
9. fib (2013) Model code for concrete structures 2010. Ernst & Sohn
10. CIB (2014) Guide for the structural rehabilitation of heritage buildings. CIB Publication n. 335. Available at http://site.cibworld.nl/dl/publications/pub335.pdf
11. ISO 2394 (2015) General principles on reliability for structures
12. ISO 13822 (2010) Bases for Design of structures—Assessment of existing structures
13. Schueremans L, Verstrynge E (2008) Use of reliability methods for evaluating safety of historic structures. In: D'Ayala D, Fodde E (eds) Structural analysis of historical constructions. Taylor & Francis Group, London, pp 1321–1329

Building Regulations and Existing Buildings: Setting the Requirements on Structural Safety in Portugal

João Branco Pedro and Vitor Campos

Abstract This paper deals with the application of building regulations to the rehabilitation of existing buildings, taking a recent Government initiative to deregulate construction works in existing buildings as a case-study. To contextualize the analysis, the Portuguese housing stock is briefly described and the way building regulations apply to existing buildings in Portugal is reviewed. The deregulation initiative was presented by the Portuguese Government as a pragmatic policy measure to stimulate urban rehabilitation but sparkled wide opposition from a number of professional and scientific associations, who questioned both its strategic sense and its technical convenience. The structural safety professional and scientific community was particularly active in this debate. Focusing on them, the analysis of the process shows that: (i) the opposing positions were founded on two equally reasonable priorities (i.e., to stimulate the rehabilitation of existing building stock versus to improve structural safety), (ii) both parties were unable to support their positions with hard facts and argued mostly around their respective priorities and (iii) the debate ended in a deadlock. Main conclusions are that to overcome the deadlock more applied research is needed to bring forward appropriate solutions to rehabilitate existing buildings, along with a better dissemination of this practical knowledge in the building and real estate sectors. An informed and enlightened debate involving all relevant stakeholders, including the consumers' associations and the public, should also be actively pursued. A final important conclusion is that increased attention should be given to sustainability and resilience when accessing social and economic efficiency in the process of designing and updating building regulations. Conducting these processes in the future will require a broader and more comprehensive approach and a more extended reference time frame that have often been adopted in the past.

Keywords Building regulations · Existing buildings · Structural safety · Portugal

J.B. Pedro (✉) · V. Campos
Laboratório Nacional de Engenharia Civil, Av. do Brasil 101, 1700-066 Lisbon, Portugal
e-mail: jpedro@lnec.pt

V. Campos
e-mail: vcampos@lnec.pt

H. Cruz et al. (eds.), *Historical Earthquake-Resistant Timber Framing in the Mediterranean Area*, Lecture Notes in Civil Engineering,
DOI 10.1007/978-3-319-39492-3_21

Introduction

Building regulations set minimum requirements for safe, healthy, functional, energy-efficient and accessible buildings for everyone who lives and works in and around them. The general purpose of building regulations is to enhance social and economic efficiency of the built environment. Issuing building regulations is a public responsibility in any modern state and an instrument to promote sustainable territorial development [1].

The application of building regulations to existing buildings is often constrained by pre-existing site and building conditions to an extent that makes it not always viable to achieve the same overall level of performance of new buildings, within the given technical and economical envelope of each operation. Therefore, setting building regulations for application to existing buildings often requires a balance between cultural, social, environmental, technological and economic factors. This balance is sometimes delicate to achieve as it depends on values predominating in society at each given period. To determine the break-even point, the opinion of different stakeholders should be actively taken into account [2]. How building regulations deal with existing buildings is therefore a subject for research and debate, involving both the professionals and the public. This is especially important in a time when urban rehabilitation becomes a high priority in the political agendas.

On the basis of this rationale, this paper analyses the positions expressed by different stakeholders when the requirements on structural safety for existing buildings were recently discussed in Portugal. To contextualize the analysis, the current condition of the Portuguese housing stock is described and the way Portuguese building regulations apply to existing buildings is reviewed. The paper then proceeds to explore the priorities of different stakeholders in a recent Government initiative that changed the technical requirements applying to construction works in existing buildings, focusing on the debate around the requirements on structural safety. To conclude, recommendations for future action in this field are presented.

The Portuguese Housing Stock

According to the last census (2011) [3], the Portuguese housing stock is relatively new and in good condition (Fig. 1): 86 % was built after 1945 and 71 % does not require repairs. However, if we focus on the buildings prior to 1946 a different scenario arises: 42 % of the stock is in need of minor or medium repairs and 18 % needs major repairs or are severely deteriorated. Buildings prior to 1946 represent only 14 % of the housing stock but account for 58 % of the buildings in need of major repairs or being severely deteriorated.

In 2011, nearly half of the buildings had are in forced concrete structure (49 %) and roughly a third of the buildings had a structure of masonry walls with slab

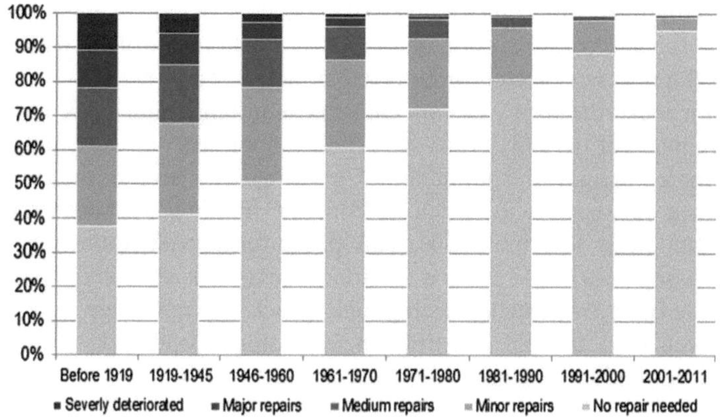

Fig. 1 Maintenance condition of residential buildings per construction date [3]

(32 %) [3]. However, the type of structure varies significantly with the construction period. Buildings prior to 1946 are predominantly made of masonry walls with timber floor and roof structures, while the buildings after that date increasingly incorporated concrete in their structural elements.

Maintenance condition of buildings varies significantly according to the type of structure (Fig. 2) [3]: 79 % of the buildings that incorporated concrete in their structural elements do not need repairs whereas that percentage decreases to 40 % for the buildings made of masonry walls, rubble stone or adobe, without concrete slab. In fact, 68 % of the buildings in need of major repairs or rated as severely deteriorated were built with masonry walls, rubble stone or adobe, without concrete slab.

Regarding seismic resistance, the buildings prior to 1946 have varying degrees of structural safety, many being expected poor or very poor performance levels in

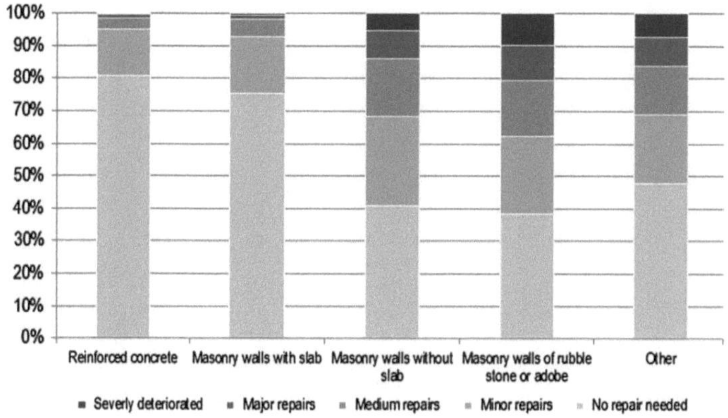

Fig. 2 Maintenance condition of residential buildings per type of structure [3]

case of earthquake. Buildings after 1952 had to comply with increasingly demanding seismic safety requirements and regulations, so their performance in case of earthquake is expected to be better or significantly better.

Based on these data we can conclude that buildings built prior to 1946, predominantly made of masonry walls with timber floor and roof structures, are those requiring more extensive rehabilitation. The technology used in their construction has also been superseded, as structural concrete framing has become the rule in almost all buildings after the 1950s. This poses further difficulties when rehabilitation of this segment of the stock is in order, due to lack of skilled workers able to deal with the older construction technologies, lack of knowledge on the compatibility between new and old construction technologies and lack of technical and economic information to support de interventions.

The Portuguese Building Regulations

During the last 20 years the Portuguese regulatory framework for building has evolved positively towards better safeguarding quality of construction [4]. 88 % of the legal documents concerning building regulations that were in force in 2006 had been approved after 1986 (Fig. 3) [5]. Major regulatory improvements were made in the fields of energy-efficiency, accessibility and technical installations. This was partly due to: (i) the evolution of the Portuguese economy and its integration in the European and global markets, (ii) the improvement of the knowledge basis and (iii) the direct influence of European directives and regulations.

While better safeguarding quality of construction, this evolution has also brought some problems with it. One of the problems regards the sharp increase in the number of regulations applying and the frequent lack of coherence and uniformity of criteria between them [5]. This has a disturbing effect, increasing the costs of

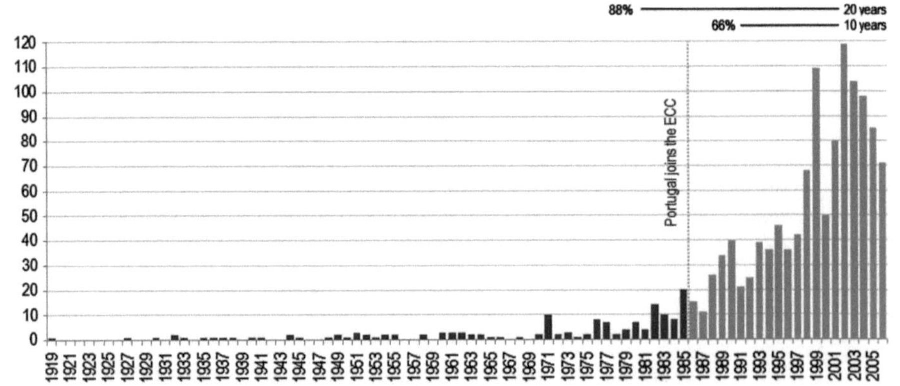

Fig. 3 Legal documents concerning building regulations in force in 2006, per date of approval [5]

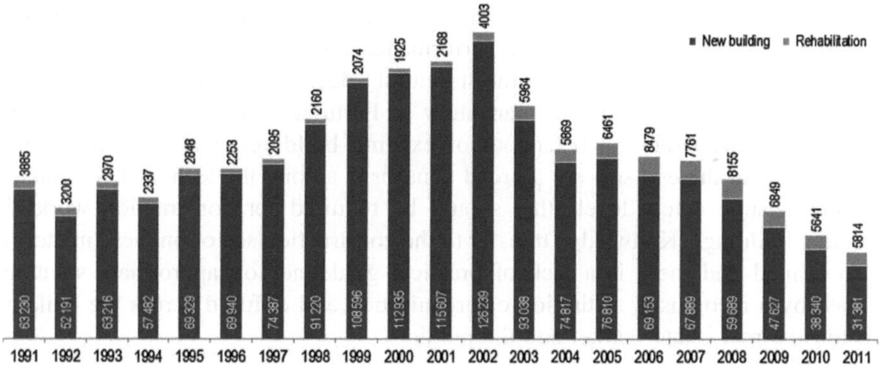

Fig. 4 New building and rehabilitation in Portugal: number of dwellings concluded per year 1991–2011 [3]

context in the building sector and thus affecting its economic efficiency. Another problem is that current Portuguese building regulations were designed with the focus on new buildings [6]. New building has largely dominated the Portuguese real estate market since the mid-20th century and rehabilitation only accounted for an almost insignificant share of the total number of dwellings concluded annually: 5, 6 % in average between 1991 and 2011 (Fig. 4) [3]. As public policy and the market are now switching towards the rehabilitation of the existing building stock, the mismatch between the existing regulatory framework and the object of the interventions becomes evident and is creating difficulties.

Regarding their application to existing buildings, current Portuguese building regulations adopt a *general protection principle*: buildings do not have to comply with regulations issued after their date of construction and, therefore, construction works carried out on an existing building do not necessarily have to improve on the building's performance levels but only have to avoid down-grading them.

However, the law states that this principle does not apply (i) if building regulations explicitly set out specific requirements for certain activities to occur in buildings (even if these activities were already taking place in the building prior to rehabilitation), or (ii) if the regulations explicitly require that certain complementary works are carried out to improve on safety or habitability whenever construction works take place in existing buildings [7, 8].

Portuguese building regulations on structural safety do not require that complementary works are carried out to improve on safety whenever construction works take place in existing buildings and their scope does not explicitly state that they apply to existing buildings. Therefore, the general protection principle applies.

One common difficulty directly associated with the general protection principle is to decide which performance levels should be considered: the building's actual performance levels or the original performance levels. This doubt allows for a wide margin of subjectivity in the application of structural safety regulations and is a cause for occasional differences between promoters and permit-issuing authorities [9].

These differences often arise when deciding on which methodology and criteria shall be used to assess structural safety performance levels and to prove that planned construction works do not reduce the structural safety conditions.

The Eurocodes are still not mandatory in Portugal. There has been extensive research on the structural behaviour of existing buildings, namely their seismic resistance, but there is still not a broad consensus among the technical community on the seismic safety levels that should be required for construction works in existing buildings. Knowledge transfer to the construction sector on these matters is also limited and there is a lack of practical guidelines on appropriate solutions issued by a recognised institution or organisation and diffused across the building sector, the real estate market and the public.

The Special Legal Regime for Buildings Rehabilitation

In 2012–2014, the Portuguese Government prepared and adopted a special legal regime aimed at dealing with the alleged mismatch between the current regulatory framework and the rehabilitation of the existing building stock [10]. This special regime applies to the rehabilitation of residential buildings that are more than 30 years old or that are located in the are as the municipalities designate for urban renewal. The special regime is designed tobe in force until 2020.

The Government approach has been to deregulate, sharply reducing the requirements for construction works on existing buildings by exempting construction works from complying with a number of requirements on habitability, accessibility, acoustics, energy efficiency, gas installation and telecommunication infrastructures [10].

In this context, additional requirements on structural safety were not favoured. The new regime opted to reiterate the general safeguard already laid down in the Portuguese legal framework: "*interventions in existing buildings* ***cannot decrease*** [...] *the structural and seismic safety of the building*" [10].

The new special regime was forwarded by the Government as a pragmatic policy measure aimed to eliminate unnecessary and inappropriate regulatory constraints, thus reducing the uncertainty and the cost of interventions and promoting a proactive policy of urban rehabilitation. The argumentation was built around five main ideas [9]:

(1) many buildings need repair;
(2) Portuguese building regulations do not take into account the technical and constructive specificity of old buildings;
(3) the performance levels set by current building regulations is excessive given the characteristics of some of the existing buildings and the financial resources currently available;

(4) the reduction of cost resulting from the reduction of the technical requirements will enable viable interventions in all urban areas and make housing more affordable;
(5) rehabilitation shall be encouraged taking into account its broader economic and social benefits.

The approval of this special regime was involved in controversy, sparkling wide opposition from a number of professional and scientific associations, who questioned both the strategic sense of the policy measure and its technical convenience. The structural safety professional and scientific community was particularly active in the debate. Detractors of the special regime within the structural safety community considered that improving structural safety conditions of buildings is a standing priority and saw rehabilitation as an opportunity to achieve this strategic goal [11–13]. They argued that:

(1) most of the existing buildings do not ensure minimum levels of safety in case of an earthquake;
(2) the added cost of measures to improve structural stability of buildings is small;
(3) rehabilitation without improving resilience to risk of earthquake conveys a false impression of safety to users and it is a questionable investment from the point of view of national economy.

The view of the Government prevailed at the end, with the promise that a specific legal document on structural safety of existing buildings would be subsequently prepared. That did not happen to this day, over almost two years after the approval of the special regime. However, a specific provision on this matter was entered in the National Strategy for Housing (2015) referring to actions to improve the seismic safety of buildings when rehabilitation works are carried out [14].

It should be added that the debate on the vulnerability of older buildings to earthquake, particularly those located in central urban areas of regions more prone to significant seismic activity (e.g., Lisbon and Algarve), has been going on for decades, without major practical results. Some municipalities have improved their risk assessment and contingency plans, but national regulations on the seismic safety of buildings still do not deal specifically with the issue and continue to organize around the requirements for new building.

A move forward from the above mentioned deadlock seems thus to depend either from a new Government electing seismic risk management as a political agenda issue (likely in the broader framework of cities' policy and urban rehabilitation policy) or from the ability of organizations stemming from society, namely the structural safety community and the consumers' associations, to push forward this concern in a way that it becomes a relevant public issue.

Meanwhile, the current four main areas of difficulty will likely persist [9]: (i) to decide on which regulatory requirements the project must meet; (ii) to prove that planned construction works do not reduce the structural safety conditions; (iii) to know which reinforcement measures are appropriate for each structural type and actual building conditions; and (iv) to estimate their cost.

Conclusions and Perspectives for Future Action

The discussion of the special regime turned into a clash between opposing positions, founded on two seemingly equally reasonable priorities: to stimulate the rehabilitation of the existing building stock *versus* to improve the structural safety of existing buildings.

The debate ended in a deadlock. Both parts argued with their reasons and objectives, mainly of strategic nature, but proved unable to support their points of view with indisputable facts:

(1) on the Government side, lack of scientific and empirical data proving a positive direct correlation between the changes in the building regulations and the reduction of costs and increase of dynamics in urban rehabilitation;
(2) on the structural safety community side, irrefutable scientific and empirical data as well as practical information on the technical solutions demonstrating their effective benefits in terms of safety and cost.

The representatives of the real estate sector and the building sector welcomed the simplification of the requirements, as they always do when deregulation is in order. The consumers' organisations and the public were mainly unaffected by the debate and were not moved by the concerns put forward by the structural safety community, which may be seen as a tell-tale of the lack of their communication ability outside the restricted professional and scientific circles.

An evolution from this point will likely be possible in two circumstances: if a new Government elects the issue as relevant for the political agenda (in the broader framework of the cities' policy and urban rehabilitation policy) or if the structural safety community is able to persuade the Government and the society by supporting its arguments with hard facts and practical tools, such as:

(1) technical guidelines and recommendations for structural rehabilitation of existing buildings according to their characteristics and conditions;
(2) a cost-benefit analysis of these solutions;
(3) a practical programme of capacity building of the construction sector to implement the prescribed solutions.

In either case it seems right to conclude that more applied research on structural rehabilitation of existing buildings according to their characteristics and conditions is needed along with training and a better dissemination of the practical results within the construction and the real estate sectors. This should be supported by an informed and enlightened debate among relevant stakeholders (i.e., government bodies, research and technical community, building and real estate sectors, consumers and citizens associations).

Finally, it is important to bear in mind that sustainability and resilience have become increasingly important factors when accessing social and economic efficiency. In fact, sustainability and resilience were the cornerstones of the argumentation from the structural safety community side, though they were never outspoken in those terms. They should be taken in higher consideration in the design and update of building regulations and properly weighted against the more short-term policy objectives (e.g., the arguments supporting the Governments' deregulation measures). Conducting these processes in the future will thus require a broader and more comprehensive approach and a different, more extended, reference time frame than those that have often been adopted in the past.

References

1. Pedro JB, Meijer F, Visscher H (2010) Technical building regulations in EU countries: a comparison of their organization and formulation, 2010 CIB world congress. Salford Quays (England). Available at https://www.researchgate.net/publication/260980505
2. UN Habitat (2003) The Habitat agenda goals and principles, commitments and the global plan of action, UN Habitat Agenda. Available at http://unhabitat.org/wp-content/uploads/2014/07/The-Habitat-Agenda-Goals-and-Principles-Commitments-and-the-Global-Plan-of-Action-2003.pdf
3. Instituto Nacional de Estatística & Laboratório Nacional de Engenharia Civil (2013) O parque habitacional e a sua reabilitação—análise e evolução 2001–2011. The housing stock and its rehabilitation—analysis and evolution 2001–2011. INE & LNEC, Lisboa. Available at http://www.ine.pt/xportal/xmain?xpid=INE&xpgid=ine_publicacoes&PUBLICACOESpub_boui=165231362&PUBLICACOESmodo=2
4. Pedro JB, Meijer F, Visscher H (2009) The Portuguese building regulation system: a critical review. Int J Law Built Env 1(2):156–175. Available at https://www.researchgate.net/publication/242021483
5. Pedro JB, Campos V (2015) Reflexões sobre a codificação das Normas Técnicas da Construção. Reflections on the codification of the building regulations. LNEC, Lisboa. Available at https://www.researchgate.net/publication/276269079
6. Pedro JB, Coelho AL, Pinto A, Pina dos Santos C, Viegas J, Grandão Lopes J, Patrício J, Campos V (2015) Regulamentação técnica da construção nas obras em edifícios existentes: análise da aplicação e sugestões de melhoria. Technical building regulations in the works in existing buildings: analysis of their application and suggestions for improvement. LNEC, Lisboa
7. Decree-Law no. 555/99 (1999) Legal framework for urban development and construction works. Official J (PT), Ser 1(291):8912–8942. Last amendment by Law no. 136/2014, 9 of September. Consolidated version available at http://www.pgdlisboa.pt/leis/lei_mostra_articulado.php?nid=625&tabela=leis
8. Decree-Law no. 307/2009 (2009) Legal framework for urban regeneration. Official J (PT) Ser 1(206):7956–7975. Last amendment by Law no. 136/2014, 9 of September. Consolidated version available at http://www.pgdlisboa.pt/leis/lei_busca_assunto_diploma.php?buscajur=estrutural&artigo_id=&pagina=1&ficha=1&nid=1197&tabela=leis&diplomas=&artigos
9. Pedro JB, Campos V (2015) Memorando: o RERU e a segurança estrutural. Memo: the RERU and structural safety. LNEC, Lisboa
10. Decree-Law no. 53/2014 (2014) Exceptional and temporary legal framework for the rehabilitation of existing buildings or fractions. Official J (PT) Ser 1(69):2337–2340. Available at https://dre.pt/application/file/25344956
11. Regime Excecional para a Reabilitação Urbana e Risco Sísmico (2014) Special legal regime for urban rehabilitation and seismic risk. Constr Mag. Available at http://www.construcaomagazine.pt/scid/webCM/defaultArticleViewOne.asp?articleID=1712&categoryID=780
12. GECoRPA Grémio do Património (2014) Comunicado de imprensa: o regime excecional para a reabilitação urbana face ao risco sísmico. Press release: the special legal regime for urban

rehabilitation and seismic risk. Available at http://www.gecorpa.pt/Upload/SalaImprensa/comunicado_RERU.pdf

13. Lei ignora risco sísmico (2014) Law ignores seismicrisk. Correio da Manhã. Available at http://www.cmjornal.xl.pt/nacional/sociedade/detalhe/lei-ignora-risco-sismico.html
14. Resolution of the Council of Ministers no. 48/2015 (2015) Approves the housing national strategy for 2015–2031. Official J (PT) Ser 1(136):4826–4850. Available at https://dre.pt/application/file/69809543

Seismic Assessment Procedures for Flexible Timber Diaphragms

Ivan Giongo, Gianni Schiro, Roberto Tomasi, Dmytro Dizhur and Jason Ingham

Abstract A review of the available procedures and guidelines for the in-plane assessment of existing timber diaphragms in unreinforced masonry (URM) buildings is provided. Most of the provisions that concern flexible wood floors and that are valid for URM, can as a matter of fact be adopted for historic masonry buildings reinforced with timber frames. Recommendations from the most recent international standards like ASCE 41-13 and NZSEE 2015 were thoroughly analysed. Particular attention was given to producing a step-by-step practice-oriented guideline that provides background information which helps understand better the procedure frame work and also gives suggestions for alternative interpretations when multiple choices are possible.

Keywords Flexible timber floors · Diaphragm in-plane behaviour · Seismic assessment

I. Giongo (✉) · G. Schiro · R. Tomasi
Department of Civil, Environmental and Mechanical Engineering,
University of Trento, Trento, Italy
e-mail: ivan.giongo@unitn.it

G. Schiro
e-mail: gianni.schiro@unitn.it

R. Tomasi
e-mail: roberto.tomasi@unitn.it

D. Dizhur · J. Ingham
Department of Civil and Environmental Engineering,
University of Auckland, Auckland, New Zealand
e-mail: ddiz001@aucklanduni.ac.nz

J. Ingham
e-mail: j.ingham@auckland.ac.nz

H. Cruz et al. (eds.), *Historical Earthquake-Resistant Timber Framing in the Mediterranean Area*, Lecture Notes in Civil Engineering,
DOI 10.1007/978-3-319-39492-3_22

Introduction

At present ASCE 41-13 "Seismic Evaluation and Retrofit of Existing Buildings" [1] is the standard that addresses the flexible timber diaphragm issue most extensively. Brief origin of the procedure outlined in ASCE 41-13 [1] is discussed. ASCE 41-13 [1] is the updated version of the ASCE 41-06 "Seismic Rehabilitation of Existing Buildings" [2] which is mainly based on FEMA 356 "Prestandard and Commentary for the Seismic Rehabilitation of Buildings" [3]. FEMA 356 updates and converts FEMA 273 "NEHRP Guidelines for the Seismic Rehabilitation of Buildings" [4] and into mandatory language. Consequently, the roots of the "diaphragm assessment procedure" currently adopted by the ASCE 41 go back to the late '90s when FEMA 273 [4] was drafted. The research background is mainly represented by the experimental program carried out by ABK and published in the Topical Report TR03 in 1981 [5]. The influence of the experimental outcome from [5] is still noticeable even in the most recent standards.

A section of the paper deals with New Zealand provisions on flexible timber diaphragms as a new approach has been recently introduced. The roots of the diaphragm assessment procedure contained in NZSEE 2006 [6] are derived from ASCE and FEMA and consequently the focus will be on NZSEE 2015 [7] which updates Chap. 10 of [6]. The experimental background for the changes introduced by Ref. [7] lays mainly in the research work recently carried out at the University of Auckland (see Giongo et al. [8] for detailed reference).

American Standard ASCE 41-13

Seismic input for URM buildings. According to ASCE 41-13 [1], the distribution of seismic forces for unreinforced masonry (URM) buildings with flexible diaphragms is determined by the in-plane behaviour of the diaphragms. In fact, when referring to Linear Static Procedures (valid for demand/capacity ratios DCR $\leq$ 3 or less if *m-factor* is smaller; *m-factor* is the component demand modification factor to account for expected ductility) the pseudo seismic force has to be calculated for each span at each level by using a formula that depends on the diaphragm period. A diaphragm is classified as flexible if the maximum horizontal deformation is more than twice the average story drift. The diaphragm period T_d can be estimated by applying the Rayleigh's quotient for a fixed end flexural beam:

$$T_d = (3.07\Delta_d)^{0.5} \tag{1}$$

where Δ_d is the maximum deflection [m] due to a lateral force equal to the weight tributary to the diaphragm (corresponding to the inertial force generated by an acceleration of 1 g). Once the period has been determined, the pseudo seismic force is obtained from:

$$V_d = C_1 C_2 C_m S_a W_d \quad (2)$$

where V_d is the pseudo lateral force; C_1 is a *modification factor to relate expected maximum inelastic displacements to displacements calculated for linear elastic response*; C_2 is a *modification factor to represent the effect of pinched hysteresis shape, cyclic stiffness degradation and strength deterioration on maximum displacement response*; C_m is a factor equal to 1; S_a is the response spectrum acceleration; W_d is the effective seismic weight of the diaphragm (including the weight of the out-of-plane walls). Both C_1 C_2 and S_a are directly dependent on the natural period evaluated by using Eq. 1. Such method of force distribution can be applied only if diaphragm deflections do not exceed 6 in (approximately 150 mm). No amplification effects due to the diaphragm location from the ground are considered as wall deformations are assumed to be negligible when compared to diaphragm deformations. To calculate Δ_d, the default shear stiffness values (G_d) given in Table 1 can be employed as the *diaphragm shall be considered to remain elastic under the prescribed lateral forces* [1]. G_d represents the equivalent shear stiffness of a structural component and should not be confused with the shear modulus of a material. Δ_d can be obtained as following:

$$\Delta_d = \beta \frac{W_d L}{G_d B} \quad (3)$$

where L is the diaphragm span (dimension orthogonal to the loading direction) and B is the diaphragm width. β is a coefficient that is related to the static scheme

Table 1 Default expected strength values for wood diaphragms [1]

Diaphragm type		Shear stiffness (G_d)		Expected strength (Q_{CE})	
		[lb/in]	[KN/m]	[plf]	[KN/m]
Single straight sheathing		2000	350	120	1.75
Double straight sheathing	Chorded	15,000	2627	600	8.76
	Unchorded	7000	1226	400	5.84
Single diagonal sheathing	Chorded	8000	1401	600	8.76
	Unchorded	4000	701	420	6.13
Straight sheathing over diagonal sheathing	Chorded	18,000	3152	900	13.13
	Unchorded	9000	1576	625	9.12
Double diagonal sheathing	Chorded	18,000	3152	900	13.13
	Unchorded	9000	1576	625	9.12
Panel overlay on sheathing unblocked	Chorded	9000	1576	450	6.57
	Unchorded	5000	876	300	4.38
Panel overlay on sheathing blocked	Chorded	18,000	3152	–	–
	Unchorded	7000	1226	–	–

assumed (β = 1/4 for a shear beam under point load; β = 1/8 for a shear beam under uniformly distributed load; β = 5/32 for a shear beam under parabolic load).

Alternatively, ASCE 41-13 [1] provides formulations to calculate diaphragm midspan deflection at yield, which in case of single straight sheathing diaphragms, double straight sheathing diaphragms, single diagonal sheathing diaphragms, diagonal sheathing with straight sheathing or flooring above diaphragms and double diagonal sheathing diaphragms is:

$$\Delta_y = \frac{5v_yL^3}{8EAB} + \frac{v_yL}{4G_d} + \frac{\sum(\Delta_cX)}{2B} \tag{4}$$

where Δ_y is the deflection at yield [in.]; A is the area of the diaphragm chords section [in.^{2}]; E is modulus of elasticity of diaphragm chords [lb/in.^{2}]; $\sum(\Delta_cX)$ is the sum of individual chord-splice slip values on both sides of the diaphragm, each multiplied by its distance to the nearest support; v_y is the shear per unit length at yield in the direction under consideration [lb/in.]. G_d is expressed in [lb/in.], while the length quantities L and B are in [in.] and [ft] respectively [1]. However some consistency issues are present unless in the first term of the formula v_y is expressed in [lb/ft] and L in [ft]. For more clarity please see Appendix A of ATC guidelines [9]. Equation 4 could be rewritten:

$$\Delta_d = \frac{5W_dL^3}{192EAB} + \frac{W_dL}{8G_dB} + \frac{\sum(\Delta_cX)}{2B} \tag{5}$$

where Δ_d [m], W_d [kN], G_d [kN/m], L [m], B [m], Δ_c [m], X [m], A [mm^2] and E [Gpa].

It appears that to comply with the static scheme proposed by Eq. 4 a factor β = 1/8 should be adopted for Eq. 3. In the former version of ASCE 41 [2] instead:

$$\Delta_y = \frac{v_yL}{2G_d} \tag{6}$$

which made it straight forward to employ Eq. 3 with β = 1/4 as Eq. 3 can be directly derived from Eq. 6. Obviously using the three-term formula instead of Eq. 3 gives higher values of Δ_d which generally lead to smaller pseudo lateral forces. In case of existing timber diaphragms constituted by wood structural panels (wood structural panel sheathing, wood structural panel overlays on straight or diagonal sheathing) Δ_y is calculated as:

$$\Delta_y = \frac{5v_yL^3}{8EAB} + \frac{v_yL}{4Gt} + \frac{\sum(\Delta_cX)}{2B} + \alpha Le_n \tag{7}$$

where G is the modulus of rigidity of wood structural panels; t is the effective thickness of wood structural panel for shear; α is a parameter depending on the nailing pattern (α = 0.188 for constant nailing; α = 0.376 for variable nailing) and

e_n is the nail deformation at yield. Equation 7 reflects the assumption of the girder analogy where the sheathing (web) is solely responsible for the shear transfer and the chord elements (flange) are supposed to resist all the bending forces. This results in the deformability contribution of each component being in series with one another as the sheathing and the chords behave independently. Conversely in Eq. 3 the shear deformability contribution (second term of the formula) is not related to a shear modulus as in Eq. 7 but to an equivalent shear stiffness that has been derived from actual deflection values obtained from experimental results [5]. This means that the chord stiffening contribution and also its deformability are already included in the G_d values. In fact different G_d values are provided by the standard for chorded and unchorded diaphragms (see Table 1). The first term of Eq. 4 might therefore appear as redundant, unless the G_d values for chorded diaphragms refer to specimens with "non-deformable" chords (negligible deformation).

Diaphragm verification. Once the pseudo lateral seismic force is determined the diaphragms need to be checked. Because in URM buildings diaphragms transfer the inertial load from out-of-plane loaded walls to the in-plane resisting elements, they are to be considered as primary elements. In order to evaluate the diaphragm capacity it is necessary to distinguish between deformation controlled and force controlled actions, depending on the component behaviour. Generally, components that exhibit behaviours like type 3 curve (Fig. 1a) are classified as force-controlled, while components that exhibit behaviours like type 1 and 2 curves can be classified as deformation-controlled if the elastic-plastic range (before any substantial loss of in-plane capacity) is at least twice the elastic range. Force-displacement relations for timber diaphragms can be obtained by using the modelling parameters that are provided by the standard and reported in Table 2 (Fig. 1b).

The result is that all the timber diaphragm typologies addressed by ASCE 41 are to be treated as deformation controlled (cf. *d*-parameters in Table 2). Besides, the designation of *m*-factors (see Table 2) is a sufficient requirement for the classification of a component as deformation-controlled. The safety check is:

$$mkQ_{CE} \geq \frac{V_d}{2B} \tag{8}$$

where k is the knowledge factor (range 0.75–1.00) and Q_{CE} is the expected diaphragm strength (mean value) which is provided by ASCE 41 as illustrated in

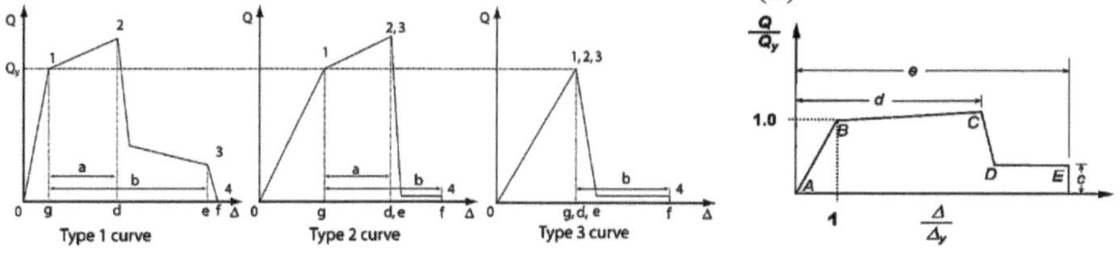

Fig. 1 Component force versus deformation curves [1]

Table 2 Modelling parameters and acceptance criteria for linear and nonlinear procedures [1]

Diaphragms[a]	Chord	Linear procedures		Nonlinear procedures				
		L/B	m-factors primary LS	L/B	Modelling parameters			m-factors primary LS
					Δ/Δ_y		Resid. s.	
					d	e	c	
Single straight sheathing	Yes	≤ 3	2	≤ 2	2.5	3.5	0.2	2.5
	No	≤ 3	1.5	≤ 2	2	3	0.3	2
Panel overlay on sheathing	Yes	≤ 3	2.5	≤ 3	3	4	0.3	3
		4	2	4	2.5	3.5	0.3	2.5
	No	≤ 2.5	2	≤ 2.5	2.5	3.5	0.4	2.5
		3.5	1.5	3.5	2	3	0.4	2

[a]For the diaphragm types not reported here, please refer to Tables 12.3 and 12.4 in [1]

Table 1. Q_{CE} values are consistent with the experimental shear capacity observed by Ref. [5] where two symmetrical point loads have been applied at one third of the specimen span (load direction parallel to the joists; L/B = 3). The related bending capacity (M_{res}) has therefore to be:

$$M_{res} = (mkQ_{CE})BL/3 \tag{9}$$

which ensures that the bending capacity required to resist the lateral seismic force is automatically guaranteed if Eq. 7 is satisfied, even when a parabolic inertial load distribution is adopted as prescribed by ASCE 41 (Fig. 2).

The following statement is contained in ASCE 41-13 [Table C7-1]: "*If the diaphragm carries lateral loads from vertical-force-resisting elements above the diaphragm level, then M and V shall be considered force-controlled actions*". This seems to be the case when the diaphragm joists (oriented parallel to the seismic action) are supported on the out-of-plane walls. In fact the diaphragms transfer to the in-plane resisting walls the inertial load of the vertical-force-resisting walls that are subjected to out-of-plane earthquake action (it should be noted though, that m-factors are available for the out-of-plane assessment of masonry walls). For force controlled actions, the following inequality needs to be satisfied:

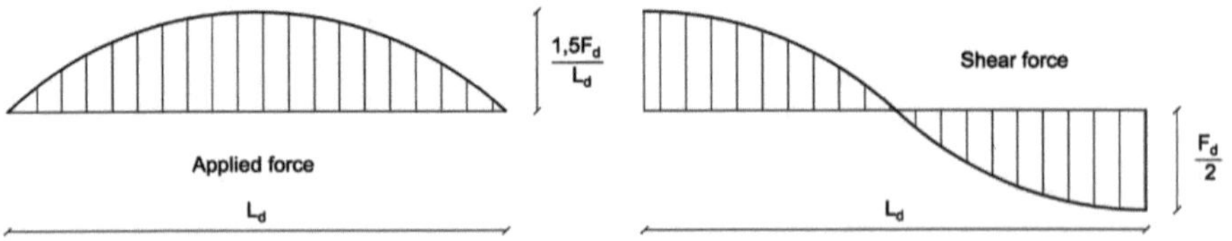

Fig. 2 Plausible force distribution in a flexible diaphragm [1]

$$kQ_{CL} \geq Q_{UF} \rightarrow 0.85kQ_{CE} \geq \frac{C_m S_a W_d}{2JB} \tag{10}$$

where Q_{CL} is the lower bound estimate of the component strength (the relation $Q_{CL} = 0.85 \cdot Q_{CE}$ can be assumed); Q_{UF} is the force-controlled action; J is the force-delivery reduction factor (≥ 1) taken as the small DCR of the components in the load path delivering force to the diaphragm. Approximately J is equal to 2.0 for high level of seismicity, 1.5 for moderate seismicity and 1.0 for low seismicity.

Non-Linear Procedures. For Non-Linear Procedures the acceptance criteria for timber diaphragms are provided for each performance level as acceptable deformation ratios (Table 2). The diaphragms are to be modelled using full backbone curves as obtained through the parameters provided in Table 2. The yield capacity Q_y (point B, Fig. 1b), required to define the force-deformation response, can be taken as the expected strength Q_{CE} (point C, Fig. 1b) divided by 1.5. The diaphragm deformation capacity corresponds to the deformation limit multiplied by the knowledge factor. In case of force-controlled actions (diaphragm carrying lateral loads from vertical-force-resisting elements), diaphragms can be modelled as linear elastic components. The maximum analysis forces shall be compared to the diaphragm lower-bound strength multiplied by the knowledge factor ($0.85 \cdot k \cdot Q_{CE}$). Alternative modelling parameters and acceptance criteria based on experimental cyclic testing are permitted by the standard.

New Zealand Standard NZSEE

Seismic input and diaphragm demand. As in [1], the in-plane resisting system of URM structures with flexible diaphragms is regarded as rigid and the building response is governed by the dynamic behaviour of the diaphragms (no amplification factor). The diaphragm period is based on the formula proposed by Wilson et al. [10] where the Rayleigh's quotient is applied to a shear beam under parabolic load distribution:

$$T_d = 0.7 \cdot \left(\frac{W_d L}{G'_{d,eff} B} \right)^{0.5} \tag{11}$$

where $G'_{d,eff}$ is the effective shear stiffness of diaphragm. Experimental support to the approach adopted in Eq. 11, can also be found in [11, 12] where dynamic snap-back testing was performed on full-scale vintage diaphragms. The effective shear stiffness $G'_{d,eff}$ can be determined as follows:

$$G'_{d,eff} = \alpha_w \cdot \frac{A_{net}}{A_{gross}} \cdot G_d; \quad \alpha_w \cong 1 + \left(\frac{t_l^3}{H_l^3} + \frac{t_u^3}{H_u^3} \right) \frac{L^2}{B} \frac{E_m}{G_d} \frac{A_{gross}}{A_{net}} \tag{12}$$

where A_{gross} is the diaphragm plan area included penetrations; A_{net} is the diaphragm net area (the direct relation between the penetration size and diaphragm stiffness is based on numerical results from Wilson [13] and is valid up to penetration areas equal to 10 % of the gross area); α_w is a parameter to account for the out-of-plane stiffness of face-loaded walls; t_l is the thickness of the walls below the diaphragm; t_u is the thickness of the walls above the diaphragm; H_l is the height of the walls below the diaphragm; H_u is the height of the walls above the diaphragm; E_m is Young's modulus of masonry. α_w is based on a simplified analytical approach where a one-way vertically spanning behaviour is adopted. The stiffness contribution of the out-of-plane walls is obtained by assuming cracked walls responding as cantilevers above and below the diaphragm [8, 14].

Like the ASCE standard, also the New Zealand standard provides reference values for diaphragm shear stiffness, which are reported in Table 3. It can be noted that an orthotropic behaviour is attributed to existing diaphragms, with the floors loaded parallel to the joists being stiffer than the floors loaded orthogonal to the joists. Data in Table 3 refer to single straight sheathed diaphragms which are the most representative of New Zealand existing timber diaphragms. In case of other floor typologies (double straight sheathing, single diagonal sheathing, double diagonal sheathing and straight sheathing above diagonal sheathing), stiffness multipliers derived from ASCE default properties are provided.

Existing diaphragm conditions are accounted for through a condition rating consisting of three classes: good, fair and poor. Unlike the k-factor previously mentioned, which is related to the uncertainty level associated with the data/info collection process, the condition rating classes depend entirely on diaphragm conditions (refer to Table 10.7 in [7]). Diaphragm mid-span displacement demand

Table 3 Shear stiffness values [kN/m] for straight sheathed vintage timber floor diaphragms [7]

Direction of loading	Joist continuity	Condition rating	Shear stiffness (G_d)[a]
Parallel to joists	Continuous or discontinuous joists	Good	350
		Fair	285
		Poor	225
Perpendicular to joists[b]	Continuous joist, or discontinuous joist with reliable mechanical anchorage	Good	265
		Fair	215
		Poor	170
	Discontinuous joist without reliable mechanical anchorage	Good	210
		Fair	170
		Poor	135

[a]Values may be amplified by 20 % when the diaphragm has been renailed using modern nails and nail guns

[b]Values should be interpolated when there is mixed continuity of joists or to account for continuous sheathing at joist splice

is determined assuming an ideal scheme of a shear beam under parabolic load distribution (as per [1]) where a shear factor for rectangular section (6/5) is adopted:

$$\Delta_d = \frac{3}{16}\frac{C(T_d)W_dL}{G'_{d,eff}B} \tag{13}$$

where $C(T_d)$ is the horizontal design action coefficient for the diaphragm.

Diaphragm verification. Once the diaphragm in-plane displacement demand has been defined, it has to be checked against the diaphragm deformation capacity. Three different acceptance criteria govern the diaphragm performance [8]. The first criterion limits diaphragm distortion so that the most severely loaded connections (located at the diaphragm lateral supports where the nail couple rotation is maximum) exhibit nail slips not detrimental to the connection strength capacity.

$$\Delta_d \leq \varphi\frac{Le_n}{2s} \tag{14}$$

where e_n is the maximum permissible nail slip; s is the nail spacing within the board-to-joist joint and φ is a coefficient that depends on the deformed shape observed for single straight sheathed diaphragms when loaded parallel (φ = 5/8) or orthogonal (φ = 1) to the joist direction. The actual deflection limit values suggested by the standard, reflect the average geometrical/mechanical properties of board-to-joist connections in New Zealand vintage diaphragms. By assuming a permissible nail slip of 6 mm, a nail spacing of 100 mm, Eq. 14 becomes:

$$\Delta_d \leq \frac{L}{33}(perpendicular)\Delta_d \leq \frac{L}{53}(parallel). \tag{15}$$

Despite one might argue than selecting an admissible nail deformation equal to the nail slip at the peak strength could be overly conservative when Life Safety is concerned, it should be noted that the distortion criterion is generally (for New Zealand diaphragms) less severe than the other criteria. The second criterion addresses the out-of-plane stability of face-loaded walls. The critical diaphragm in-plane displacement that onsets the out-of-plane collapse of the supported walls ensues from experimental results [15] and is equal to 50 % of the wall thickness:

$$\Delta_d \leq 0.5t_w \tag{16}$$

where t_w is the thickness of the supported walls (for cavity walls t_w is the thickness of the inner wythe). The third criterion is taken from NZS 1170.5 [16] where a 2.5 % limit is set for the inter-storey drift. Under the hypothesis of negligible deformation of the in-plane resisting wall system (as accepted for determining the diaphragm demand) and rigid foundations, the third criterion can be written as:

Table 4 Strength values for existing materials [6]

Materials	Strength values [KN/m]	Φ^a
Roofs with straight sheathing and roofing applied directly to the sheathing	6	0.7
Roofs with diagonal sheathing and roofing applied directly to the sheathing	15	
Floors with straight tongue and groove sheathing	6	
Floors and roofs with sheathing and existing plaster renailed to the joists or rafters	Add 2 KN/m to the values for timber framed walls	

[a]Φ is the strength reduction factor

$$\Delta_d \leq 0.025H_l + \Delta_{d,level\ immediately\ below}. \tag{17}$$

In both Eqs. 16 and 17, diaphragms located on levels adjacent to that of the diaphragm being assessed, are assumed to remain undeformed. The final displacement capacity of the diaphragm is the minimum among the three values provided by the acceptance criteria.

Diaphragm strength capacity shall also be assessed. One should refer to the *probable strength capacity,* which is: "*best estimate of the actual strength obtained from available information and/or testing*". Mean values from test results can be employed as probable strengths. When test results are not available the values reported in Table 4 (shear strength at yield of the diaphragm assembly) can be used, together with strength reduction factors equal to 1 [7].

$$Strengthvalue \geq \frac{C(T_d)W_d}{2B}. \tag{18}$$

Alternatively, formulae to directly calculate the diaphragm strength can be found in Appendix 11B of [6]. As evidenced by Oliver [17], the choice of using the procedure available in Appendix 11B instead of the default strength values, leads to significantly smaller capacities.

Further Remarks on Default Stiffness and Strength Values

As already mentioned, the G_d values provided by ASCE standard (Table 1) derive from experimental evidence. In particular they refer to the equivalent secant shear-stiffness at yield (based on estimated value which for straight sheathed diaphragms corresponds to approximately 15 mm) obtained from quasi-static cyclic tests carried out on specimens which had been previously subjected to testing. Stiffness values given in Table 3 are also experimentally derived but relate to equivalent secant shear-stiffness measured at a displacement of 100 mm, due to the critical condition of out-of-plane loaded URM walls and also maximum admissible inter-storey drift. Despite the default shear strength value for straight sheathed diaphragms from NZSEE (yield strength = 6 KN/m, Table 4) is significantly higher than that suggested by ASCE (peak strength = 1.75 KN/m, Table 1), it appears to be consistent with the experimental data available for vintage timber diaphragms in New Zealand (Giongo et al. [11] observed no damage up to a shear load of approximately 6 KN/m).

Conclusions and Future Work

Timber diaphragm assessment procedures as recommended by two widely recognized standards for seismic assessment of existing buildings were considered. The main approach-differences can be synthesized as: (1) diaphragm demand is based on the elastic shear stiffness in ASCE [1] while NZSEE [7] employs the secant stiffness at 100 mm midspan deflection; (2) unlike ASCE, NZSEE regards diaphragm in-plane behaviour as orthotropic; (3) different static scheme are assumed for the evaluation of the diaphragm period; (4) direct limitation to diaphragm deflection is prescribed only by NZSEE for linear procedure. An extensive diaphragm modelling work was undertaken by the authors, the results of which are currently being analysed and will be reported in the near future. Preliminary results seem to suggest that factors like the aspect ratio and the diaphragm size might have a significant influence on the diaphragm response and on the definition of diaphragm properties. Therefore, in order to compare and correctly evaluate different diaphragm assessment procedures, the specific background conditions behind each procedure should be accounted for (e.g. details about the experimental background, assumptions on diaphragm/building behaviour that are based on aspects common in the country where the standard has originally been applied).

Acknowledgments The authors gratefully acknowledge the Italian RELUIS consortium for the support to the research.

References

1. ASCE/SEI 41-13 (2013) Seismic evaluation and retrofit of existing building. American Society of Civil Engineers, Virginia
2. ASCE/SEI 41-06 (2006) Seismic rehabilitation of existing building. American Society of Civil Engineers, Virginia
3. FEMA 356 (2000) Prestandard and commentary for the seismic rehabilitation of buildings. American Society of Civil Engineers
4. FEMA 273 (1997) NEHRP guidelines for the seismic rehabilitation of buildings. American Society of Civil Engineers
5. ABK—TR—03 (1981) Methodology for mitigation of seismic hazards in existing unreinforced masonry buildings: diaphragm testing. ABK
6. NZSEE (2006) Assessment and improvement of the structural performance of buildings in earthquakes. New Zealand Society for Earthquake Engineering, Wellington

7. NZSEE (2015) Assessment and improvement of the structural performance of buildings in earthquakes—section 10 revision seismic assessment of unreinforced masonry buildings. New Zealand Society for Earthquake Engineering, Wellington
8. Giongo I, Wilson A, Dizhur DY, Derakhshan H, Tomasi R, Griffith MC, Quenneville P, Ingham J (2014) Detailed seismic assessment and improvement procedure for vintage flexible timber diaphragms. NZSEE Bull 47:97–118
9. Guidelines for the design of horizontal wood diaphragms (1981). ATC
10. Wilson A, Quenneville P, Ingham J (2013) Natural period and seismic idealization of flexible diaphragms. Earthq Spectra, 1003–1019
11. Giongo I, Dizhur DY, Tomasi R, Ingham JM (2015) Field testing of flexible timber diaphragms in an existing vintage URM building. J Struct Eng 141 (United States)
12. Giongo I, Dizhur DY, Tomasi R, Ingham JM (2013) In-plane assessment of existing timber diaphragms in URM buildings via quasi-static and dynamic in-situ tests. Adv Mater Res SHATIS 778:495–502
13. Wilson A (2012) Seismic assessment of timber floor diaphragms in unreinforced masonry buildings. Doctoral dissertation, The University of Auckland, New Zealand
14. Derakhshan H, Griffith MC, Ingham JM (2013) Out-of-plane behavior of one-way spanning URM walls. ASCE J Eng Mech, 409–417
15. Griffith MC, Vaculik J, Lam NTK, Wilson J, Lumantarna E (2007) Cyclic testing of unreinforced masonry walls in two-way bending. Earthq Eng Struc, 801–821
16. NZS 1170.5 (2004) Structural design actions, part 5: earthquake actions—New Zealand. Standards New Zealand, Wellington
17. Oliver SJ (2010) A design methodology for the assessment and retrofit of flexible diaphragms in unreinforced masonry buildings. SESOC J, 19–49

Part IV
Diagnosis

The Role of Terrestrial LiDAR for Mechanical and Safety Evaluation of Historic Timber Floors

Clara Bertolini-Cestari, Antonia Spanò, Stefano Invernizzi, Elisabetta Donadio, Tanja Marzi and Giulia Sammartano

Abstract The paper describes an activity conducted by a multi-disciplinary research group (architecture, civil engineering, technology, geomatics) for the mechanical evaluation of a timber floor belonging to a historic residential building located in the city center of Torino. The floor was made of beams of solid wood and wooden boards and presented a mid-span deflection. Since the floor structure was hidden by a suspended ceiling, the difficulty was to determine with precision all the geometrical characteristics of the main beam and its creep with the aim of verifying its mechanical resistance and stiffness. Through an accurate application of laser-scanner technique, it was possible to carry out the precise assessment of the structure.

Keywords Timber floor · LiDAR · Safety

C. Bertolini-Cestari (✉) · A. Spanò · E. Donadio · T. Marzi · G. Sammartano
Department of Architecture and Design, Politecnico di Torino,
Viale Mattioli 39, 10125, Torino, Italy
e-mail: clara.bertolini@polito.it

A. Spanò
e-mail: antonia.spano@polito.it

E. Donadio
e-mail: elisabetta.donadio@polito.it

T. Marzi
e-mail: tanja.marzi@polito.it

G. Sammartano
e-mail: giulia.sammartano@polito.it

S. Invernizzi
Department of Structural, Geotechnical and Building Engineering,
Politecnico di Torino, C.so Duca degli Abruzzi 24, 10128 Torino, Italy
e-mail: stefano.invernizzi@polito.it

H. Cruz et al. (eds.), *Historical Earthquake-Resistant Timber Framing in the Mediterranean Area*, Lecture Notes in Civil Engineering,
DOI 10.1007/978-3-319-39492-3_23

Introduction

One of the main difficulties in assessing this timber floor structure was certainly due to the suspended false ceiling which was hiding the structure above, together with the fact that the flat was occupied by the inhabitants living there during the intervention. The present study reports the geometrical survey, the diagnostic and the preliminary structural analysis.

Thanks to the LiDAR (Light Detection and Ranging) survey it was possible to prove that the central beam of the floor, which is the most loaded, is affected by inelastic deformations that are not compatible with any of the values prescribed by the standards. The diagnosis of the timber beam was executed with drill resistance measurement (Resistograph®), which revealed the good quality of the wood (Larch). On the contrary, the homogeneity of the material was poor, and voids at the extrados of the beam midspan have been detected. The accounting for the detailed geometry obtained with laser survey and void location provided by Resistograph® analysis allowed for a proper simulation and assessment of elastic and inelastic deformation.

The structural assessment was conducted according to the Italian standards, calculating both the maximum bending stresses and the maximum initial and long-term displacements, under the effect of dead weights and serviceability loads. As a result, although the strength limit was verified, the displacements were well above the limit prescribed by the standard. Finally, it is worth noting that in a seismic event, vertical acceleration spectra must be considered together with horizontal ones, which directly emphasize the beam deflections. In this view, a retrofitting of the beam is necessary in order to improve the bending stiffness both with respect to static and dynamic loading.

Terrestrial LiDAR Survey and Application for Close-Range and Indoor Environment

Nowadays, geomatic multisensor tools are well-known methods for the knowledge and documentation of multiscale objects [1]. Two of the most widespread techniques to acquire 3D information are LiDAR (active sensor) and digital photogrammetry (passive sensor). These techniques allow to acquire high precision data, suitable for several applications, and to generate accurate 3D models (point clouds and textured meshes). These models allow to study not only the geometry, but also to assess material degradation and pathologies. In addition, the structural model can be obtained from the identification of the generating profiles, together with valuable information about past construction techniques.

In recent years, terrestrial laser scanning has become a standard method of data acquisition for many applications in close range domain. An enormous variety of working fields can be studied with such technique, but one of the most important and extensive is the modelling of buildings, which is used for 3D city models, architecture, archaeology or, more generally, Cultural Heritage [2]. Many applications have demonstrated the high potentialities of such technique also when applied to wooden structures and beams [3]. 3D laser technologies give the possibility to record a big amount of high resolution data also in very articulated and complex objects such as timber structures, allowing to detect shape anomalies such as subsidence or other pathologies in detail, supporting structural assessment.

It is important to highlight that historical timber roof structures are often rough, dusty and dark. This last condition can cause a difficulty for photogrammetric method, often used together with TLS, whereas laser sensor is unaffected from semi darkness. Some tests have proved that the recorded laser intensity values show somewhat higher values for night-time measurements [4]. This is determined by the fact that, contrary to the *passive sensors* (photogrammetry), laser scanner uses an encoded light (*active sensor*) that is almost indifferent to the environment's light conditions in which it operates, allowing to better achieve also shaded portions.

3D generated models allow to understand the geometry of such structures, their state of maintenance and stability, representing a support for further analysis, for example in the conservation, restoration and consolidation phase. A 3D model of a wooden system allows to clearly read the whole structure in its complexity. At first, wooden structures can be very difficult to read and to separate into their different parts, since more elements are involved. However, from detailed 3D models it is possible to analyse single parts and understand their hierarchical structure [5].

For all the reasons explained, we evaluated the laser scanning technique as the best procedure to create a 3D model in this case study. The goal of the survey was to elaborate a 3D accurate model understandable in each of its different element, on which it was possible to measure the geometrical characteristics of the main beam and its creep with the aim of verifying the mechanical resistance and stiffness.

Data Acquisition and Initial Point Cloud Elaboration

The geometric survey was carried out carefully in order to acquire the visible structure as complete as possible (Fig. 1).

The acquisition phase has shown a series of initial difficulties, based on the following reasons:

- the location of the scans acquisition and the choice of the targets placement were more difficult due to the large number of objects and furniture present in the room (Fig. 2);
- the limited breadth of the breach on the suspended ceiling has limited the visibility of the structure itself.

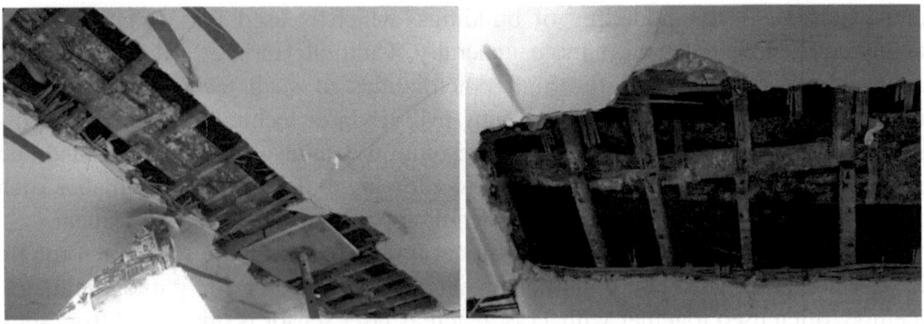

Fig. 1 Images of the timber beam through the broken ceiling: at its maximum deflection in the middle of the room (*right*); at its connection with the wall, in which there is a thickening element between beam and secondary beams (*left*)

Fig. 2 Instruments used in the survey: total station *Leica TCR850 power* (ext. *right*), and terrestrial laser scanner *Laser Focus3D Faro-Instruments* (*right*). The *Focus3D* during the acquisition in the indoor location (*left*); one of the six *targets* used as a similar control points, helpful to register accurately each point cloud to another (ext. *left*)

The acquisition phase has been featured by strengthened procedures:

- Several edges of the room ceiling and the *Resistograph* test points positioned on the beam have been measured with a total station. These measures allowed an initial estimation of the structure depression and provided useful coordinates for the scans registration.
- 6 targets have been placed all around the room, in well visible places so that they were visible from all the scan positions (Fig. 2). This operation is fundamental since they represent homologous points, useful in the clouds registration phase.
- The LIDAR survey has been realized with the terrestrial Laser Scanner Focus3D —CAM 2 (Fig. 2). Three scans have been executed in position that have been chosen in order to optimize the complete data recording of at least three faces of the beam. In particular, they have been placed not directly under the beam, but slightly laterally, so that it was possible to acquire more data diagonally between the timber structure and the suspended ceiling.

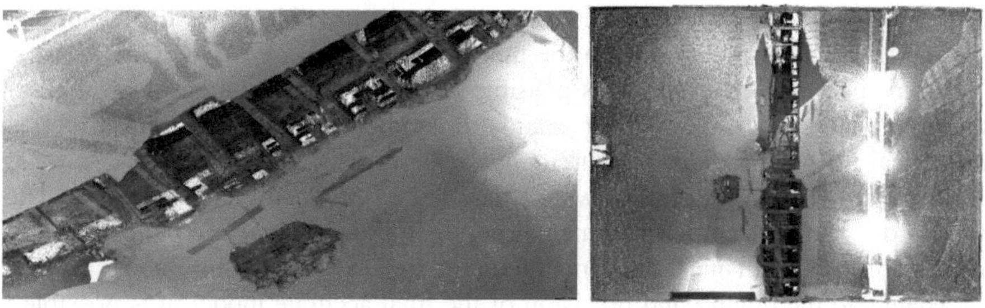

Fig. 3 The registered *point cloud* of the intrados ceiling: view of the break ceiling for diagnostic evaluation of the beam (*left*) and bottom-up view of beam and secondary beams (*right*)

- The obtained clouds have been post-processed using the FARO software SCENE, for adjusting them in the same reference system, through reference targets and measured points (Fig. 3).

3D Model Generation, Optimization, and Extraction of 3D Information for 2D Representation

The metric survey produced a highly detailed 3D point cloud made up of by millions of points which represented both the ceiling and the visible timber structure. The following step was the generation and optimization of a mesh surface, after a decimation and interpolation of the cloud (Fig. 4). This operation creates continuous models which are able to describe in detail geometrical characterizations; in our study it modelled the ceiling and the wooden structure. In this processing phase, we used a software of point cloud managing and processing, 3D meshing, CAD surface reconstruction, 3D comparison, which is often used in a wide range of 3D survey application: *3DReshaper* (by Technodigit).

The processed model can be visualized in a 3D digital space and some volumetric operations on the geometrical surface can be automatically run: an unlimited number of horizontal and vertical profile sections can be extracted wherever the

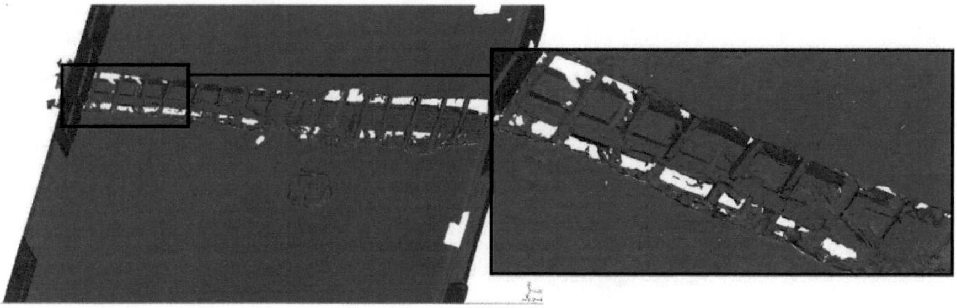

Fig. 4 View of the *mesh* of the ceiling and the beam, and zoom on the model that exactly represents the real configuration of the wooden structure

user needs them. With specific software, it is possible to extract unlimited two-dimensional views also from the point cloud.

From such extracted data, it is possible to elaborate two-dimensional representations (plans and sections) for standard architectonical views and to obtain numerical information and measurements.

In particular, several profiles were examined by producing cross sections. They were processed in specific fundamental points in order to evaluate punctual ceiling depression and mechanical deformation on the timber beam at those points (Fig. 5).

Thanks to the quality and precision of the final accurate model, it is possible to evaluate at any point the dimension of structural elements, their state of deterioration and pathologies in order to verify the mechanical resistance (Fig. 6).

Another potentiality offered by the model analysis and numerical evaluation of surfaces, has been the exploration of the depression trend of the suspended wattle ceiling. This representation (Figs. 6 and 7) has been obtained by progressive planar section of the surface along the z axis (*isolines curves* representation).

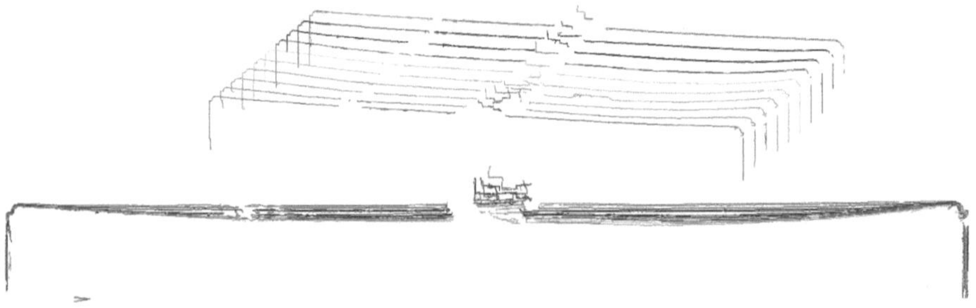

Fig. 5 Progression of cross section through ceiling and wooden structure, in a 3D spatial view (*top*) and a planar view (*down*) in which it is visible the depression of the wattle dropped ceiling surface

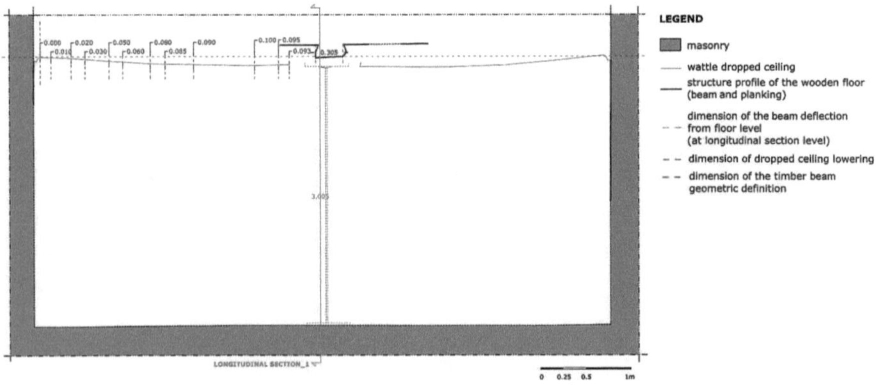

Fig. 6 CROSS SECTION_4: transversal section of the wooden beam and evaluation of the increasing depression of the wattle dropped ceiling

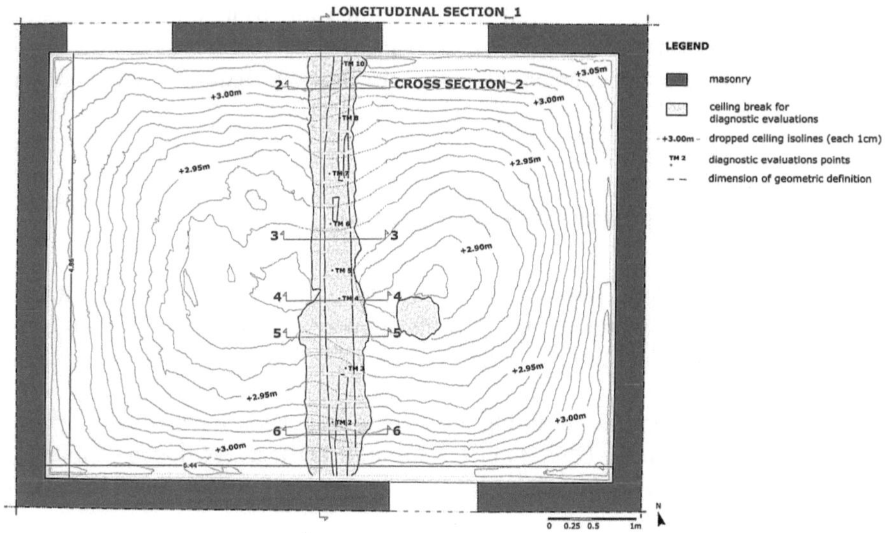

Fig. 7 CEILING PLAN of the room, in a bottom-up view, with indication of longitudinal and cross sections: isolines, derived from progressive planar sections each 1 cm (indicated each 5 cm step), show the depression of the dropped ceiling

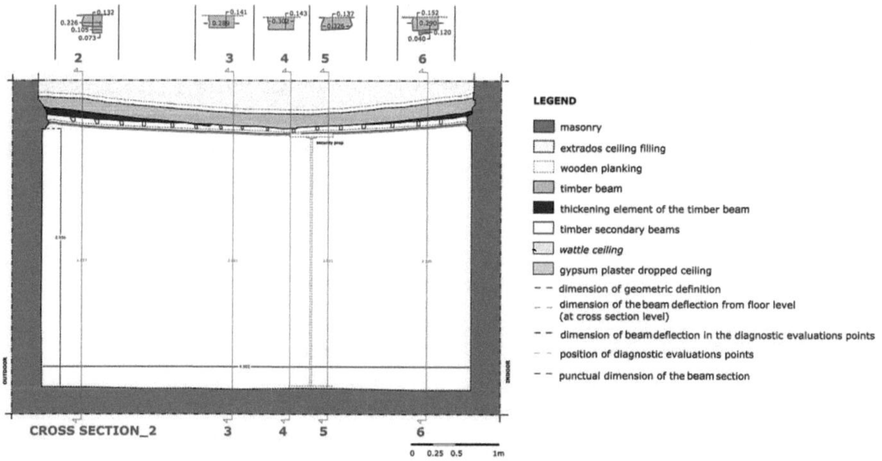

Fig. 8 LONGITUDINAL SECTION of the room with indication of multiple cross section and geometrical characterization of the beam section in those points

The bottom-up view of the ceiling plan below shows the position of *Resistograph* test points and the planar development of the ceiling lowering. This depression was caused by the beam creep and deep deflection (that even reaches the measure of 24 cm max) (Fig. 8), and it was not clearly visible by the human eye.

All of these results would have been quite unfeasible using only a total station, with which the user has to manually measure preselected points that could never be assimilated with the big amount of data automatically acquired with the laser scanner.

Assessment of the State of Conservation

According to the criteria foreseen by the Italian UNI 11118 (Cultural heritage—Wooden artefacts—Criteria for the identification of the wood species), the first activity requires the identification of the type of wood. The recognized type was oak (*Quercus*).

The procedures carried out during the on-site inspection for resistance grading followed the guidelines UNI 11119 (Cultural heritage—Wooden artefacts—Load-bearing structures—On site inspections for the diagnosis of timber members).

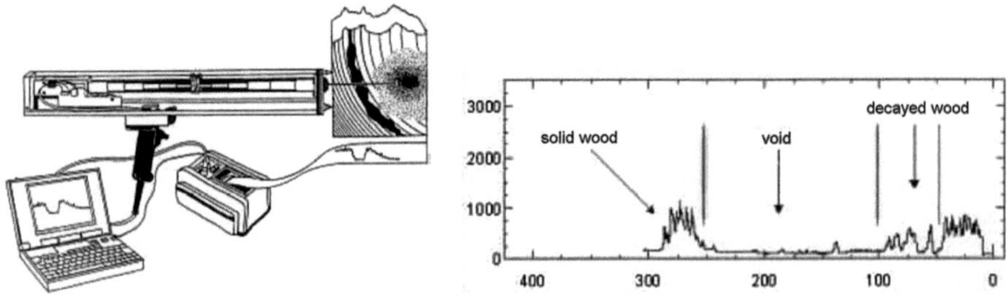

Fig. 9 Schematic example of the resistance drilling instrument and test that it can perform

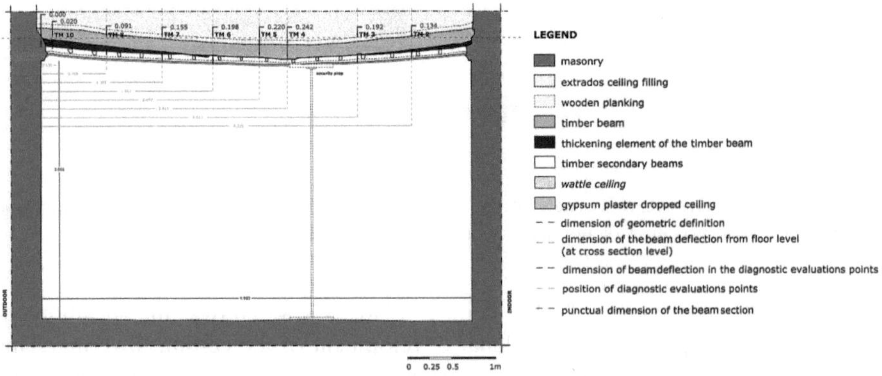

Fig. 10 LONGITUDINAL SECTION of the room with positioning of diagnostic evaluation points by resistance drilling test: its dimensions from the wall and its progressive deflection

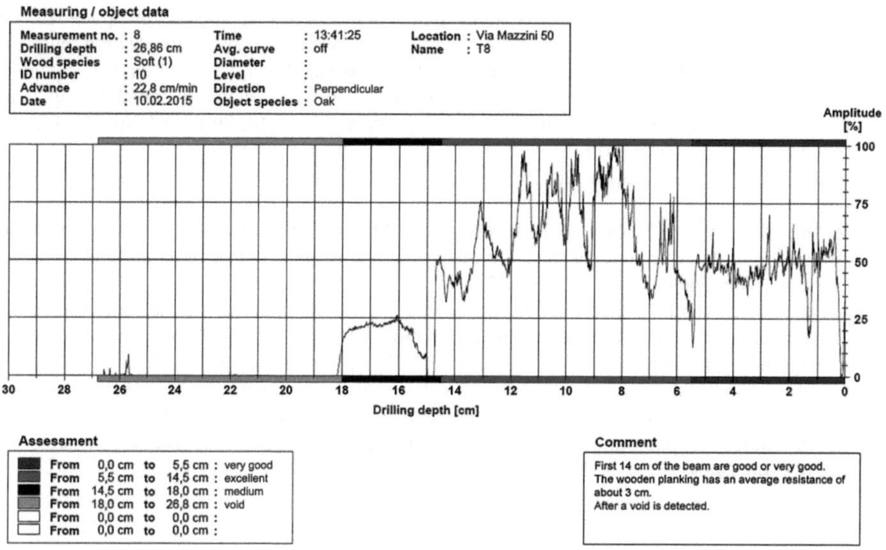

Fig. 11 Results of resistance drilling test conducted in the middle of the beam. Wood has a good state of conservation and resistance to penetration. After 18 cm of penetration the instrument meets a void (corresponding to the creep)

The evaluation of the biotic and abiotic degradation was carried out according to the standards of the UNI 11130 (Cultural heritage—Wooden artefacts—Terms and definitions concerning wood deterioration).

Afterwards a NDT diagnosis was performed with the instrument Resistograph®, which evaluates the resistance that the wood opposes to the penetration of a drill (drilling hole diameter 3 mm) (Figs. 9, 10 and 11).

Conclusions

From the survey it appears that the wooden elements observed were is in a good state of conservation. However, the creep is 24 cm, therefore the beam is not acceptable in terms of current standards. This result requires a necessary reinforcement intervention for the safety of the elements and of the building.

References

1. Overviews on methods and applications for 3d modelling arised since early 2000s. Oosterom P, Jansen E, Stoter J (2006) Large-scale 3D data integration: challenges and opportunities. Taylor & Francis. Patias P, Santana M (2012) Introduction to heritage documentation. In: CIPA

heritage documentation best practices and applications, Series 1, 2007–2009, pp 9–13. An example of a totally laser scanning application for evaluation and conservation purposes proposed from authors: Bertolini Cestari C, Chiabrando F, Invernizzi S, Marzi T, Spanò A (2014) The thin concrete vault of the Paraboloide of Casale, Italy. Innovative methodologies for the survey, structural assessment and conservation interventions. In: Structural faults and repair, London, UK, 8–10 July 2014
2. ISPRS (2012) ISPRS Technical Commission V. http://www2.isprs.org/commissions/comm5.html
3. Bertolini-Cestari C, Chiabrando F, Invernizzi S, Marzi T, Spanò A (2013) Terrestrial laser scanning and settled techniques: a support to detect pathologies and safety conditions of timber structures. Adv Mater Res 778:350–357
4. Voegtle T, Schwab I, Landes T (2008) Influences of different materials on the measurements of a terrestrial laser scanner (TLS). In: The international archives of the photogrammetry, remote sensing and spatial information sciences, vol XXXVII, Part B5, pp 1061–1066
5. Balletti C, Berto M, Gottardi C, Guerra F (2013) Ancient structures and new technologies: survey and digital representation of the wooden dome of SS. Giovanni E Paolo in Venice. ISPRS annals of photogrammetry, remote sensing and spatial information sciences, vol II-5/W1, pp 25–30

Traditional Timber-Framed Infill Structure Experimentation with Four Scales Analysis (To Connection from a House Scale)

Yannick Sieffert, Florent Vieux-Champagne, Stéphane Grange, Philippe Garnier, Jean-Charles Duccini and Laurent Daudeville

Abstract The aim of this paper is to contribute toward a better understanding of the seismic behavior of timber-framed infill structure. For this purpose, the results of a multi-scale experimental program are presented. The paper presents also the feasibility to use a DIC analyse on a full-scale of a house tested on a shake table.

Keywords Earth · Seismic-resistance · Earthquake · Timber · Shake-table · DIC

Y. Sieffert (✉) · F. Vieux-Champagne · S. Grange · L. Daudeville
Université Grenoble Alpes, 3SR, 38000 Grenoble, France
e-mail: yannick.sieffert@3sr-grenoble.fr

F. Vieux-Champagne
e-mail: florent.vieux-champagne@3sr-grenoble.fr

S. Grange
e-mail: stepahne.grange@3sr-grenoble.fr

L. Daudeville
e-mail: laurent.daudeville@3sr-grenoble.fr

Y. Sieffert · F. Vieux-Champagne · S. Grange · L. Daudeville
CNRS, 3SR, 38000 Grenoble, France

F. Vieux-Champagne · P. Garnier
CRAterre, AE&CC Research Unit, National School of Architecture of Grenoble, 38000 Grenoble, France
e-mail: philippe.garnier@grenoble.archi.fr

J.-C. Duccini
FCBA, French Technological Institute for Forestry, Cellulose, Timber Construction and Furniture, 33000 Bordeaux, France
e-mail: jean-charles.duccini@fcba.fr

H. Cruz et al. (eds.), *Historical Earthquake-Resistant Timber Framing in the Mediterranean Area*, Lecture Notes in Civil Engineering,
DOI 10.1007/978-3-319-39492-3_24

Introduction

Timbered masonry structures are popular worldwide mainly due to the reduced cost of its construction thanks to the use of local resources easy available and know-how, as well as for their aesthetics and/or ability to achieve greater resistance to seismic hazard (see [1]). Traditional timbered-framed structures with infill have indeed exhibited a remarkable behavior during recent major earthquakes (Turkey in 1999, Greece in 2003, Kashmir in 2005 and Haiti in 2010), during which they have often suffered little damage. At contrary, the seismic resistant behavior of new construction made of masonry blocks or concrete has generally been bad or disastrous leading to widespread destruction and loss of life. Poor design, bad construction and affordability are reasonable explanations for many reinforced concrete structure [2, 3].

Unfortunately, vernacular structures were forgotten by local people and by decision-makers because the lack of knowledge of their behaviors and the lack of building code and standards to design them.

But recently, some research project have been conducted in order to enhance the seismic resistant behavior knowledge of this kind of traditional timber frame structures. This paper presents the experimental study of a Haitian timbered masonry structure called "Kay peyi" in an attempt to analyze its behavior under seismic loads and hence offer scientific answers regarding this set-up [4]. This work is based on specific houses [5, 6], as shown in Fig. 10, adapted from Haiti heritage. This kind of structure is already built in Haiti as part of different Haitian reconstruction projects (i.e. Misereor, SC/CF projects in collaboration with the CRAterre laboratory and local partners, such as the Haitian NGO "GADRU", Entrepreneurs du Monde).

This structure is studied at four different scales, encompassing the connection (scale 1), the elementary cell (scale 2), the shear wall (scale 3) and extending to the house (scale 3). This multi-scale approach is illustrated in Fig. 1 and used in [7] for the seismic vulnerability analysis of timber frame structure using industrial products such as wood-products panels to brace the structure.

Scale 1: Connections

At the first scale, type-1 connections are studied in order to determine the local influence of the nail number under two loading directions (Fig. 2). Figure 3 presents the results of normal and tangential tests for each configuration (from two to eight nails). Note that "A" is written for average curve obtained with three distinct monotonic test and "Env" is written for the envelope curve of the associated cyclic test. The number following the letter indicates the number of nails.

The resistance of the connection is nearly proportional to the number of nails.

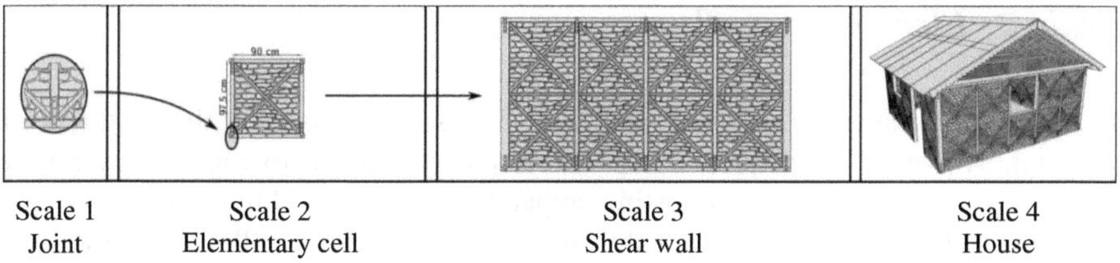

Fig. 1 The four scales of this experimental study

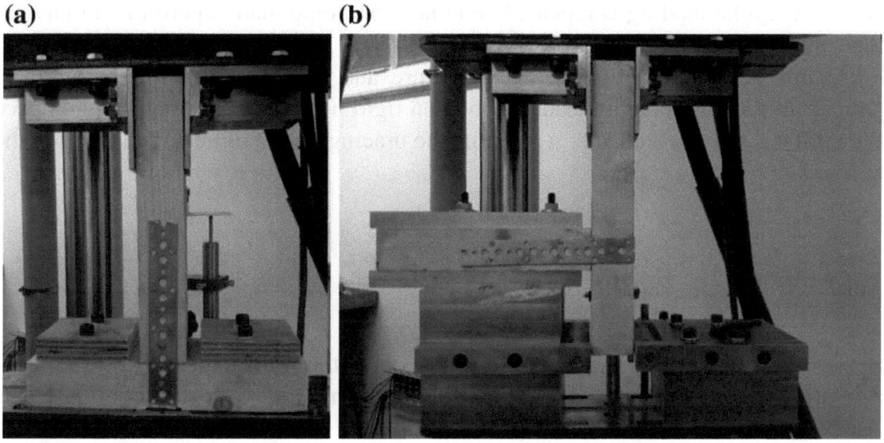

Fig. 2 Tests on the connections. **a** Normal test. **b** Tangential test

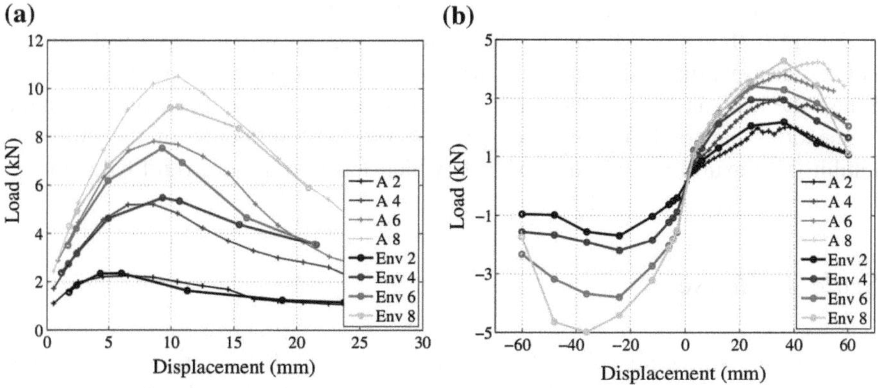

Fig. 3 Load-displacement diagram at the connection scale. **a** Normal test results. **b** Tangential test results

Scale 2: Elementary Cell

Different infill materials are used worldwide. Thus, the analysis of their influence is relevant. However, it is quite complicated and expensive to make a parametric analysis at the wall scale. A testing apparatus, shown in Fig. 4 has thus been designed to test an intermediate scale, such as an elementary cell, that remains part of the wall.

At the scale of elementary cell, two parameters have been studied: the influence of the bracing (with and without) and the influence of the infill material (stone, adobe and empty). We present below only the result with Andrew's cross (Fig. 5). Reversed cyclic loading is repeated and one monotonic load is performed to assure their reliability.

At this scale, the influence of the nature and presence of infill seems to be similar. In fact, the bracing leads to obtain a rigid behavior, like a lattice and hence the infill has only two effects: it prevents the bracing form failure and it modifies the stiffness.

Scale 3: Shear Wall

Scale 3 is examined to highlight the influence of the infill material [8]. Two configuration are tested, one with infill stone and one empty. In the both case, the Andrew's crosses are present. Figure 7 shows the load-drift diagram for reversed cyclic (C) and monotonic (M) tests on both filled (F) and empty (E) walls. This figure proves the significant effect of infill on wall behavior: the maximum load is almost twice for a filled configuration (44.3 kN for the monotonic) than for an empty one (26.9 kN for the monotonic) (Figs. 6 and 7).

Figure 6 shows that flexural deformation is present in the empty wall whereas lattice deformation is shown in the wall with infill, leaving the infill undamaged. When the wall is empty, the Andrew's crosses are free to bend and the nailed connections cannot limit this and are push out. The influence of the infill is to inhibit the flexural deformation of cells which leads to a more rigid behavior.

Scale 4: House

The dynamic structural behavior of timber-framed infill structure is studied at the house scale in order to complete the previous scales which have been studied on the basis of a quasi-static experimental campaign. For this purpose, a representative ground motion of Haiti's 2010 earthquake [9] has been designed (Figs. 8 and 9) and a shake table has been used (Fig. 10).

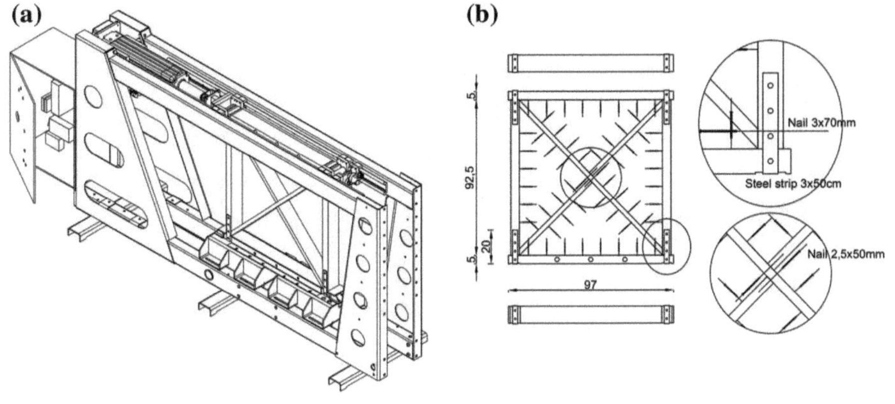

Fig. 4 Elementary cell layout. **a** Shear test on an elementary cell. **b** Elementary cell dimensions

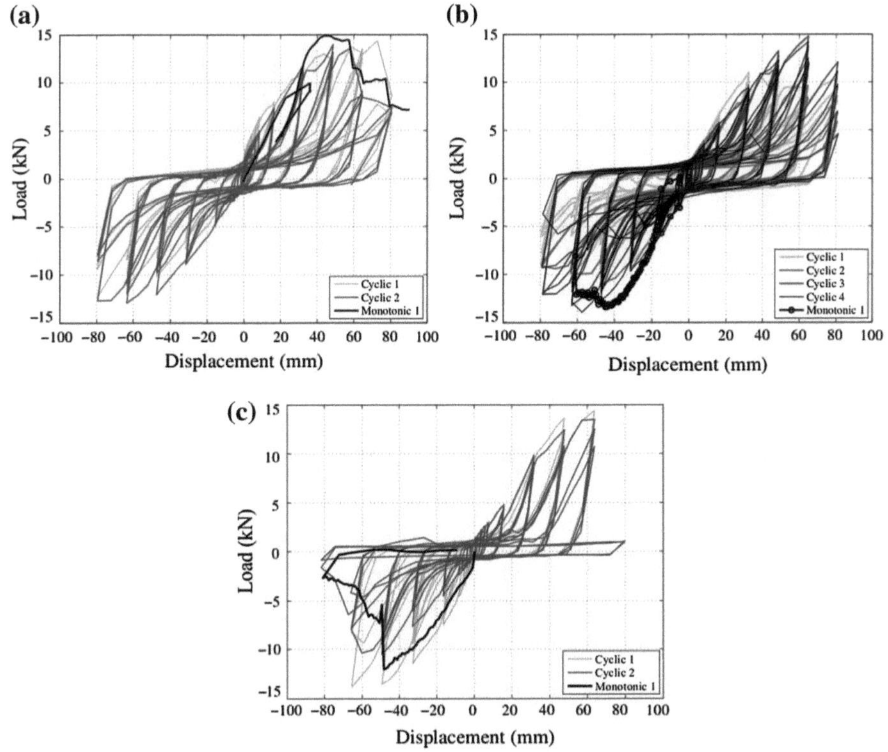

Fig. 5 Cyclic and monotonic curves for braced cells filled with stone, adobe or empty. **a** Infill with stone. **b** Infill with adobe. **c** Empty

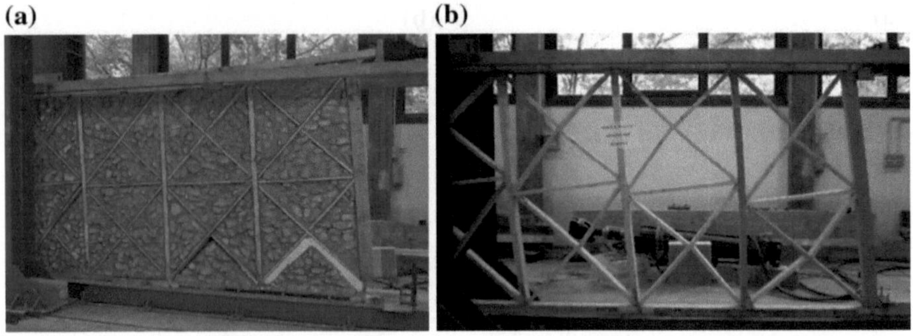

Fig. 6 Wall deformation at the end of the monotonic test. **a** Filled wall deformation. **b** Empty wall deformation

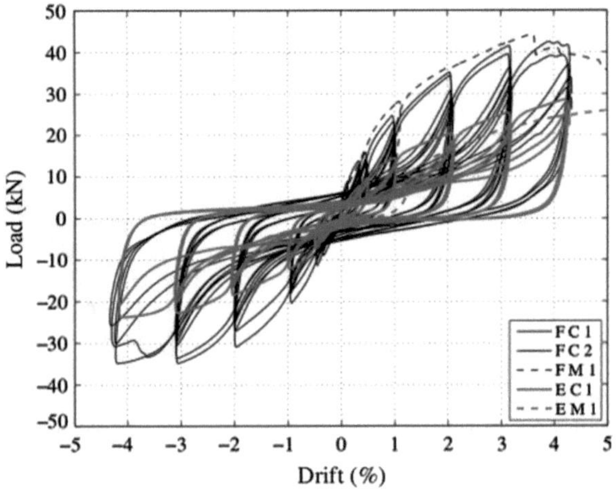

Fig. 7 Load-drift diagram—wall both filled and empty

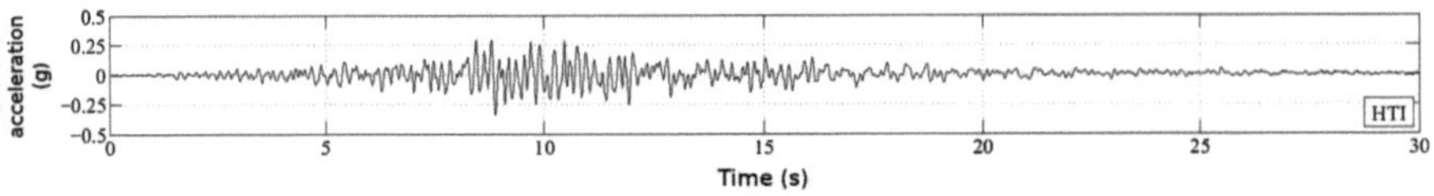

Fig. 8 Time-history of the ground motion Haiti

The dynamic tests were performed on the uniaxial earthquake simulation facility in the Mechanical Laboratory of the French Institute FCBA. The shake table is composed of a 6 × 6 m^2 aluminum platform moved by a 250 kN servo-hydraulic actuator.

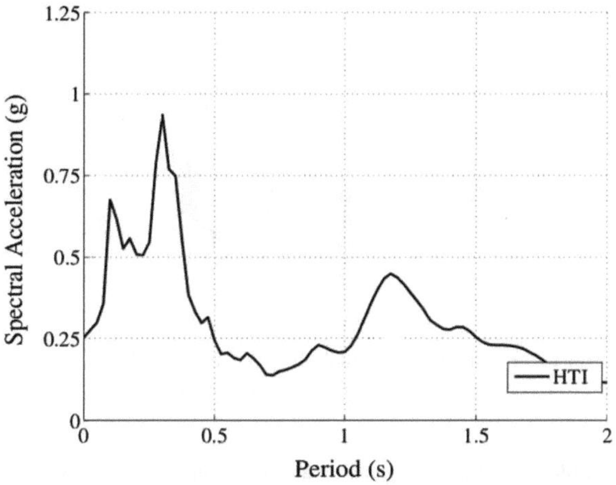

Fig. 9 Absolute acceleration response spectrum for 5 % damping of the ground motion Haiti

Fig. 10 The house builds on the shake table

Before the seismic tests, a modal analysis was conducted by measuring three axis acceleration responses at each wood member intersection of the structure while subjected to a white noise excitation. This process yielded: the first three modal shapes (Fig. 11), natural frequencies, and damping ratios (Table 1). The first two modes are similar and represent the transverse wall racking mode (only the first one is shown on the Fig. 11a). The small frequency difference can be attributed to the various positions of openings in the shear wall as well as the intrinsic differences in the workmanship of the two walls (weight, geometric dimensions, etc.). The second primary mode is the shear wall vibration (Fig. 11b).

Under the dynamic tests, the monitoring of the house is made by classical local transducers (LVDT: Linear Variable Differential Transformer, DWDS: Draw Wire Displacement Sensor and accelerometer) but also with non-contact measurement for the full-field displacement of one shear wall (parallel of the shake table axial displacement) with high-speed camera (Phantom v611 with a resolution of 2560 × 1600 pixels and a frequency of 150 frame per second) associated DIC

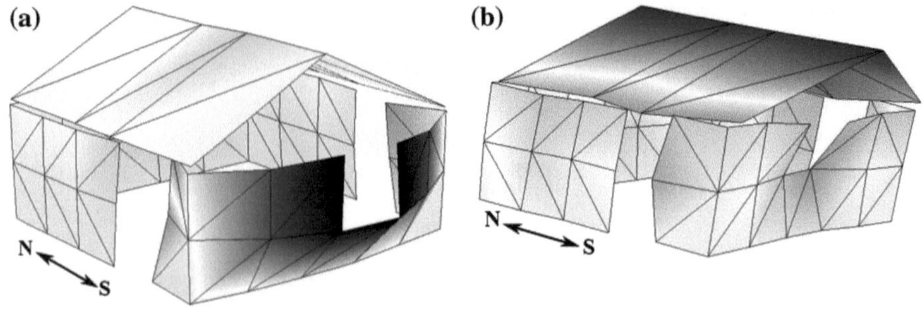

Fig. 11 Fundamental modal shape of the structures. **a** First modal shape of the first transverse wall. **b** Third modal shape of the shear wall

Table 1 Natural frequencies and the equivalent viscous damping ratio

Pole	Element	Frequency (Hz)	Damping ratio (%)
1a	Transverse wall	5.2	5.2
1b	Transverse wall	5.9	5.7
2	Shear wall	10.9	5.3

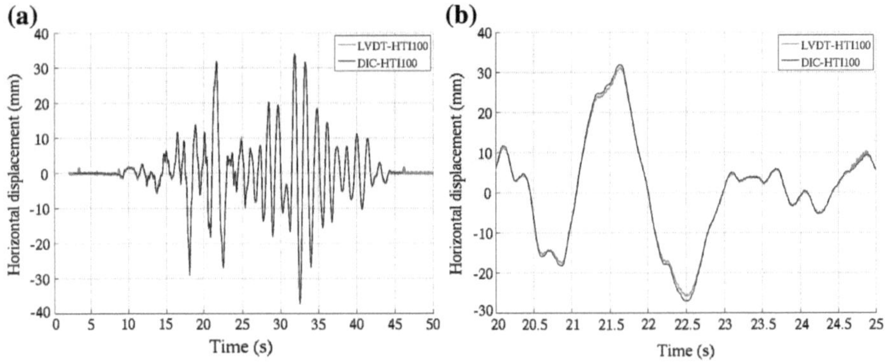

Fig. 12 Comparison between DIC and LVDT for the shake table displacement. **a** All signal. **b** Zoom near peaks

(Digital Image Correlation) method. This paper presents some results obtained on this dynamics tests to validate the method implemented for DIC measurements with the software Tracker [10, 11]. Near 4000 pixels are tracking to describe the global response of the shear wall and an average of 3 pixels are used to follow the shake table displacement. Figure 12 shows a perfect acquaintance of the shake table displacement recorded by DIC and LVDT as the differences are less of 5 %. It is important to note that the LVDT signal is more noised that the one obtained with DIC technique.

Fig. 13 Deformed configuration

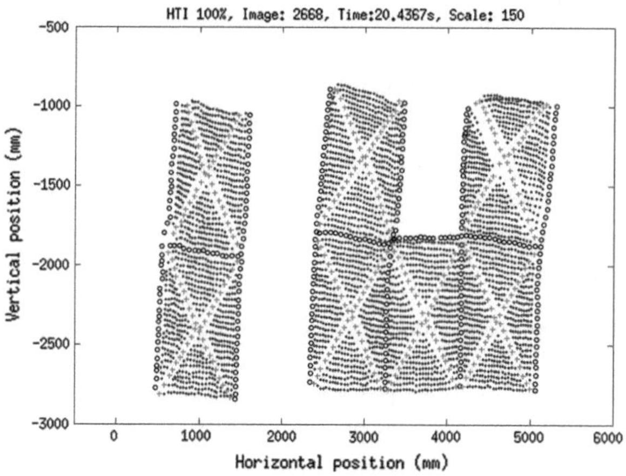

Fig. 14 Vector displacement

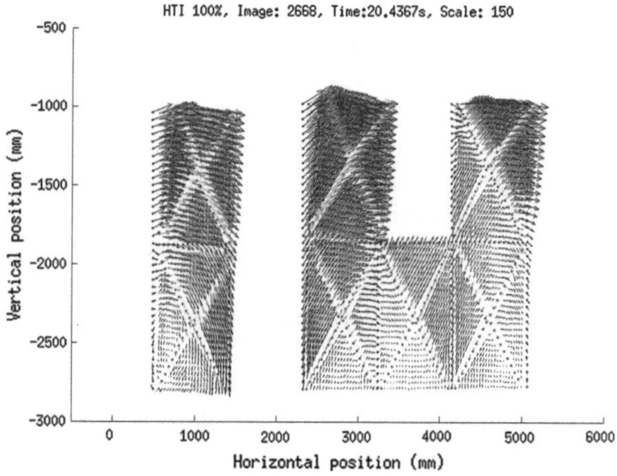

Figures 13 and 14 carry out the deformed configuration and the vector displacement when the maximum relative displacements in the horizontal direction on the top right post is archived. This maximum relative displacement is 1.67 mm then an amplification factor of 150 is used. Note that these two figures give very good results in regards of the very low relative displacement and the number of pixels (2.16 mm/px) of the camera which needs the use of subpixel process (0.216 mm/px) for the DIC precision. Black color represents the beam or the post, red color represents the Andrew's crosses and blue color represents the infill material.

One main conclusion is that the powerful of the DIC allows to see, with the help of an amplification factor, the separation displacement in the same cell of the infill material and the flexural behavior of the timber elements and of the entire timber frame. These are impossible to catch with naked eye.

Conclusion

The paper has presented a multi-scale experimental program, during which a connection, an elementary cell, a shear wall and a house have been tested to analyse the behavior of a timber-framed structure filled with stones and earth.

At the scale 1, both normal and tangential tests were conducted on the connection. The resistance of the connection is nearly proportional to the number of nails.

At the scale 2, tests were completed in order to analyze the influence of infill. These results show that at this scale, infill exert no influence on the lateral load capacity.

At the scale 3, shear wall tests demonstrated the benefit of infill and the high ductility of these filled timber structures.

At the scale 4, the house displayed a good resistance when submitted to a ground motion equivalent to the January 12th Haiti earthquake. No damage was observable which provide direct proof of the seismic-resistant behavior of timber-framed structures with infill.

This paper has presented a field measurement of displacement with the help of a high-speed camera. The innovation in this paper is to prove the feasibility of this measurement into a large sample which have real dimension of a house's wall (9.3 m^2) and under a seismic loading applied by a shake table. To validate the DIC analyses, a comparison was made from the displacement recorded by LVDT. The difference between the two measurements was less than 5 %. The powerful of the DIC method give new perspective in the engineering structure analysis.

References

1. Dutu A, Ferreira J, Guerreiro L, Branco F, Goncalves A (2012) Timbered masonry for earthquake resistance in Europe. Mater Constr 62(308):615–628
2. Doğangün A (2004) Performance of reinforced concrete buildings during the may 1, 2003 bingöl earthquake in turkey. Eng Struct 26:841–856
3. Langenbach R (2008) Learning from the past to protect the future: armature crosswalls. Eng Struct 30:2096–2100
4. Vieux-Champagne F, Sieffert Y, Grange S, Polastri A, Ceccotti A, Daudeville L (2014) Experimental analysis of seismic resistance of timber-framed structures with stones and earth infill. Eng Struct 69:102–115

5. Vieux-Champagne F, Sieffert Y, Grange S, Belinga Nko'o C, Duccini J, Faye C, Daudeville L (2015) Experimental analysis of a shake table test of a timber-framed structures with stones and earth infill. Eng Struct (submissed)
6. Sieffert Y, Vieux-Champagne F, Grange S, Garnier P, Daudeville L (2016) Full-field measurement with a digital image correlation analysis on a shake table test of a timber-framed structure filled with stones and earth. Eng Struct 123:451–472
7. Humbert J, Boudaud C, Baroth J, Hameury S, Daudeville L (2014) Multi-scale modelling of wood shear walls. I: calibration and validation underquasi-static reversed-cyclic loading. Eng Struct 65:52–61
8. Poletti E, Vasconcelos G (2015) Seismic behaviour of traditional timber frame walls: experimental results on unreinforced walls. Bull Earthq Eng 13(3):885–916
9. St Fleur S, Courboulex F, Bertrand E, Deschamps A, Mercier de Lepinay B, Prepetit C, Hough S (2013) A combined source and site-effect study of ground motions generated by an earthquake in Port au Prince (Haiti). EGU General Assembly 2013, held 7–12 April 2013 in Vienna, Austria
10. Combe G, Richefeu V (2013) Tracker: a particle image tracking (pit) technique dedicated to nonsmooth motions involved in granular packings. In: Powders and grains 2013: proceedings of AIP conference on micromechanics of granular media, vol 1542(1), pp 461–464
11. Richefeu V, Combe G, Viggiani G (2012) An experimental assessment of displacement fluctuations in a 2d granular material subjected to shear. Géotechnique Letters 2:113–118

Characterization of Timber Masonry Walls with Dynamic Tests

Ana Gonçalves, Paulo Candeias, Luís Guerreiro, João Ferreira and Alfredo Costa

Abstract Most of the Lisbon 18th century timber-framed masonry "Pombalino" buildings currently need seismic rehabilitation due to: (i) natural degradation with aging; (ii) need for adaptation to new serviceability requirements, generally involving higher loads and consequent structural changes; (iii) former interventions with elimination or damage of structural elements, affecting seismic resistance and (iv) noncompliance with the present seismic codes. The research presented in this paper aimed at experimentally evaluating the seismic vulnerability of the "Pombalino" buildings and at proposing a strengthening technique to reduce it by reinforcing their timber framed "Frontal" walls. The experimental program was based on extensive dynamic testing on prototype representative of the current characteristic of "Frontal" walls. The results of the dynamic tests carried out in the LNEC 3D shaking table are presented regarding the two types of tests performed: (a) seismic tests, in which the seismic action was applied with increasing amplitude in the direction of the walls; (b) dynamic identification tests, aiming at evaluating the decrease of the mechanical properties of the models. A comparison of the performance of the non-strengthened and strengthened tested prototypes is also presented.

A. Gonçalves (✉) · L. Guerreiro · J. Ferreira
Instituto Superior Técnico (IST), Universidade de Lisboa, CEris,
Avenida Rovisco Pais, 1, 1049-001 Lisboa, Portugal
e-mail: goncalves.amn@gmail.com

L. Guerreiro
e-mail: luisg@civil.ist.utl.pt

J. Ferreira
e-mail: joao.gomes.ferreira@cg.tecnico.ulisboa.pt

P. Candeias · A. Costa
Laboratorio Nacional de Enganria Civil (LNEC), Av. Do Barsil 101,
1700-066 Lisboa, Portugal
e-mail: pcandeias.lnec@gmail.com

A. Costa
e-mail: alfcc.lnec@gmail.com

H. Cruz et al. (eds.), *Historical Earthquake-Resistant Timber Framing in the Mediterranean Area*, Lecture Notes in Civil Engineering,
DOI 10.1007/978-3-319-39492-3_25

Keywords "Pombalino" buildings · "Frontal" walls · Timber-masonry walls · Shaking table tests

Introduction

Timber framed masonry wall buildings are seen all over Europe, especially in seismic regions, given its adequacy to resist earthquake [1–4, 6, 7]. The "pombaline" buildings, named after the Marquis of Pombal, present a structure with a maximum of four storeys, composed by arcades at the ground floor, masonry facade walls and internal timber framed masonry walls. These walls together with the floors' timber beams, form the 3-D cage that constitutes the seismic resistant structure. The Marquis of Pombal ordered their construction after the 1755 earthquake that destroyed Lisbon, aiming at providing the city with seismic resistant buildings [8].

In some recent earthquakes (Turkey 1991, Greece 2003, India 2005) half-timbered building showed a better performance than unreinforced masonry buildings [5]. In fact, this construction system is pointed out by several authors as being one of the most efficient earthquake resistant structure in the world [5, 6], but their popularity is not only due to their seismic performance, but also to their low cost and the strength they offer.

In traditional half-timbered building, it is usual that half-timbered walls are also placed in parallel to external gravity load bearing masonry walls. In unreinforced masonry structures, the capacity of the building is good in terms of vertical loads, since masonry has generally an adequate compressive strength. When subjected to horizontal loads, shear and flexural capacity of such structures are quite low being, particularly vulnerable to out-of-plane failure mechanisms. When a bracing system is added, the lateral resistance increases, improving the seismic behavior of the structure. The internal half-timbered walls represent the lateral load bearing walls of the structure, since due to their lightness and higher flexibility than traditional masonry walls, they are expected to perform better under seismic loads. The timber-framed masonry walls are connected to the external masonry walls by means of the timber floor beams, which are connected both to the timber-framed and to the external masonry walls. This system can be also beneficial to reduce the out-of-plane vulnerability of masonry walls [9].

To improve the understanding of "frontal" walls structures, a research initiative was started at IST (Lisbon, Portugal), which included the shaking table tests described in this paper and the quasi-static cyclic testing [9–11]. It consisted of unidirectional dynamic testing of a two-storey wall with and without masonry and reinforcements (metal plate), which was constructed at real scale. The objective of this shaking table test was to observe the behaviour of such mixed structures when subjected to seismic loading and to collect high-quality data that can be used for the validation of numerical models.

Shaking Table Tests Setup

Model Definition

The shaking table tests were performed at LNEC (Laboratório Nacional de Engenharia Civil, Lisbon) at NESDE department (Earthquake Engineering and structural dynamics Division). The tests were performed unidirectional along the shaking table lateral axis, since the purpose of this tests was to study the plan response of the "frontal" walls.

The LNEC shaking table has, in plan, $4.6 \times 5.6\ m^2$ and the model geometry was according to those dimensions. Another factor that was important in the design of the models is the simulation of the facade effects over the "frontal" walls and is presented as innovative feature on this tests.

The "frontal" walls models reproduce the behavior of the two floors, with 6 m height, 3 m length with wall and a distance between walls of 2.7 m, with 12 cm of wall thickness. Each wall has two panels of $3 \times 3\ m^2$, with nine Saint Andrew's crosses (Fig. 1). The walls are connected of the timber floors at 3 and 6 m height. Saint Andrew's crosses are made up vertical, horizontal and diagonals timber elements. Cross-halving joints were used in the connections between vertical and horizontal timber elements and between the crossing diagonals. The connections between the diagonals and their end nodes were obtained by simply sawing them at 45°. All connections were reinforced with iron nails. The wood used in the experimental study was stone pine (pinus pinea) class C18 [12]. The wood sections were $16 \times 12\ cm^2$ (base and top beam) and $8 \times 12\ cm^2$ (all other elements).

The choice of the type of masonry was very important since there are several types of infill, including mortar with bricks or mortar with tiles or even mortar

Fig. 1 Models: **a** MODEL 0; **b** Model 1; **c** Model 2

mixed with small stones. In this study, mortars with bricks were used. The masonry consists of bricks and cement-lime-sand mortar with a volume ratio of 1:2:6 (hydrated lime: portland cement 32.5 N: sand). Although ancient mortars were solely composed of lime and sand, cement was added in these cases to ensure a faster cure. The drying time was two months.

As previously stated, this test aimed to study the behavior of the "frontal" walls when subjected to seismic action, where the reproduction of the facade effect is fundamental. The facade was simulated by a metal structure, which was based on the weight distribution along the vertical direction of the walls. The metallic facade is connected to the front wall through the beams of St. Andrew cross, in a total of six points on each wall. The metallic structure consists of 12 masses of 600 kg, distributed in pairs with several degrees of freedom to have capacity for deformation and redistribution of the weights between the floors. The articulated structure was designed by using pin connections between the masses of the metallic structure.

Three models, representing the "frontal" walls, were built: "Model 0" consisting of timber frames without masonry infill (Fig. 1a); "Model 1" consisting of a timber frames with masonry infill (Fig. 1b) and "Model 2" consisting of a timber frames with masonry and reinforced cross-halving joints (Fig. 1c).

Test Setup

The main objectives of the shaking table tests are to assess the seismic performance of the "Pombalino" buildings and to validate the efficiency of strengthening techniques. This was carried out through two experimental procedures: (a) the response evaluation of the model based on the results obtained in the seismic tests (e.g. crack patterns and displacements) with increasing seismic amplitude applied at the base; (b) the definition of the seismic vulnerability curves based on the decrease of the natural frequencies of the model with the application of seismic tests with increasing amplitude.

The methodology for defining the seismic vulnerability curves through experimental tests is, usually, based on the identification of the dynamic properties of the model (natural frequencies, mode shapes and damping ratios) along a series of seismic tests with increasing input excitations [13, 14]. Dynamic identification tests at the shaking table (forced vibration tests) were carried out, aiming at characterizing the initial dynamic properties. The tests for the characterization of the dynamic properties were done applying a series of accelerations at the base of the model, in one direction.

The input signals used in the dynamic identification tests (Fig. 2a) correspond to a "white noise", with large frequency range and low amplitude, aiming at avoiding further damage in the model. The duration of the signals were 320 s (intense phase) and the maximum amplitude is about 0.2 g in the longitudinal directions. Software such LNEC-SPA [13] was used to generate artificial accelerograms.

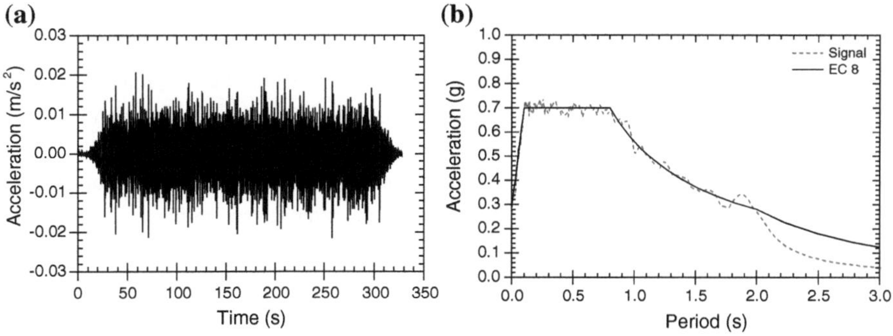

Fig. 2 Dynamic identification tests: **a** the input signals (white noise); **b** response spectrum defined by Eurocode 8

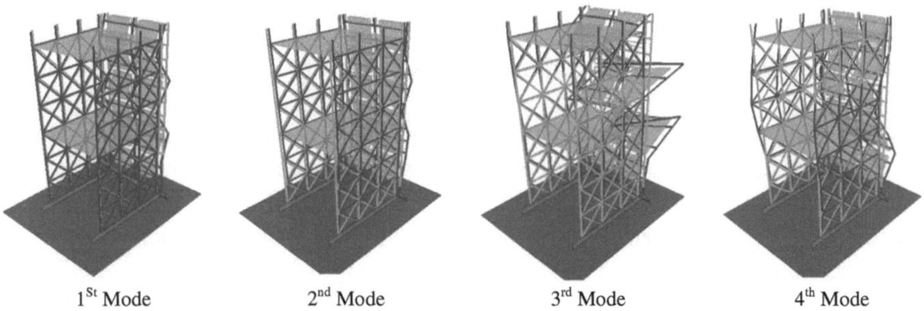

Fig. 3 Mode shape of the *Model 0*

The signal used in the seismic tests was performed at the LNEC 3D shaking table by imposing accelerograms compatible with the design response spectrum defined in Eurocode 8 [15], for a 2.3 zone (far earthquake in Lisbon) and soil type D, with a 5 % damping coefficient. Figure 3b shows the response spectrum, corresponding to the earthquake signal for a return period of 475 years.

The shaking table tests of the model were composed by five seismic tests with amplitudes equal to an earthquake with a return period of 73 years, 225 years, 475 years, 975 years and 2000 years. Between each seismic test, a dynamic identification test was realized with a total of six tests (Table 1).

Instrumentation

Besides the construction of the model, the preparation of the tests involved the following tasks: (a) selection and preparation of the facility to simulate the seismic scenario; (b) selection and setup of the devices to measure the response of the

Table 1 Shaking table tests carried

Number	Identification	Description
1	DI 1	Dynamic identification test before the first seismic test
2	Earthquake 73 years	Seismic test with earthquake for 73 years of return period
	DI 2	Dynamic identification after seismic test
3	Earthquake 225 years	Seismic test with earthquake for 225 years of return period
	DI 3	Dynamic identification after seismic test
4	Earthquake 475 years	Seismic test with earthquake for 475 years of return period
	DI 4	Dynamic identification after seismic test
5	Earthquake 975 years	Seismic test with earthquake for 975 years of return period
	DI 5	Dynamic identification after seismic test
6	Earthquake 2000 years	Seismic test with earthquake for 2000 years of return period
	DI 6	Dynamic identification after seismic test

model; (c) preparation of the acquisition systems to record the input and output signals.

The model was instrumented with 28 accelerometers, 8 displacement transducers and 4 optical acquisitions. The accelerometers were installed on the two "frontal" walls and on the metal wall. Accelerometers were placed in the model in several levels, located on the horizontal frame on the front wall and on each mass of the metal structure.

Results and Discussion

In this paper the dynamic properties of the three models are defined (Model 0, Model 1 and Model 2).

The dynamic identification tests aimed at estimating the dynamic properties, namely the frequencies, mode shapes and damping ratios of the model. Several methods can be used to identify the dynamic properties through input/output vibration tests, such as the Peak Picking, Circle Fit, Rational Fraction Polynomial or Complex Exponential.

In the dynamic identification tests, the input and output acceleration signals were measured at the shaking table and at the mock-ups, respectively, with sampling frequency equal to 250 Hz. The signals were firstly processed aiming at obtaining the dynamic properties of the model, namely by removing the DC components (0 Hz) and by filtering using a lowpass Fourier filter with a cutting frequency equal to 40 Hz [17, 18]. The *welch* method was used to calculate the power spectrum. In order to smoothen it and reduce the dispersion of the FRF (Frequency Response Function), 2^{10} (1024) samples per frame were used filled with a minimum number of zeros (padding) and *Hanning windows* with overlap equal to 2/3. This set of operations were carried out in the software MATLAB [19]. The functions used for this calculation were: pwelch (Welch's Power Spectral Density Estimate), cpsd, (Cross Power Spectral Density) and mscohere (Magnitude Squared Coherence).

Once the dynamic properties of the model were obtained, the variation of the frequency of the modes can be used to quantify the evolution of the damage, by taking into account the fundamental relation between natural frequency, mass and stiffness of system with a single degree of freedom [20].

The dynamic behaviour of the model depends on several modes, which are associated to different types of stiffness. Therefore, for simplicity, the damage indicator $d_{i,n}$ for each frequency i and step n (DI n) is assumed to be linearly proportional to the ratio between the corresponding frequencies for the $n(fi, n)$ and the first ($f_{i,0}$) dynamic identification tests, as presented below:

$$d_{i,n} = 1 - \frac{fi, n}{fi, 0} \quad (1)$$

The *Model 0* correspond to the structure only with timber frame and metal wall (Fig. 1a). In the first dynamic test four modes of vibration were identified with frequency between 3.84 and 16.98 Hz (Fig. 3), which correspond to three longitudinal modes and one combined mode. For frequencies higher than 18 Hz, it was difficult to identify the modes due to the coherence between the signal. The frequency of the first longitudinal mode is equal to 3.84 Hz. The longitudinal modes are associated to the global behaviour of the model and were clearly identified in the FRF's. The second and third longitudinal modes are equal to 11.09 and 16.98 Hz, respectively. The second mode presents a frequency of 4.59 Hz and is a combined mode (longitudinal + torsion). Table 2 presents the vibration modes and Fig. 3 its configuration modes.

The *Model 1* correspond to the structures with masonry infill on the timber frame (Fig. 1b). In the dynamic tests, five modes of vibration were identified between 4.9 and 19 Hz, which correspond to three longitudinal and two combined modes (longitudinal with tortion, Fig. 4). The frequency of the first longitudinal mode is equal to 4.90 Hz. The frequencies of the other longitudinal modes are equal to 14.08 and 18.75 Hz for the second and third longitudinal modes, respectively. The combined modes are equal to 5.71 Hz and to 12.35 Hz (Table 3).

The evolution of modal frequencies throughout the testing is illustrated in Fig. 5, where the frequency of the vibration mode decreases. For the first mode there is a decrease of 39 % between the first and the last testing. In the third mode the decrease is 19 % and in the fourth and fifth mode there is a decrease of 12 and

Table 2 Frequency mode of the *Model 0*

Mode	1st	2nd	3rd	4th
Frequency (Hz)	3.84	4.59	11.09	16.98
Mode shape	Longitudinal	Combined	Longitudinal	Longitudinal

Note Combined = Longitudinal + torsion

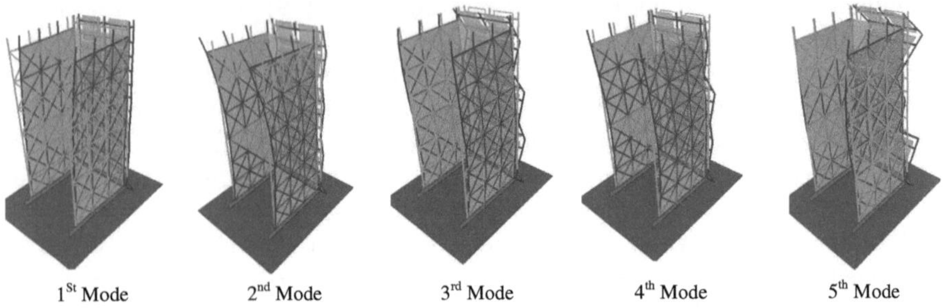

Fig. 4 Mode shape of the Model 1

Table 3 Frequency mode of the *Model 1*

Mode	1st	2nd	3rd	4th	5th
Frequency (Hz)	4.90	5.71	12.35	14.08	18.75
Mode shape	Longitudinal	Combined	Combined	Longitudinal	Longitudinal

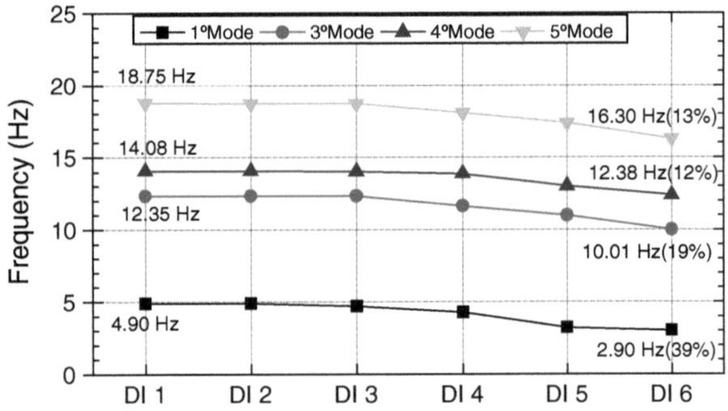

Fig. 5 Variation of frequencies

13 %, respectively. The Fig. 6 represents the evolution of modal damping throughout the different characterizations, concluding that the damping increases, especially in modes where the damage is significant.

The Model 2 is a strengthened model, reinforced with metal plate on the cross-halving joints and braced laterally. In the dynamic tests five mode shapes were estimated, ranging from 5.74 to 20.28 Hz, corresponding to longitudinal modes (Table 4; Fig. 7). The frequency of the first longitudinal mode is equal to 5.74 Hz. The frequencies of the second and third longitudinal modes are equal to 14.23 and 20.28 Hz, respectively. These modes were very similar, only diverging in the inflection point location on the walls (Fig. 7; Table 4).

The Fig. 8 represents the evolution of the Model 2 damage for each dynamic identification. It is observed a frequencies deterioration between 3.2 and 14 %. The change on the frequency value over the test is smaller than in the Model 1, showing

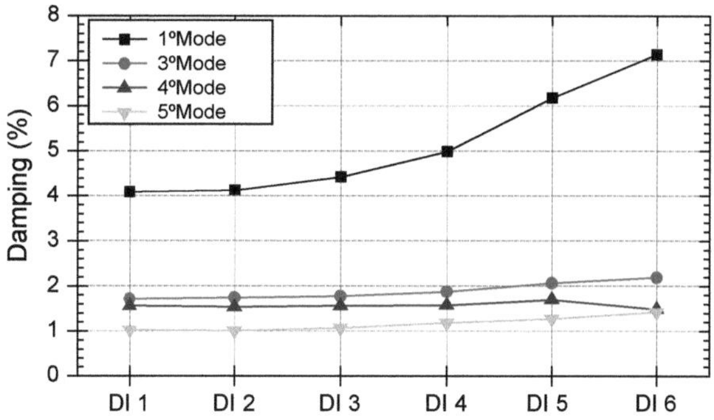

Fig. 6 Damping during the test

Table 4 Frequency mode of the *Model 2*

Mode	1st	2nd	3rd	4th
Frequency (Hz)	5.74	14.23	15.19	20.28
Mode shape	Longitudinal	Longitudinal	Longitudinal	Longitudinal

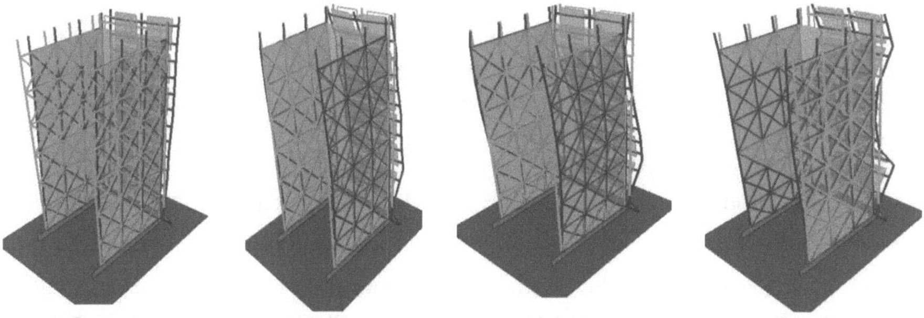

Fig. 7 Mode shape of the *Model 2*

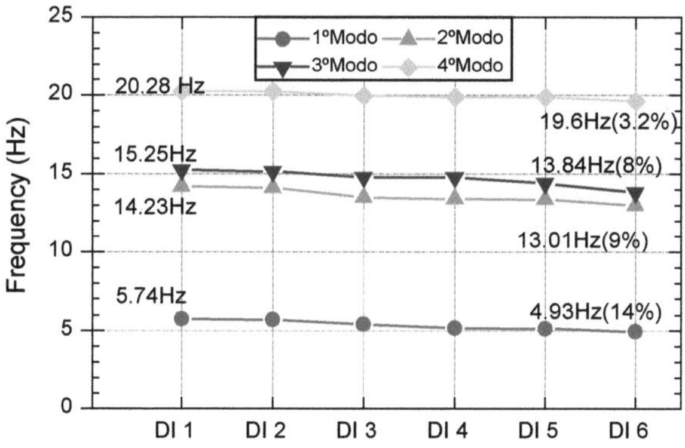

Fig. 8 Variation of frequencies

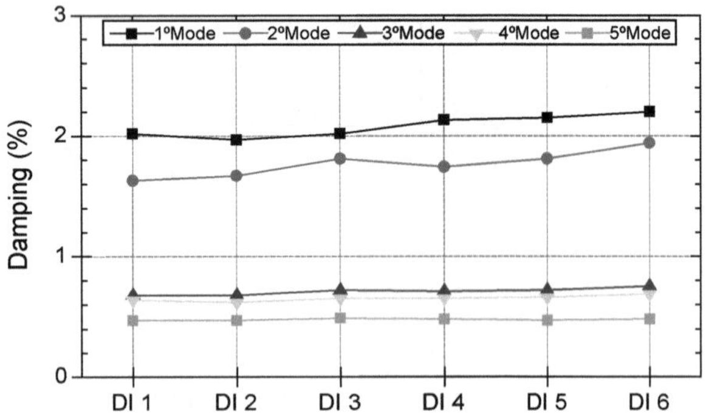

Fig. 9 Damping during the test

there is damage mitigation due to the reinforcement plates. The Fig. 9 shows the development of the modal damping throughout the test, showing a damping increase with increasing seismic intensity and damage.

Conclusion

This paper intended to provide useful information on the dynamic behaviour of traditional timber-framed masonry walls, typical of the Portuguese *Pombalino* buildings and thus intending to contribute to the state of the art on the matter, given the scarce results available in literature.

The experimental analysis was carried out on three distinct models (Model 0, Model 1, and Model 2). From the experimental campaign carried out it was possible to observe that:

- The masonry increases the rigidity of the model, having the disadvantage of increasing the mass of the structure. In the first modal identification test, it was observed that in all modes the frequency is lower in the model without masonry.
- The damage in Model 1 is highly concentrated at the connections between floors resulting in the large deformation of the walls that presented a maximum damage of 39 %.
- The modal frequencies increased in the reinforced model due to a stiffness increase.
- By analysing the evolution of frequencies during the dynamic test, it is shown that the strengthened model has a good behaviour to seismic test, presenting a maximum damage of 14 %.

Acknowledgments The financial support of the Foundation for Science and Technology (FCT) throughout the research project PTDC/ECM/100168—REABEPA is acknowledged. The authors also acknowledge HCI for his invaluable help, and the researchers team of the Earthquake Engineering and Structural Dynamics Division (NESDE), for their help and support in the laboratory works.

References

1. Diskaya H (2007) Damage assessment of 19th century traditional timber framed structures in Istanbul. In: From material to structure—mechanical behaviour and failures of the timber structures, ICOMOS IWC—XVI international symposium, 11th–16th November
2. Dogangun A, Tuluk IO, Livaoglu R, Acar R (2006) Traditional wooden buildings and their damages during earthquakes in Turkey. Eng Fail Anal 13
3. Dutu A, Ferreira J, Guerreiro L, Branco F, Gonçalves A (2012) Timbered masonry for earthquake resistance in Europe [online]. Instituto de Ciencias de la Construcción Eduardo Torroja del Consejo Superior de Investigaciones Científicas
4. Gülkan P, Langenbach R (2004) The earthquake resistance of traditional timber and masonry dwellings in Turkey. In: 13th world conference on earthquake engineering Vancouver, B.C., Canada, 1–6 August, Paper No. 2297
5. Langenbach R (2007) From 'Opus Craticium' to the 'Chicago Frame': earthquake-resistant traditional construction. Int J Architectural Heritage 1:29–59
6. Makarios T, Demosthenous M (2006) Seismic response of traditional buildings of Lefkas Island, Greece. Eng Struct 28:264–278
7. Redondo E, Ros R (2003) Wooden framed structures in Madrid domestic architecture of 17th to 19th centuries. In: Huerta S (ed) Proceedings of the first international congress on construction history, Madrid, 20th–24th Jan 2003. Instituto Juan de Herrera, Madrid
8. Ferreira J, Texeira MJ, Dutu A, Branco F, Gonçalves A (2014) Experimental evaluation and numerical modelling of timber framed walls. J Exp Tech 38(4):45–53
9. Gonçalves A, Ferreira J, Guerreiro L, Branco F (2014) Seismic retrofitting of timber framed walls. Materiales de Construcción, CSIC, vol 64(316)
10. Gonçalves A, Ferreira J, Guerreiro L, Branco F (2015) Experimental characterization of timber framed masonry walls cyclic behavior. Struct Eng Mech, Int J 53(2)
11. Goncalves A (2015) Structural rehabilitation of masonry walls in Pomnalino buildings (in Portuguese). Ph.D. thesis, University of Lisbon, Instituto Superior Técnico, Lisbon, Portugal
12. NP EN 338 (2003) Madeira para Estruturas - Classes de Resistência. IPQ, Lisboa
13. Degée H, Denoël V, Candeias P, Campos-Costa A, Coelho E (2007) Experimental investigation on the seismic behaviour of North European masonry houses. In: 7th national congress on seismology and earthquake engineering (Sísmica 2007), Porto, Portugal (In Portuguese)
14. Bairrão R, Falcão M (2009) Shaking table tests of two different reinforcement techniques using polymeric grids on an asymmetric limestone full-scaled structure. Eng Struct 31 (6):1312–1330. doi:10.1016/j.engstruct.2008.04.039
15. EN 1998-1 (2004) Eurocode 8: design of structures for earthquake resistance—general rules, seismic actions and rules for building. European Committee for Standardization
16. Mendes P, Oliveira S (2008) Análise dinâmica de estruturas. Utilização integrada de modelos de identificação modal e modelos de elementos finitos. ITB ADE 16, Laboratório Nacional de Engenharia Civil, Lisboa
17. Carvalho E, Oliveira C, Fragoso M, Miranda V (1998) Regras gerais de reabilitação e econstrução de edifícios correntes afectados pela crise sísmica do Faial, Pico e S. Jorge iniciada pelo sismo de 9 de Julho de 1998. Materiais de Construção e Estruturas, Relatório 100/98, Laboratório Regional de Engenharia Civil, Região Autónoma dos Açores, Lisboa
18. Bendat J, Piersol A (2000) Random data: analysis and measurement procedures. Wiley Inter Science, ISBN: 0471317330
19. MathWorks (2014) MATLAB the language of technical computing, version 9, MathWorks Inc. http://www.mathworks.com (consultado em Novembro de 2013)
20. Chopra A (2001) Dynamics of structures: theory and applications to earthquake engineering, 3rd edn. Prentice Hall

Assessment and Diagnosis of Two Collar Timber Trusses by Means of Visual Grading and Non-destructive Tests

Isabel D. Gomes, Fotis Kondis, Hélder S. Sousa, Jorge M. Branco and Paulo B. Lourenço

Abstract The main objective of the present work is to evidence the importance of the inspection and diagnosis phase on the assessment of existing timber structures through visual grading and non-destructive testing (NDT) of its elements, especially for old structures with historical importance. To that aim, two collar timber trusses were inspected and assessed. These trusses were part of an old roof with over than one hundred years old that was decommissioned and then reassembled in laboratory conditions in the University of Minho (Portugal). The timber trusses were then tested on full scale at laboratory conditions until failure. The first part of this work presents the visual grading procedure complemented with NDT. During that part, all the elements of the trusses were visually inspected on each 40 cm segments with consideration of its geometric characteristics (exterior cross-section, wane and deformation), its defects (cracks, knots and decay) and critical sections. Non-destructive techniques, such as impact penetration, ultrasound and drilling resistance tests, were carried out in order to characterize the level of conservation of the elements at its present condition. The combination of the results of all NDTs allowed for a more substantiated grading of the segments and consequently of the elements. The results are presented through damage maps indicating the main pathologies of the structure and its residual cross-section, as well as correlations between different NDTs. In the second part of this work, a comparative analysis between the two trusses is made considering the prior results of the NDTs and the

I.D. Gomes · H.S. Sousa (✉) · J.M. Branco · P.B. Lourenço
ISISE—Department of Civil Engineering, University of Minho, Guimarães, Portugal
e-mail: sousa.hms@gmail.com

I.D. Gomes
e-mail: a61836@alumni.uminho.pt

J.M. Branco
e-mail: jbranco@civil.uminho.pt

P.B. Lourenço
e-mail: pbl@civil.uminho.pt

F. Kondis
School of Architecture, National Technical University of Athens, Athens, Greece
e-mail: fotis.kondis@gmail.com

H. Cruz et al. (eds.), *Historical Earthquake-Resistant Timber Framing in the Mediterranean Area*, Lecture Notes in Civil Engineering,
DOI 10.1007/978-3-319-39492-3_26

performance of the trusses on the full scale tests. To that aim the results of the visual grading and NDTs are correlated to the sections were failure of the segments was visible.

Keywords Timber trusses · Visual grading · NDTs · Damage maps

Introduction

The Mediterranean Region presents several examples of historical heritage that must be preserved, in particular timber structures that are often neglected in comparison to other structures and materials. Timber roof structures are one of the most characteristic and defining typologies of these structures. There are several configurations of timber roofs ranging from simple purlins and boards roof to complex tridimensional trusses. Often, these roofs are neglected in terms of assessment and preservation due to its inaccessibility or just because they are hidden behind vaults or other interior roof sub-structures.

As wood is a natural material, during its life it is subjected to biological degradation associated to the surrounding environmental conditions. This situation, allied to poor or inexistent maintenance practices, often leads to structures that are seriously decayed or damaged and inevitably must be repaired or even replaced. This results that the assessment of the structural capacity of existing timber structures and its residual bearing capacity are essential to determine the need and level of possible intervention actions [1, 2].

For heritage and other relevant structures, this information must be obtained through methods that will not damage the elements, thus visual inspection allied with non-destructive techniques (NDTs) must be considered. Commonly used non or semi-destructive tests, among others, are impact penetration, controlled drilling resistance and ultrasound tests [1, 3–5].

In [6], an accurate geometric and numerical evaluation of the timber elements of two king post trusses were made resulting from the data obtained in the diagnostic phase (using visual inspection and NDTs), and from full scale tests. Also, in [7], the damages on a queen post truss were identified through NDTs and the mechanical performance of each timber element was assessed by full scale loading tests in elastic field. The results from the diagnostic phase were then used to calibrate numeric and predictive models, where it was noticeable that the presence of decay is a crucial parameter influencing the performance of the truss, as also seen in [8].

With respect to the seismic response, the contribution of a timber truss roof is often not considered for numerical modelling of historical structures, due to the weak connection between the timber truss and the supporting walls [9]. Therefore it is of utmost importance to assess the conditions of the connections and supports. In [10], it is stated that detailing and condition of connections are important parameters for the seismic response. In the same study, it was stated that, based on observations and analysis, a low vulnerability timber roof should comprise: (i) no unrestrained thrusts, (ii) an effective connection with the supporting walls, (iii) a comparable response capability in the different connections, and (iv) suitably reinforced carpentry connections to avoid disassembling and brittle behaviour. Within that scope, in [11], full-scale experimentation on a roof truss was made to determine its performance and to calibrate numerical models for the connections.

In the referenced works, the diagnostic phase was essential to understand and predict the overall structural performance of the timber trusses. In the present work, two collar trusses are assessed according to visual grading complement with NDTs. These results are then discussed and correlated according to the results of full scale tests made to the trusses.

Materials and Methods

The Collar Trusses. The timber trusses were part of a 18th century building situated in Coimbra, Portugal. The timber structure was comprised of a three pitched roof including four trusses with a span of 14.2 m. As a consequence of the significant deterioration and damage, the trusses were brought down and replaced (more information in [12]). The trusses were dismantled and brought to University of Minho (Portugal) where they were assessed by visual inspection and NDTs. After being reconstructed two of the trusses (further on identified as trusses T1 and T2) were tested in full scale (Fig. 1).

Inspection and Testing. In a first phase, the cross sections dimensions of the trusses' elements were measured at each 40 cm segments. Visual inspection, according to [13], was made to each of the segments in order to obtain a detailed grading of the element along its length as well as an overall grading of each structural element, attending to its critical sections. For visual grading of a single element, the used norm considers three classes (I, II and III) regarding onsite diagnosis. The timber element is from a given class if it fulfills all the imposed requirements and considered as non-classifiable (NC) if it does not fit to either class.

Fig. 1 Collar truss: **a** on the building site, **b** dismantled elements; **c** reconstructed in laboratory

In this test campaign three NDTs, namely impact penetration, ultrasound and drilling resistance tests, were considered both for increasing the reliability of the visual inspection and to assess the depth of decay along the length of each element. The equipments used were, respectively, the impact hammer Pilodyn® 6J, Pundit Lab (54 kHz frequency transmission transducers) and the controlled drilling device Resistograph® 3450. The penetration impact tests were firstly considered as they are a fast and inexpensive method, providing straightforward results for an initial characterization of the surface layer of timber elements. Drilling resistance tests were made to obtain information about the interior of the cross-sections (local data), while ultrasound was made to obtain global information about each segment. Each cross-section was analyzed by impact penetration tests (3 tests per section face) and drilling resistance tests (2 tests in width and 3 in height), while the segments between sections were analyzed by impact penetration tests (5 tests per segment face) and ultrasound (1 test per segment face), as presented in Fig. 2.

After the diagnosis phase, two full-scale tests were made to the trusses, until failure was reached. In both cases failure was found in the rafters near the loading positions (Fig. 3). Also evidences of crushing were found in the supports.

Results of Inspection and Diagnosis

Visual Inspection and Grading. The most common found defects were the existence of knots and misalignment of grain direction. Decay was found mainly in the bottom part of the rafters (support area in contact with the masonry walls). Also wane was found, more due to the initial sawing process, rather than to loss of cross section integrity during the lifetime of the structure. Therefore, wane was not considered as a limiting parameter for visual grading in this study, but instead only the residual cross section without wane was considered. The results of visual grading by each segment and also regarding the percentage of elements with given class are given in Figs. 4 and 5. The class for each element was given by the consideration of the worst grading found for the segments within that element.

From the analysis of Figs. 4 and 5 it is observed that truss T2 had a higher percentage of elements in lower classes, as well as more segments in lower classes. This is particularly noticeable for the rafters supports and near the connection with the tie-beam.

Impact Penetration Tests. The results of the impact penetration tests for truss T1 are shown in the damage map presented in Fig. 6 for the case of the back face (more detailed results in [15]). The tests evidenced that one of the support areas had higher penetration depths, as well as in the connections between rafters and collar beam and in the connection between rafters and post. This was consistent with the decay found visually and also with the visual grading. The interior elements (post, collar beam and diagonals) had lower penetration depths as they are not in direct contact with the roof covering or the masonry walls, and therefore away from possible damp due to water infiltrations. The presence of decay led to significant

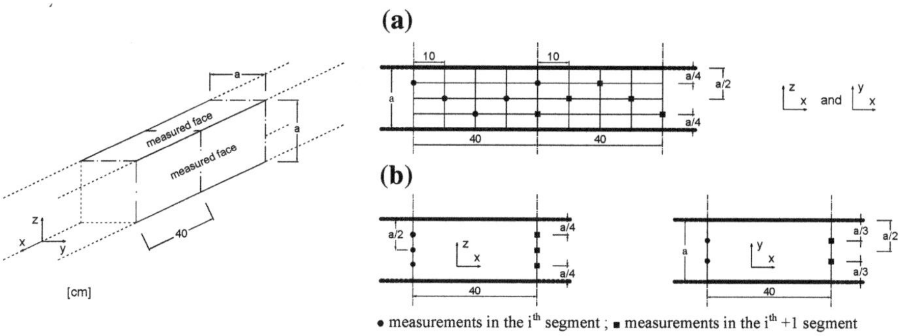

Fig. 2 Test measurement mesh: **a** impact penetration; **b** drilling resistance, adapted from [14]

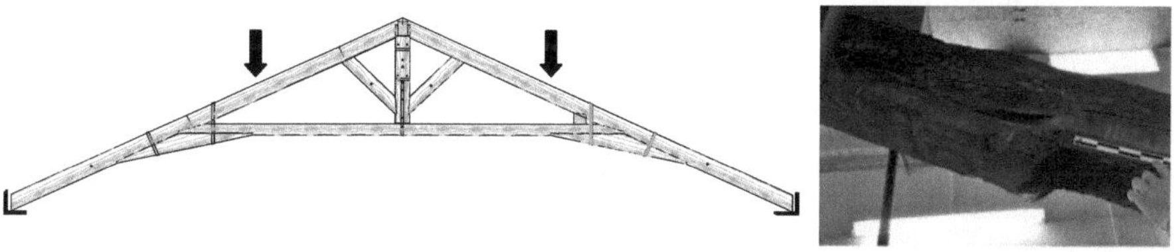

Fig. 3 Schematic of test setup and details of a failure on one of the rafters

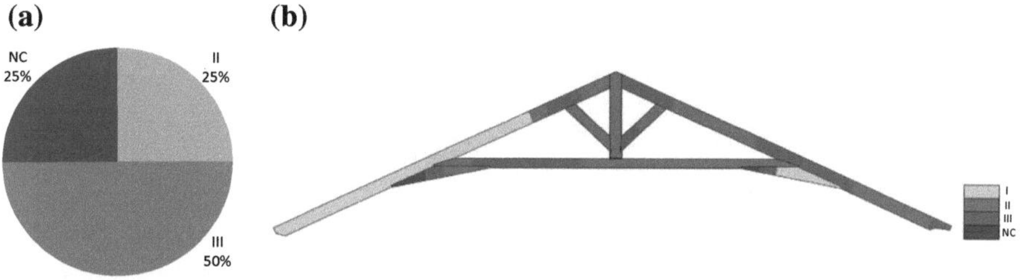

Fig. 4 Visual grading results for truss T1: **a** percentage of elements with a given grade; **b** grading by segment

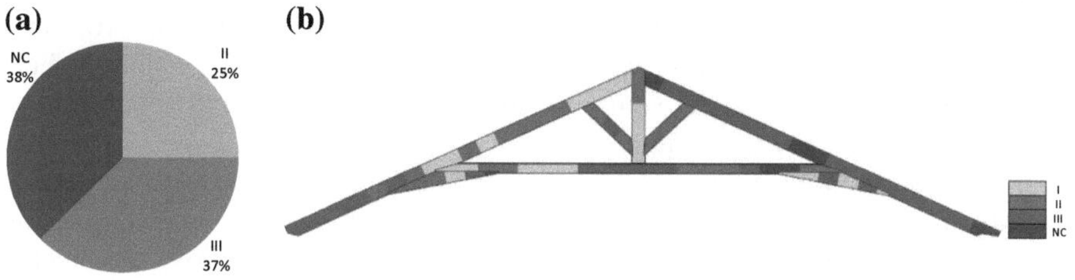

Fig. 5 Visual grading results for truss T2: **a** percentage of elements with a given grade; **b** grading by segment

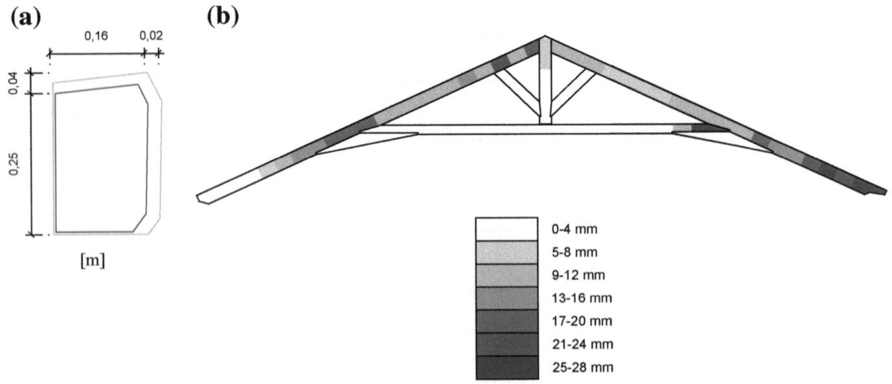

Fig. 6 Impact penetration results: **a** example of a residual cross-section; **b** damage map of the back face of truss T1

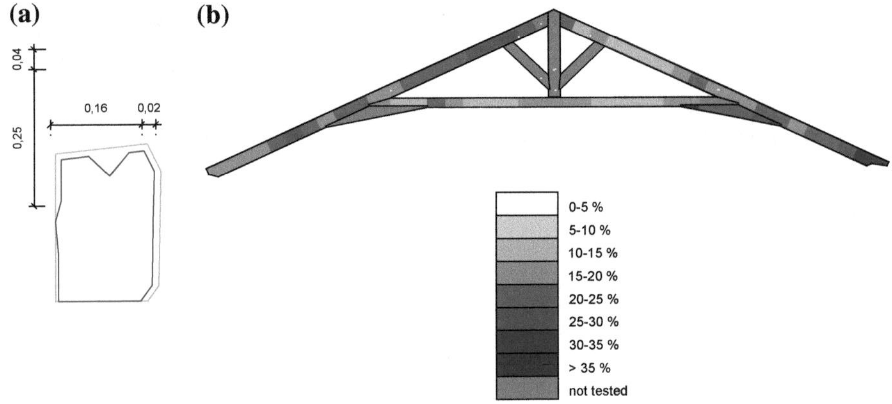

Fig. 7 Drilling resistance results: **a** example of a residual cross-section; **b** damage map with percentage of loss of effective cross-section for truss T1

variation of results even within the same segment. In the case of the rafters, the mean coefficient of variation (CoV) of the penetration depth ranged between 60 and 90 %. The mean value of penetration depth, for all elements, was 16.3 mm. Considering that reference tests made to undecayed (and with no visible defects) had a mean penetration depth value of 8.0 mm, it was considered that the mean value of decay depth was 8.3 mm.

Drilling Resistance Tests. The results of the drilling resistance tests for truss T1 are shown in the damage map presented in Fig. 7 regarding the percentage of loss of effective cross-section. Regarding the elements with higher loss of effective cross-section, namely the rafters, mean values around 20 % are found with CoV of 18 and 35 %, respectively for the north and south rafters.

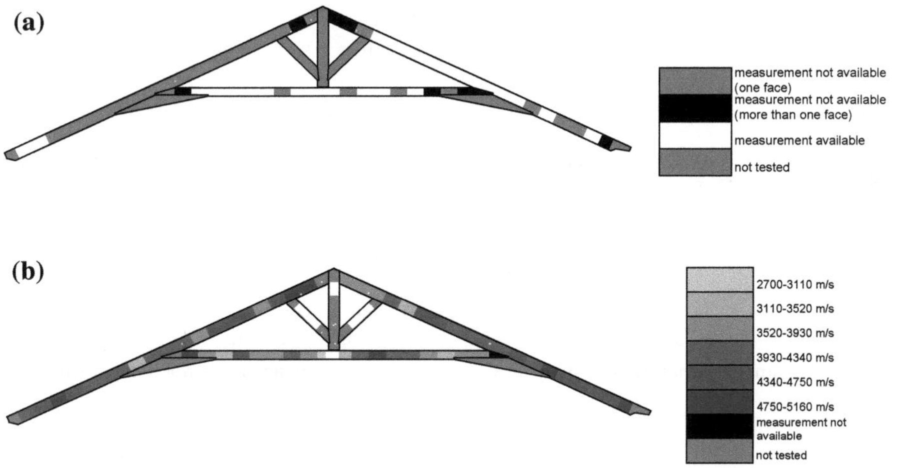

Fig. 8 Ultrasound wave velocity results map for truss T1: **a** description of segments with no achieved measurements; **b** mean wave velocity for the four faces of each segment

Ultrasounds. The results of the ultrasound tests for truss T1 are shown in the damage map presented in Fig. 8 regarding the wave velocity. In the same figure, the segments where no measurement was possible (signal loss) are also presented. The values range from 2700 to 5160 m/s. It is seen that higher wave velocities are found in the rafters supports and on the connections with the tie-beam and diagonals. The segments where measurements were not obtained, due to loss of signal, corresponded to segments with significant knots or grain misalignment, or considerable superficial decay.

Correlation Between NDTs

The different NDTs provided an important insight to the conservation condition of the timber trusses. The results of NDTs efficiently evidenced the segments where higher decay was present. Considering the overall results from the NDTs, the critical sections are found in the rafters, mainly in the supports and on the connections with the tie-beam and post. In this point, the results of the NDTs are consistent with the results of the visual grading.

Nevertheless, the correlations between quantitative results obtained by section (for impact penetration tests and drilling resistance tests) are low. This is mainly to the high variation of results and also due to the fact that impact penetration tests assess mainly the superficial layer of the element, whereas the drilling resistance tests assess the full thickness of the section. In both cases, the measurements are confined to a relative small portion of the section, therefore very dependent on the location of the test (very influenced by defects). In Fig. 9 the correlations between

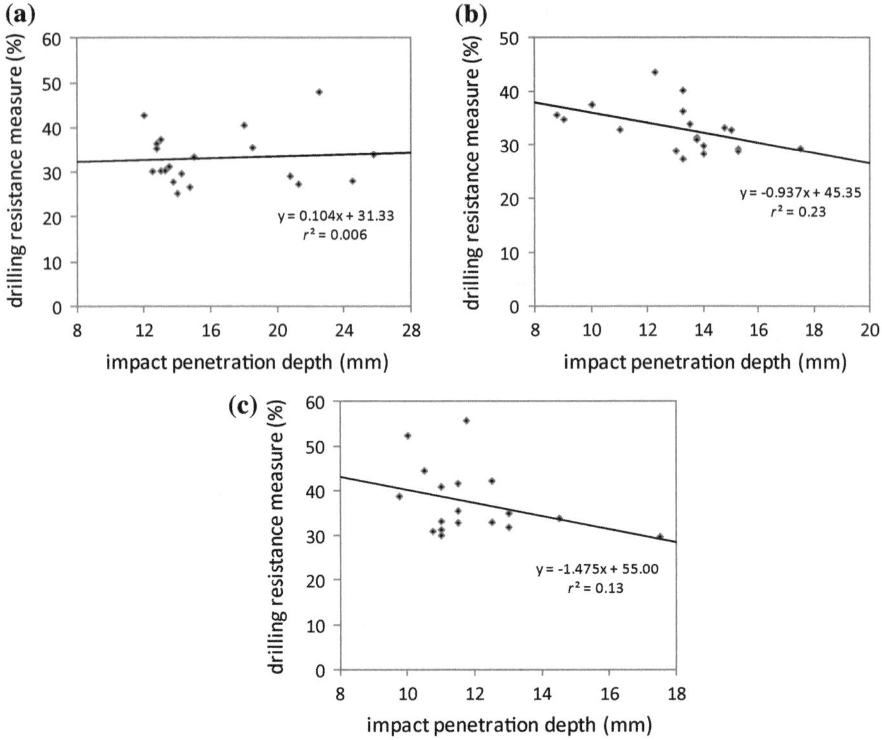

Fig. 9 Correlation between impact penetration depth and drilling resistance measure for truss T1: **a** north rafter; **b** south rafter; **c** tie-beam

mean impact penetration depth per section with resistance measurement of drilling resistance tests is presented for the case of truss T1 for both rafters and the tie-beam. More information about the correlations may be found in [15].

Comparative Analysis

The results of full-scale tests on the collar trusses in terms of load/displacement curves accounting for the loading points on the rafters are presented in Fig. 10. Truss T1 was able to sustain a higher level of load (approximately 125 and 140 kN on each rafter) compared to truss T2 (approximately 80 and 90 kN on each rafter). In terms of visual grading, truss T2 had higher percentages of lower grades and also evidenced a higher level of degradation, confirmed by NDTs, resulting in a lower level of load. It was also possible to verify that failure of the elements took place on segments that were initially considered as possible weak sections accounting their lower visual grade. Taking as example truss T2, three main failures were found during the mechanical

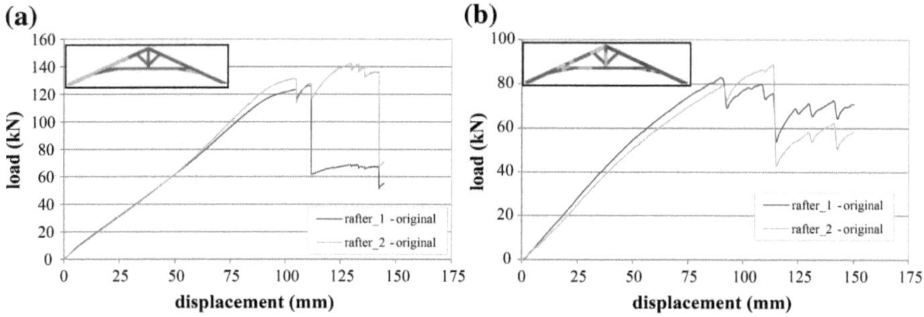

Fig. 10 Results of full-scale tests on the collar trusses: **a** truss T1; **b** truss T2

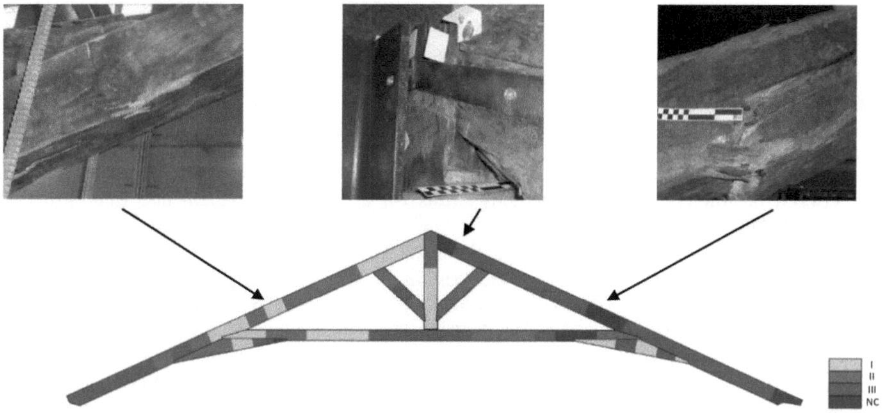

Fig. 11 Visual grading of truss T2 with details of the failure of the rafters and of the post-rafter connection

test, namely failure of both rafters on sections near the connection with the tie-beam and failure near the connection of one rafter with the post. Those failures coincided with sections with lower visual grades as presented in Fig. 11.

Conclusions

The first step of the assessment of an existing timber structure is its diagnosis accounting the results of visual inspection and non-destructive tests. This step is crucial to identify possible weak sections and thus for a better understanding of the mechanical performance of the structure. The results taken from it, are also extremely important for the calibration of more complex analysis or numerical models.

In this work, the results of visual grading and of non-destructive tests on two old collar trusses were considered for identification of their weak sections and to understand their overall state of conservation. To that aim, damage maps considering visual grading, impact penetration, drilling resistance and ultrasound results were made and analyzed. Also, correlations between tests were made, concluding that a qualitative analysis may be used to sustain the initial visual grading.

Finally, the results of the diagnosis was compared with the results of full-scale tests, and it was observed that visual grading allied with the results of non-destructive tests may efficiently identify possible weak sections, as well as serve as a means of comparison between the two tested trusses.

Acknowledgments The information obtained through training school: Assessment and reinforcement of timber elements and structures (Guimarães, Portugal) within scope of FP1101 and RILEM TC 245 RTE is acknowledged. The second author acknowledges the grant provided for his STSM within FP1101. The support of the Structural Lab from University of Minho and Augusto de Oliveira Ferreira e Companhia Lda (AOF) are greatly acknowledged.

References

1. Calderoni C, De Matteis G, Giubileo C, Mazzolani FM (2010) Experimental correlations between destructive and non-destructive tests on ancient timber elements. Eng Struct 32 (2):442–448
2. Alessandri C, Mallardo V, Pizzo B, Ruocco E (2012) The roof of the church of the nativity in Bethlehem: structural problems and intervention techniques. J Cult Heritage 13(4):e70–e81
3. Ceraldi C, Mormone V, Ermolli ER (2001) Resistographic inspection of ancient timber structures for the evaluation of mechanical characteristics. Mater Struct 34(1):59–64
4. Kasal B, Tannert T (2010) In situ assessment of structural timber. RILEM state of the art reports
5. Sousa HS, Sørensen JD, Kirkegaard PH, Branco JM, Lourenço PB (2013) On the use of NDT data for reliability-based assessment of existing timber structures. Eng Struct 56:298–311
6. Branco JM, Piazza M, Cruz PJ (2010) Structural analysis of two King-post timber trusses: non-destructive evaluation and load-carrying tests. Constr Build Mater 24(3):371–383
7. Del Senno M, Piazza M (2003) Behaviour and rehabilitation of queen post timber trusses: a case study. Advances in architectures series, pp 803–814
8. Brites RD, Neves LC, Machado JS, Lourenço PB, Sousa HS (2013) Reliability analysis of a timber truss system subjected to decay. Eng Struct 46:184–192
9. Betti M, Vignoli A (2008) Modelling and analysis of a Romanesque church under earthquake loading: assessment of seismic resistance. Eng Struct 30(2):352–367
10. Parisi MA, Chesi C, Tardini C, Piazza M (2008) Seismic vulnerability and preservation of timber roof structures. In: Proceedings, SAHC08, pp 1253–1260
11. Parisi MA, Piazza M (2002) Seismic behavior and retrofitting of joints in traditional timber roof structures. Soil Dyn Earthq Eng 22(9):1183–1191
12. Lourenço PB, Sousa HS, Brites RD, Neves LC (2013) In situ measured cross section geometry of old timber structures and its influence on structural safety. Mater Struct 46(7):1193–1208
13. UNI (2004) UNI 11119:2004 Cultural heritage—wooden artifacts—load-bearing structures—on site inspections for the diagnosis of timber members. UNI Milano
14. Sousa HS (2013) Methodology for safety evaluation of existing timber elements. Ph.D. dissertation, Department of Civil Engineering, University of Minho, Portugal
15. Gomes I (2015) Diagnóstico e ensaio de carga de uma asna tradicional de madeira. Master thesis, Department of Civil Engineering, University of Minho, Portugal (in Portuguese)

Retrofitting Assessment and Numerical Models for the Historical Timber Roof Structures of the Towers of Valentino Castle in Turin

Clara Bertolini-Cestari, Antonia Spanò, Stefano Invernizzi and Tanja Marzi

Abstract The paper analyses a reinforcement intervention conceived by one of the authors (C. Bertolini) and performed on the historical timber roof structure of the Valentino Castle in Torino (Italy) some thirty years ago. The timber roof of the towers, dating back to 17th century, is a three-dimensional frame structure with a height of about 12 m. In the transversal direction four main great trusses are present, and two small trusses are placed close to the transversal masonry walls. The pitch is strongly inclined. A ridge connects the structure in the longitudinal direction of the timber roof, in addition to five series of purlins and three orders of overlaid frames with stiffening functions. This structural complex is firmly secured to the covering planks that support the tiles of black stone. The intervention was carried out with timber elements anchored to the ancient masonry walls, in order to improve their seismic behaviour. This intervention proved to be particularly innovative, since at that time the regional area where the castle is located was not yet considered a seismic area by the national standards. Laser scanning surveying,

C. Bertolini-Cestari (✉) · A. Spanò · T. Marzi
Department of Architecture and Design, Politecnico di Torino,
Viale Mattioli 39, 10125 Turin, Italy
e-mail: clara.bertolini@polito.it
URL: http://www.polito.it

A. Spanò
e-mail: antonia.spano@polito.it

T. Marzi
e-mail: tanja.marzi@polito.it

S. Invernizzi
Department of Structural, Geotechnical and Building Engineering,
Politecnico di Torino, C.so Duca degli Abruzzi 24, 10125 Turin, Italy
e-mail: stefano.invernizzi@polito.it

H. Cruz et al. (eds.), *Historical Earthquake-Resistant Timber Framing in the Mediterranean Area*, Lecture Notes in Civil Engineering,
DOI 10.1007/978-3-319-39492-3_27

non-destructive investigations and numerical simulations are adopted in the present study, in order to assess the efficiency of the intervention and the present condition of the joints.

Keywords Earthquake-resistant · Timber roof · Assessment · Reinforcement intervention · LiDAR survey · 3D modeling

Introduction

The Valentino Caste is located in the centre of the Italian city of Turin, facing the Po riverside. The origin of the building dates back to the 17th century and owes its current structure to Princess Christine Marie of France (1606–1663), wife of Victor Amadeus I, who dwelt there from 1630 and wanted a castle in the style of the castles built in that period in France. The architecture of the castle was therefore inspired by the French principle of the pavilion system, with four towers at each angle, and a wide inner court (Fig. 1).

The timber roof of the towers is a three-dimensional hierarchical structure. More precisely, in the transversal direction, the roof is supported by four great trusses and two small trusses at the ends. The pitch is strongly inclined. In the longitudinal direction, the timber structure is composed of the ridge, five series of purlins and three orders of overlaid frames with stiffening functions (Figs. 2–3). The structural complex is firmly connected to the covering planks that support the tiles of black stone [1].

At the end of the eighties the timber structure of the tower roofs was object of new studies and analyses since, during a previous intervention in 1961, the roof tiles had been substituted with "ardesia ligure" (slate) tiles, which performed very poorly in the freeze-thaw cycles of the continental climate.

Prof. Clara Bertolini (Politecnico di Torino), who was in charge of the intervention in the eighties, adopted an innovative structural concept. In fact, the structural behaviour of the roof could not be ascribed simply to classical frame truss schemes, where the principal truss bears the weight of the independent joists and roof structure above. On the basis of this simplified hypothesis, the whole roof resulted unsafe, and should have been dismantled entirely.

Subsequently, the roof planking was considered as belonging to the structural system, together with the joists and the underlying trusses. Therefore, the structural hierarchy was in some sense reversed. As a consequence, the structure was conceived as a three-dimensional ribbed shell, quite similar to an upside down ship hull, that was very stiff with respect to both vertical and horizontal loading actions. Although seismic retrofitting was not compulsory at that time in the Turin area, special attention was devoted to giving the structure a better seismic behaviour, by introducing tying timber elements to brace the masonry tower and to connect the roof system with the underlying masonry walls in a more efficient way.

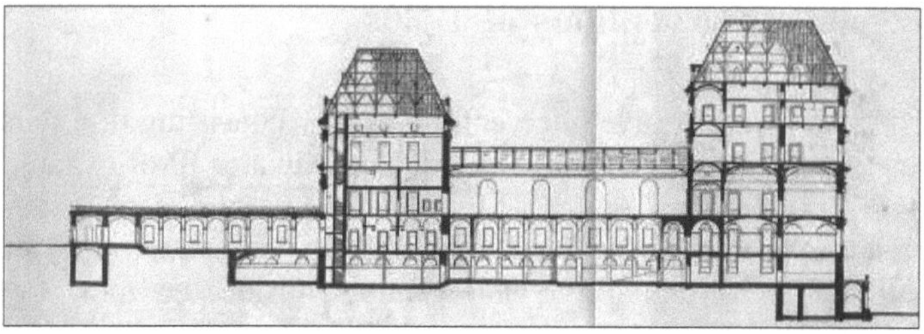

Fig. 1 Longitudinal section of timber roof structure of the Valentino castle in Turin

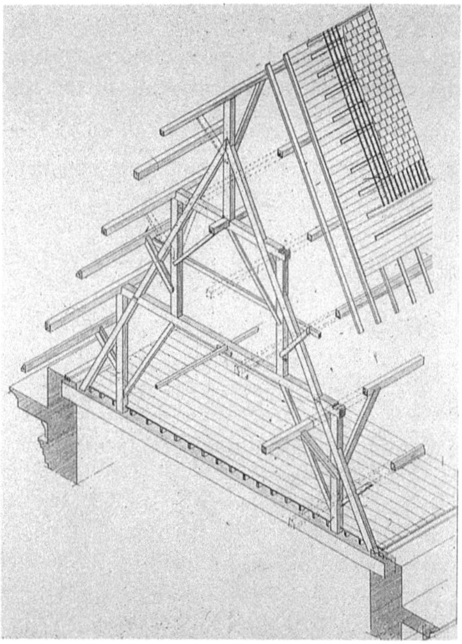

Fig. 2 Scheme of a truss and its components (drawing C. Bertolini 1986)

Fig. 3 Valentino castle in Turin today (photo T. Marzi)

Reinforcement Interventions of 1980s

In the late 1980s a conservative intervention, mainly consisting of a reinforcement of the structural system, was carried out (coordinator Prof. Clara Bertolini, Politecnico di Torino).

The restoration process took place in two separate phases, starting with the tower exterior and then covering all the interventions inside the roof. Considerable reinforcement work was carried out, as summarised hereafter. Each damaged area of the tie-beam heads was reconstructed using glue-laminated timber reinforced with fiber-glass rods. The insertion of a neoprene layer in between the wall and the glulam prosthesis guaranteed a good level of ventilation, greatly improving the durability of the intervention (Figs. 4–5).

Another phase of intervention involved the reinforcement of the support area of the timber joists and planking at the top of the masonry walls (Figs. 6–7). The

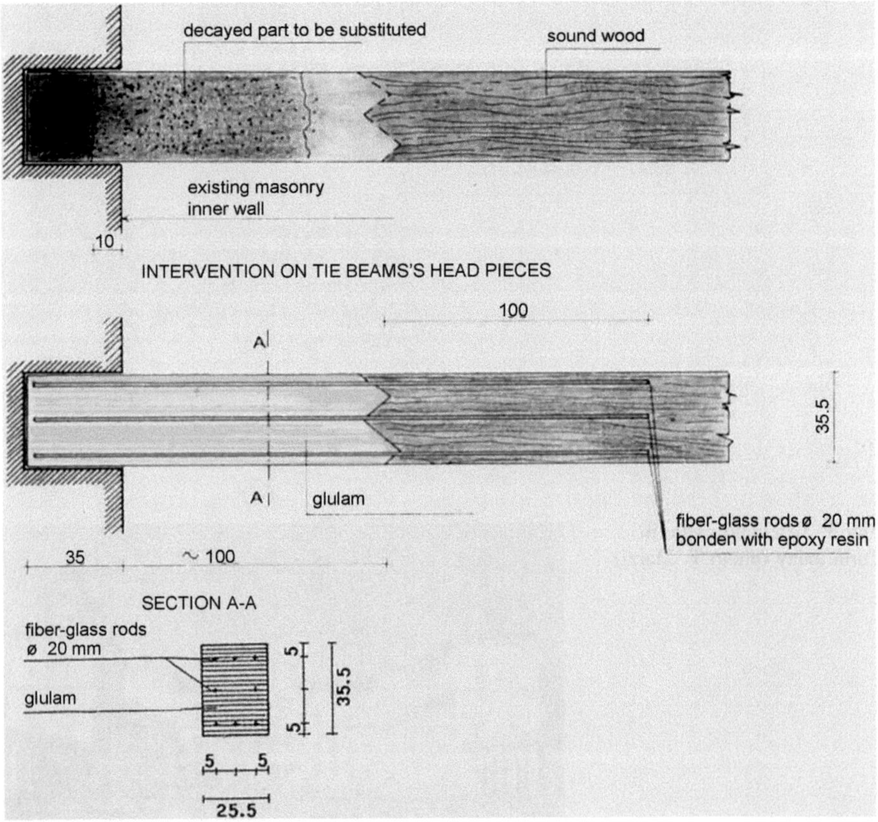

Fig. 4 Drawing of reinforcement intervention on tie beams' head-pieces using reinforced glue-laminated timber. Realization (1986)

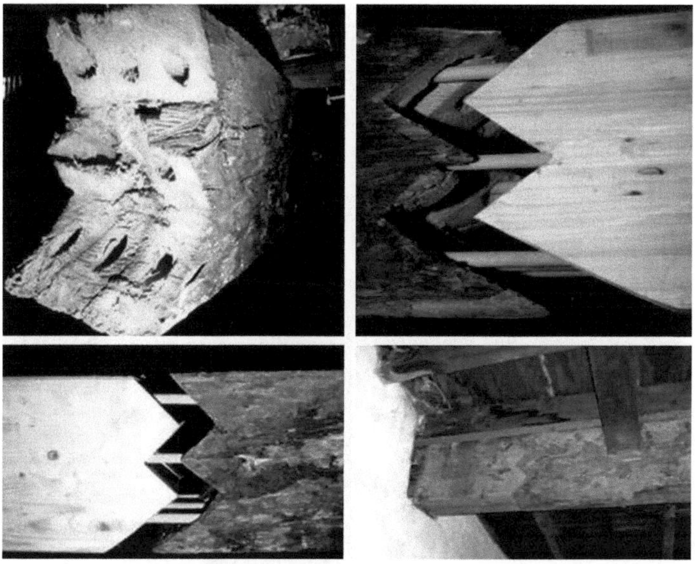

Fig. 5 Reinforcement intervention on tie beams' head-pieces using reinforced glue-laminated timber. Realization (1986)

Fig. 6 Area of the masonry wall supporting the timber carpentry before the intervention (1986)

Fig. 7 Area of the masonry wall supporting the timber carpentry after the intervention (1986)

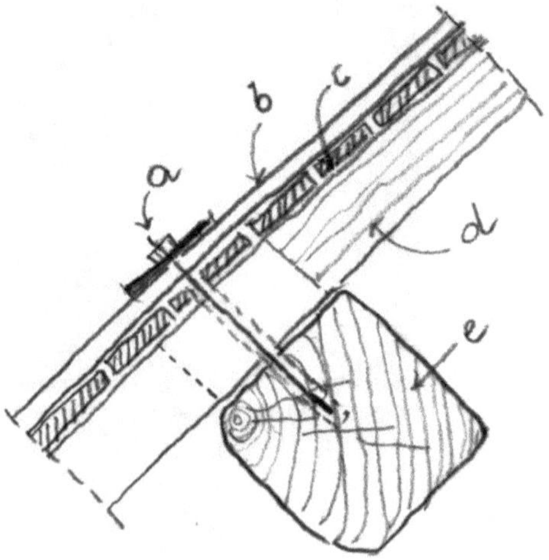

Fig. 8 Scheme of the connection between rafters, girders and double layers of planking to get the continuity of roof pitch: threaded steel pin (*a*), overlapped and crossed layers of the covering planks (*b*, *c*), purlin (*d*), girder (*e*)

Fig. 9 Restoration of covering planks

planks and tie-beams were joined with steel pins, to connect the layers of the orthogonal covering planks (similar to the current CLT technology) and to improve the "shell" behaviour (Figs. 8–9).

Seismic Improvement

The three dimensional roof structure itself, due to its lightness and stiffness, is not particularly prone to the effects of horizontal and vertical seismic acceleration spectra, provided that a correct constraint to the underlying masonry structure is

Fig. 10 Larch bracing and connection between the masonry and timber structures

guaranteed. The masonry structure, on the contrary, is prone to horizontal seismic acceleration, especially if adequate bracing of the wall tops is missing.

In the present case, the common solution of adding a horizontal concrete bracing is not the most convenient because of the added mass, and due to compatibility issues (both from the structural and restoration points of view).

A different solution was found by adding a timber (larch) bracing along the whole perimeter of the roof base, which was anchored to the masonry by means of steel connectors (Fig. 10). This choice minimised the added mass, optimising the connection between the masonry walls and the timber structure above, also providing some horizontal bracing (although it was less stiff than the corresponding concrete alternative).

Opportunities for 3D Survey of Timber Roof Structures

Some years ago the survey of a timber roof structure would have been carried out with a total-station-based polar coordinate measurement system; even though now high precision electromagnetic distance measurement (EDM) commonly exploits reflector-less devices which are very useful for inaccessible points, the method is still time-consuming and expensive in terms of labor.

The impact of new technologies and the availability of new sensors and systems is at the forefront of the scenario of Cultural Heritage documentation. Three-dimensional acquisition and modeling of structures, artefacts and interior spaces can be based on LiDAR (Light detection and ranging) systems, Photogrammetry and range cameras [2].

The innovation of techniques and their integration capabilities offer the most complete survey, modeling and visualization products, that involves the three dimensional vector reconstructions on which real world textures are draped. The descriptive abilities range from the geometrical shape of buildings and their parts or elements, dimensions, materials and even the alteration and decay of built structures, while the methods are usually high productive and automated.

In the case of an entire timber roof structure such as the one covering the tower of Valentino Castle, the darkness of the roof led to the choice of a laser scanning method, which is not affected by lighting conditions such as Photogrammetry. Furthermore, the LiDAR survey was chosen firstly because the former and main purpose related to the mechanical evaluation was the modeling of the three dimensional organization of the whole timber structures, (trusses, ridges and purlins), and then because a phase based scanner with a coaxial RGB camera was used (Laser Scanner Focus^{3D}–FaroCAM 2).

Compared with photogrammetry, and especially with the recent *Structure from Motion* processing offering a high level, low cost and accurate survey system [3], the terrestrial laser scanning equipment is still not as cost-effective as shown in the CH documentation, but the main advantages lie in the automation of the acquisition phase which enables fast, straightforward, and very accurate measurements (about 1 million points per second).

The Laser Survey Application on the Tower Roof and Future Developments

The terrestrial laser scanning (TLS) carried out in the roof of the Valentino castle tower enabled a large amount of high resolution information to be recorded, in order to model the very complex timber structures [4]. The main features are summarised below:

- a base of two topographical vertices provide a reference system for the whole survey and for the polar coordinate measurements of some thirty control points, to the record the differently positioned point clouds.
- Six well distributed point clouds were acquired in order to cover all timber elements; the scan positions were chosen in order to optimize the complete data recording of the complex structure, placing them one for each space between trusses.
- the point clouds acquired in the Valentino castle tower roof were perfectly aligned thanks to the high accuracy in the target detection, (average value of accuracy on control point coordinates about 10 mm)
- the generation of a 3D surface model (by triangulated method—mesh) which is able to create afterwards some textured continuous models or section profiles, showed a rather noisy surface, compatible with the features of historical timber roof structures that are rough, dusty, and dark. The model was meshed using the software 3D Reshaper (Technodigit).

Methods of Model Generation for the Structural Assessment and Their Automation

Generally, among laser scanning application in CH domain, the point models can be further processed in many different ways depending on the uses needed. Often delivered products are ground plans or horizontal and vertical sections, both in a two dimensional vector form or with the texture draping; other times complete 3D surface models with projected high resolution textures are increasingly required, to share visualizations and analyses with different specialists.

The points model segmentation is aimed at clustering points with similar attributes into homogeneous regions; the study of approaches to obtain this automatically is a topic theme of research interests since the introduction of some level of organization to the data is crucial and useful in order to extract significant information from the model.

The complexity of the roof timber structure of the Valentino tower required a mainly manual a segmentation setting (Fig. 11a, b). However this phase was essential for the extraction of a detailed model for the structural assessment. The original point cloud of the structures is generally too detailed for the mechanical evaluation, so a very focused strategy of the point model discretization is needed, depending on whether lines, surfaces or volume models are required. The timber structure point model was filtered and decimated in order to run a proper optimization of the interpolated *mesh* surface (Fig. 11c), and then the automatic extraction of 5 cm distance section profiles assisted the extraction of centre lines of wooden beams. (Fig. 11d, e, f).

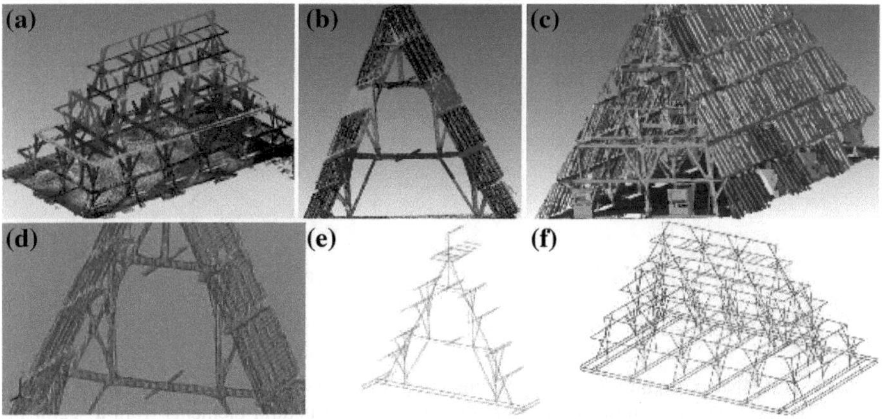

Fig. 11 **a**, **b** The points cloud segmentation. **c** The output model coming from the optimization of a mesh surface. **d** The extraction of horizontal and vertical section profiles. **e**, **f** A single truss and the complete roof structure represented by a wireframe model compose by lines beam element

Currently the conversion of laser models into new models that are properly conceived for mechanical evaluation is crucial and this issue requires a more in-depth study. It is essential to consider constructive systems and materials of historical buildings structures, since according to the method chosen for structural evaluation, various levels of detail and different geometric primitives are appropriate.

Fortunately, the numerical models of historical buildings can benefit from the recent technical developments of algorithms and tools advanced in the Reverse Engineering (RE) area of interest. The needs of RE in the industrial and design environments have highly motivated the enhancement of the so-called *as-built* measurement techniques, among which digital photogrammetry and LiDAR methods are leading [5]. Directly deriving from this, many methodologies have been suggested for the segmentation of 3D laser data, which are generally categorized into three classes: region growing, model fitting methods, and clustering of attributes [6]. Furthermore, many popular modeling software that implement cloud management, are equipped with specialized tools for the recognition and extraction of geometric primitives (planes, cylinders or other rotation surfaces).

Among many commercial solutions that manage the *region growing* algorithm by performing *feature extraction,* we tested *3DReshaper* by *Technodigit* and another tool pertaining to the sphere of BIM (Building Information Modeling) (Fig. 12) since it focuses abilities on construction system modeling and their monitoring recording. In the CH preservation sector, in fact, notable interest is directed to the investigation of benefits related to switching from a 3D content model to historic building information modeling (HBIM), aimed at supporting the conservation plans [7].

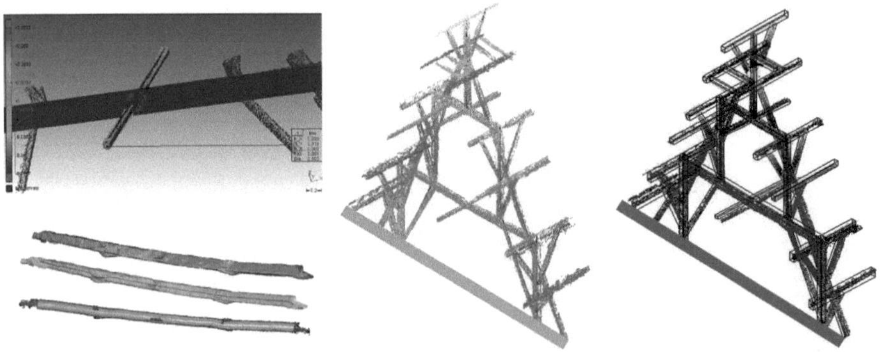

Fig. 12 Automatic recognition and approximation to cylinders of horizontal elements of the truss (using 3D reshaper) and automatic recognition of axes of structural elements from a segmented points cloud (achieved with the software scan to BIM ©IMAGINiT)

Numerical Simulations

The information acquired through the laser scanning survey served to obtain a detailed structural model. The ribbed planking was accounted for by means of an equivalent flat shell of the same stiffness, with equivalent thickness equal to 7 cm. This strategy allowed the actual behaviour of the double cross planking and rib system to be simulated, which is halfway between a "lattice" and a "shell" of the actual thickness. In addition to the dead load, due to the wood structure and to the black stone tiles (about 1750 Pa), the live load can be provided, in principle, by the snow and the wind pressure. The Italian standards, adopted for the calculation, state that the snow load can be disregarded for the present structure, since the slope of the roof is equal or higher than 60°. On the other hand, the wind thrust must be calculated according to the location and topographic configuration of the structure, and must be arranged according to three different load configurations. The wind pressure ranges between 674 Pa (compression) and 1123 Pa (depression). In order to account for different efficiency of the structural connection, two limit cases have been considered. The first model assumes that the planking system has no stiffness (Fig. 13), while the second model accounts for the planking (Fig. 14), and the three-dimensional constraint is modelled with proper periodic boundary conditions. The commercial finite element code Nòlian® (Softing srl) has been used to perform the calculation.

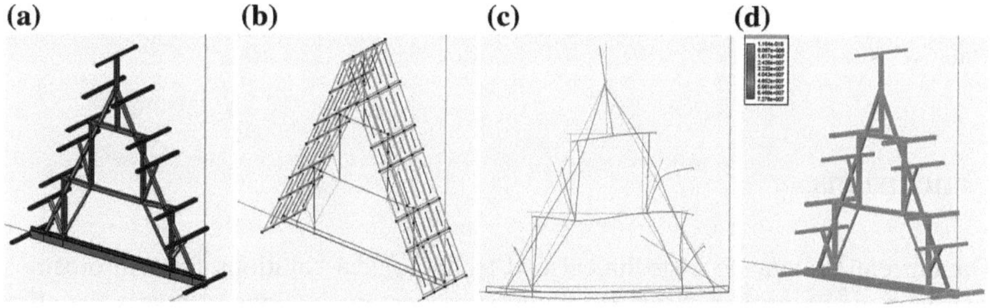

Fig. 13 Classical structural scheme: **a** solid model of the structural subsystem, **b** details of the loading masks, **c** elastic deformed shape, **d** principal compression stresses

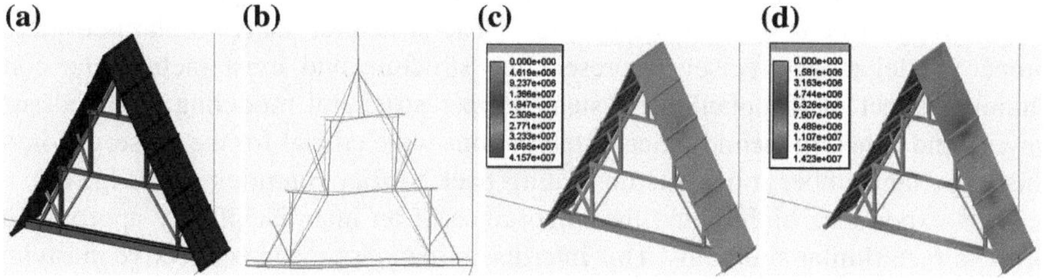

Fig. 14 Collaborating planking: **a** solid model of the structural subsystem, **b** elastic deformed shape, **c** principal tension and **d** compression stresses

Fig. 15 The timber roof structure of the north-east tower nowadays (photo C. Bertolini)

The results of the structural analysis show that if the constraining of the planking is disregarded, overestimated stresses and consequent deformations are obtained in several structural elements (especially in the median-upper part of the roof, where the second analysis provided low-intermediate levels of stress). As a consequence, the calculated levels of stress and deformation are not admissible for timber, and the roof structure would be assessed as unstable with indication to replace it for safety reasons. On the other hand, if the constraining effect of the planking is adequately accounted for, the levels of stress and deformation are admissible, even with respect to the nowadays more restrictive standard. In this way the original roof structure can be preserved as architectural heritage belonging to the Unesco World Heritage Site List.

Conclusions

The current abundance of technical and technological solutions for reinforcement interventions on timber structures requires ex-post evaluations to assess the effectiveness of interventions on historical structures. The positive assessment of the new accurate structural analysis improves and optimizes the results obtained at the time of the intervention carried out 30 years ago. This confirms the durability of an intervention performed with minimally invasive techniques due to the tri-dimensional behaviour of the structure. The structural analyses clarified that a proper model must precisely represent the structure and must include the constraining effect of the planking system. Proper structural modeling, detailed surveying and coherent technological interventions were crucial for the preservation of this important timber roof structure dating back to four centuries ago (Fig. 15).

This experience highlighted the importance of an interdisciplinary approach in order to face similar situations. This interdisciplinary approach is effective in saving resources while improving the assessment, in particular considering the different specialized investigations that are needed in any knowledge phase.

Acknowledgments The student Emilio Abbate is kindly acknowledged for his work on the HBIM model.

References

1. Bertolini Cestari C (2000) Problemas de recuperação: métodos de investigação, tecnologias de intervenção. In: Seminário Estruturas de madeira: reabilitação e inovação, Sep. 1999, pp 45–83
2. Patias P (2013) Overview of applications of close-range photogrammetry and vision techniques in architecture and archaeology. In: McGlone C (ed) Manual of photogrammetry. 6th edn. ASPRS (American Society of Photogrammetry and Remote Sensing), pp 1093–1107
3. Chiabrando F, Spanò A (2013) Points clouds generation using TLS and dense-matching techniques. A test on approachable accuracies of different tools. In: ISPRS annals of the photogrammetry, remote sensing and spatial information sciences, vol II-5/W1, XXIV international CIPA symposium. Strasbourg, France, 2–6 September 2013
4. Bertolini-Cestari C, Invernizzi S, Marzi T, Spanò A (2013) Numerical survey, analysis and assessment of past interventions on historical timber structures: the roof of Valentino castle. In: 3rd international conference on structural health assessment of timber structures, Shatis'15. Wroclaw, Poland, pp. 581–592
5. Chapman D (2013) As-built modeling of large and complex industrial facilities. In: McGlone C (ed) Manual of photogrammetry. 6th edn. ASPRS (American Society of Photogrammetry and Remote Sensing), pp 1082–1093
6. Lari Z, Habib AF, Kwak E (2011) An adaptive approach for segmentation of 3d laser point cloud. Int Arch Photogr Remote Sens Spat Inf Sci XXXVIII-5/W12:103–108
7. Oreni D, Brumana R, Georgopoulos A, Cuca B (2013) HBIM for conservation and management of built heritage: towards a library of vaults and wooden bean floors, ISPRS Ann Photogr Remote Sens Spat Inf Sci II-5/W1

Part V
Repair/Reinforcement

Development of Dissipation Panels for Seismic Retrofitting of Historic Buildings

Raquel Paula, Vítor Cóias and Luís Guerreiro

Abstract In the reconstruction of Lisbon after the devastating earthquake of 1755, an innovative structural solution was implemented. The historic timber framed masonry walls represented then a structural key element regarding seismic action. As often as not the buildings that incorporated these seismic resistant walls require interventions in respect of their original structure, re-establishing or enhancing the seismic behaviour. Innovative Dissipation Panels for seismic retrofitting are under development. The design of the Dissipation Panels was based on the role of the historic timber framed masonry walls as well as on the incorporation of the quite recent seismic protection systems, which reduce damage and limit losses more effectively than the traditional approach. The new technology was designed to improve the energy dissipation capacity of the buildings associated with seismic actions whilst preserving and respecting their original structural concept and, therefore, their authenticity. Full scale prototypes of the developed Dissipation Panels were produced and tested in order to characterize its cyclic behaviour. Experimental characterization of the dynamic behaviour of the conceived prototypes show good energy dissipation capacity.

Keywords Seismic retrofitting · Energy dissipation · Dissipation panels · Innovative technology

R. Paula (✉) · L. Guerreiro
CERIS, IST, Universidade de Lisboa, Av. Rovisco Pais, 1049-001 Lisbon, Portugal
e-mail: raquel.paula@tecnico.ulisboa.pt

L. Guerreiro
e-mail: luis.guerreiro@tecnico.ulisboa.pt

V. Cóias
Stap, S.A., R. Marquês de Fronteira, 8, 3.º Dto., 1070-296 Lisbon, Portugal
e-mail: vitorcoias@stap.pt

H. Cruz et al. (eds.), *Historical Earthquake-Resistant Timber Framing in the Mediterranean Area*, Lecture Notes in Civil Engineering,
DOI 10.1007/978-3-319-39492-3_28

Introduction

In the reconstruction of Lisbon after the devastating earthquake of 1755, an innovative structural solution was implemented. The new typology of buildings—named *Pombalino* buildings—comprised the distinctive *gaiola* system, which was designed to provide the buildings with adequate seismic behaviour. The gaiola system consisted of an interior 3D timber grid mainly formed by the timber framed masonry walls (frontal walls) and the elevated timber floors.

The historic timber framed masonry walls represented then a key structural element regarding seismic action [1]. As often as not the buildings that incorporated these seismic resistant walls have undergone uncontrolled changes and occasional modifications that decreased their seismic strength. In consequence, these buildings require interventions in respect of their original structure, re-establishing or enhancing the seismic behaviour. Moreover, rehabilitation and retrofitting in order to mitigate the effects of earthquakes on structures should also be addressed for the several other old buildings that lack seismic resistance nowadays [1–3].

Innovative Dissipation Panels for seismic retrofitting are under development. The design of the Dissipation Panels was based on the role of the historic timber framed walls as well as on the incorporation of the quite recent seismic protection systems [4], which reduce damage and limit losses more effectively than the traditional approach. The new technology was designed to improve the energy dissipation capacity of the buildings associated with seismic actions whilst respecting their original structural concept and, therefore, their authenticity.

Full scale prototypes of the developed Dissipation Panels were produced and tested in order to characterize its cyclic behaviour. The prototypes were subjected to cyclic tests performed at the Laboratory of Structures and Strength of Materials of Instituto Superior Técnico (IST). Experimental results are herein presented.

Brief Description of the Technology

During an earthquake event, large amounts of energy are input into structural systems due to ground excitation. Passive energy dissipation systems, when incorporated into a structure, absorb or consume a portion of the input energy, thereby reducing energy dissipation demand on primary structural members and minimizing possible structural damage.

Basic conception of the Dissipation Panels corresponds to an articulated supporting frame with an embodied dissipative damper. The Dissipation Panels have a similar pattern design to the notable past timber framed walls, so they can be easily integrated within/in respect of the existing structure—Fig. 1. Complementary, the addition of the damper leads to a significant increase of the energy dissipation capacity of the structure.

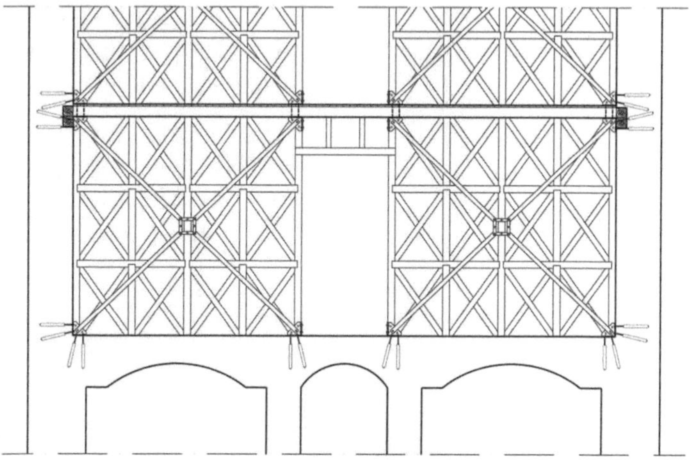

Fig. 1 Schematic representation of the dissipation panels. Installation in the existing structure

The Dissipation Panel was conceived to exploit the dissipation behaviour of the embodied damper, which for this study has hysteretic properties. First studies are hence related to an Hysteretic Damper (HD) made of steel with a very simple geometry. The Hysteretic Damper (HD), as the main and central body of the Dissipation Panel, was designed to take advantage of the plastic behaviour of steel. Testing and validation of other types of dampers are included in the scope of this research study.

Besides the increase of damping, the herein described system is aimed at enabling compatible, low intrusive and reversible retrofitting interventions thus preserving the basic characteristics of the constructions incorporating the Dissipation Panels.

Experimental Setup

In the concept design of the experimental prototypes, allowance for good cyclic behaviour was addressed. The experimental program comprehended two main parts. The first part corresponded to the mechanical characterization and hysteretic capacity assessment of the Hysteretic Dampers alone—Fig. 2. This part also included the material characterization tests, which consisted of tensile tests of the steel that was used to fabricate the Dampers. The second part of the experimental program aimed at determining the cyclic behavior of the Panels—Fig. 3—and evaluating the energy dissipation capacity as a main criterion of its efficacy. Designation adopted for the tested physical models is presented in Table 1. Note that for each complete prototype of the Panel, it was fabricated an additional central

Fig. 2 Experimental setup for testing the hysteretic dampers prototypes

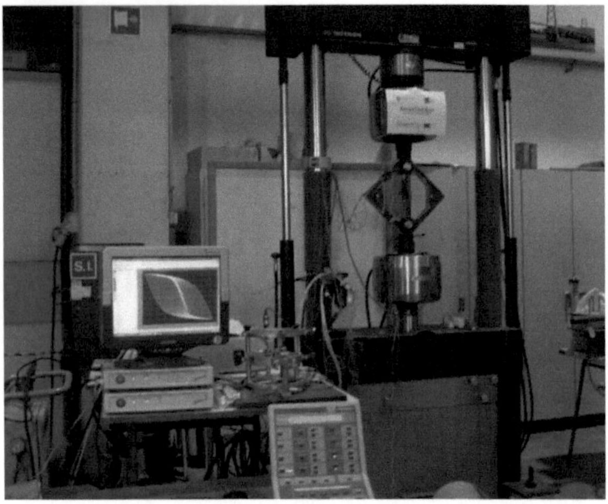

Fig. 3 Experimental setup for testing the dissipation panels prototypes. Reaction wall of the laboratory

Table 1 Experimental program and designation of tested models

Part 1 Dampers tests	Part 2 Panels tests	Notes
DHA1	P1DHA2	DHA1 damper and central damper of P1DHA2 panel were cut from the same steel plate
DHA3	P2DHA4	DHA3 damper and central dampers of P2DHA4 and P3DHA6 panels were cut from the same steel plate
	P3DHA6	

damper which was tested separately—in the first part of the experimental program as referred before.

Damper

The mechanical properties of the hysteretic dampers were obtained by means of cyclic tests that were carried out in an universal Instron testing machine as shown in Fig. 2. Physical models of the dampers were subjected to tensile-compression tests.

Dissipation Panel

Full scale models of individual prototype units of a complete Dissipation Panel were produced and tested in order to characterize the cyclic behaviour of the Panels. The Panels were subjected to cyclic quasi-static tests performed in the reaction wall at the Laboratory of Structures and Strength of Materials of Instituto Superior Técnico (IST) as presented in Fig. 3. The tests involved the application of controlled in-plane horizontal cyclic displacements at the top of the models by a 400 mm stroke mechanical actuator with 1000kN capacity that incorporated a load cell in order to measure the applied load. In order to guarantee the stability of the Panel during the test, a top guiding system that permitted its horizontal displacement was conceived as it can be observed in Fig. 3. The instrumentation of each Dissipation Panel consisted of: 8 extensometers and 2 Linear Variable Differential Transducers (LVDTs) (with 25 mm course), D1 and D2, in the central device—Fig. 4; 2 extensometers in each diagonal frame of the Panel; 1 Draw Wire Transducer at the top of the Panel (d) with a maximum measuring capacity of 500 mm. The control of the amplitude of the displacement cycles was done through this measuring device.

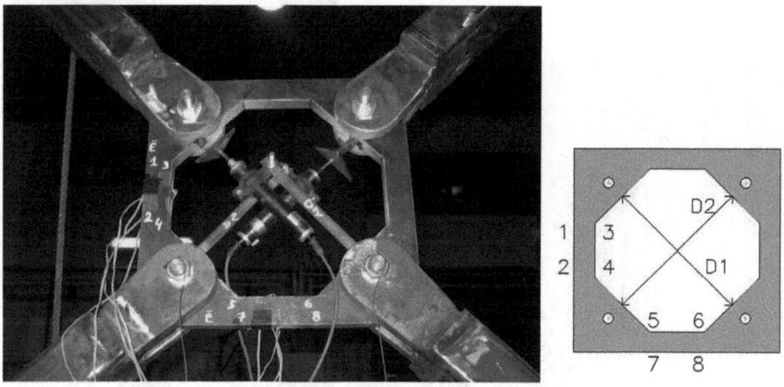

Fig. 4 Instrumentation of central hysteretic damper of the dissipation panel prototype P1DHA2. Extensometers 1–8. Transducers D1 and D2

Results and Discussion

Dampers

The experimental diagrams that resulted from testing the dampers DHA1 and DHA3 are shown in Fig. 5. The hysteretic loops presented growing displacements amplitudes from one cycle to the next cycle. As the damper plasticizes, energy dissipation occurs due to inelastic hysteretic response. Comparing the two diagrams it can be concluded that DHA3 presented more resistance than DHA1. This was associated with the fabrication process that was more accurate for the case of the DHA3 prototype regarding the interior section changes. This is also noticeable when looking at the models during the test and after failure—Fig. 6. Failure of model DHA1 occurred at an interior angle of the section. In case of model DHA3

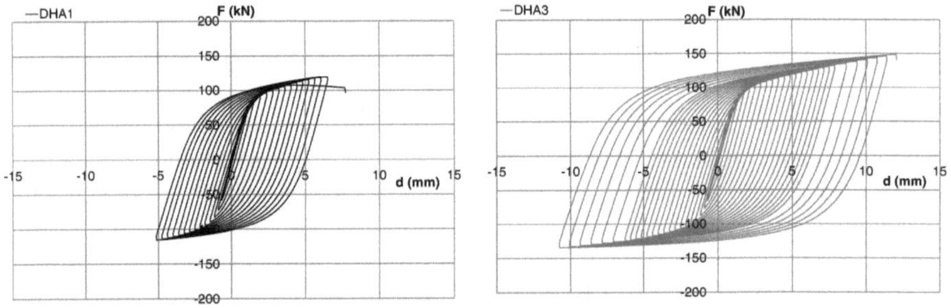

Fig. 5 Dampers (alone) testing results. Force-displacement curves

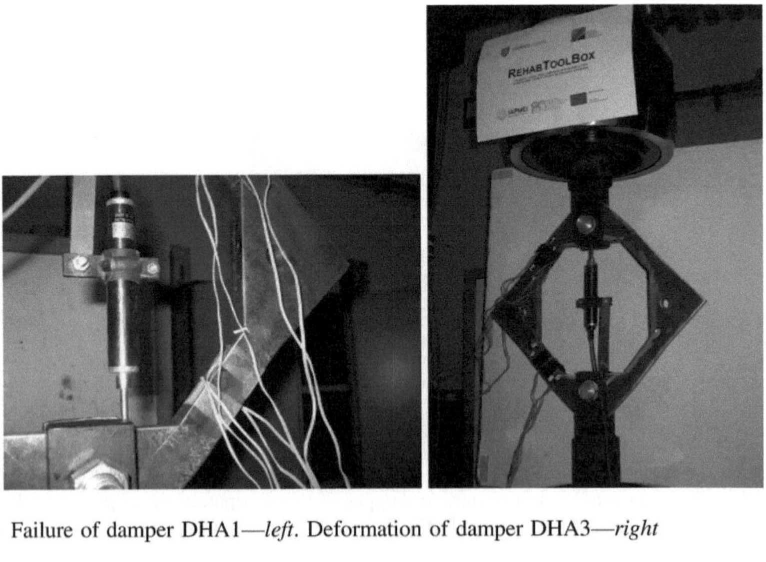

Fig. 6 Failure of damper DHA1—*left*. Deformation of damper DHA3—*right*

testing was interrupted when 'sufficient' deformation had been obtained. 'Sufficient' means enough deformation capacity to comply with the design specification for the horizontal displacement of the Panel. As it can be observed in Fig. 5, no signs of failure were noticed.

Dissipation Panels

Figure 7 presents the scheme of the horizontal displacements imposed to the prototypes of the Panels during the cyclic tests. For each model, the imposed horizontal displacements corresponded to increasing amplitudes from the beginning till the end of the cyclic test.

The experimental diagrams of the tested panels P1DHA2, P2DHA4 and P3DHA6 are shown in Figs. 8, 9, 10 and 11—on the left, the force-displacement curve of the Panel (forces and displacements—d-measured at the top of the Panel)

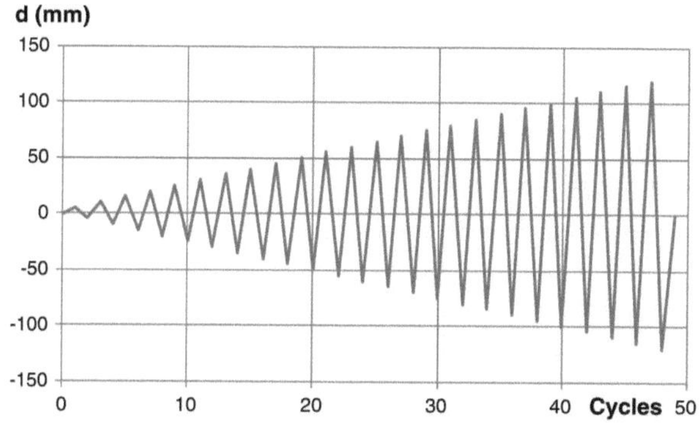

Fig. 7 Imposed horizontal displacements—cyclic tests of the dissipation panels

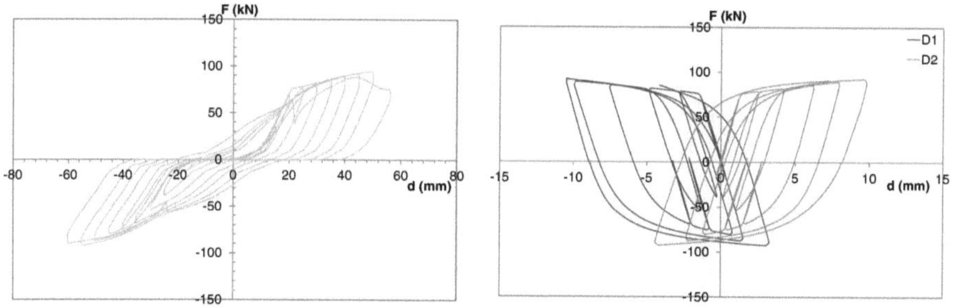

Fig. 8 P1DHA2 test results. Force-displacement curve of the panel. Force-displacement curves of the damper

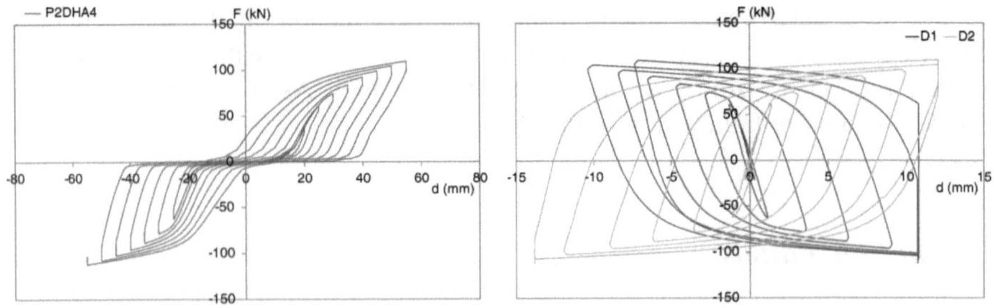

Fig. 9 P2DHA4 test results. Force-displacement curve of the panel. Force-displacement curves of the damper

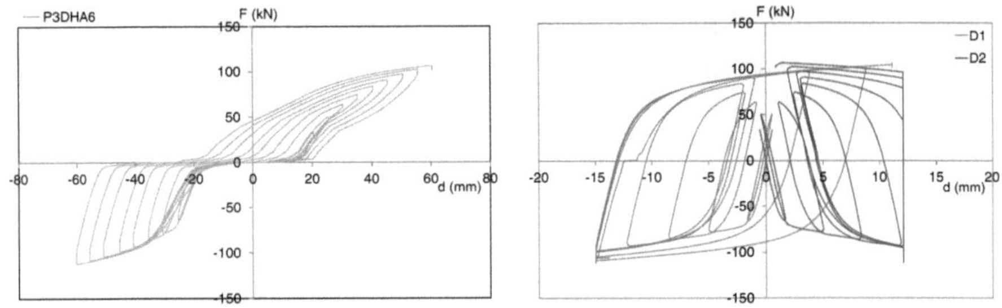

Fig. 10 P3DHA6 test results—1st phase. Force-displacement curve of the panel. Force-displacement curves of the damper

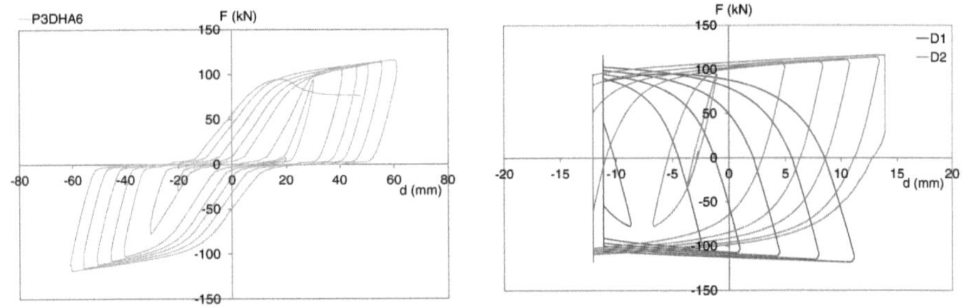

Fig. 11 P3DHA6 test results—2nd phase. Force-displacement curve of the panel. Force-displacement curves of the damper

and on the right, the force-displacement curves of the central damper of the Panel (displacements measured along the diagonals D1 and D2, as shown in Fig. 4).

For the test of Panel P3DHA6, two diagrams are presented because the cyclic test had to be done in two phases. First part was cancelled and then the test was restarted in a second part. First part had to be interrupted due to stability problems affecting the fixation of the base of the Panel to the auxiliary bottom beam that was used to execute the cyclic tests of the Panels. It should also be noted that the diagrams of the central dampers (on the right) are truncated because the maximum course of the LVDTs were exceeded.

During the tests and for all specimens, it was possible to observe the deformations of the steel dampers in each cycle and its growing amplitude before failure. In case of Panel P2DHA4 the test was interrupted before failure. Failure modes of tested models P1DHA2 and P3DHA6 are shown in Fig. 12. For both panels, failure occurred due to rupture of the central dampers which was localized at the sections where stress concentration was expected, because of the dampers variable cross section.

The experimental diagrams of the prototype Dissipation Panels show good capacity to dissipate energy, both in tension and in compression. The energy dissipated in each cycle corresponds to the area enclosed by the load-displacement curve [5]. Figure 13 presents the energy dissipated in each cycle.

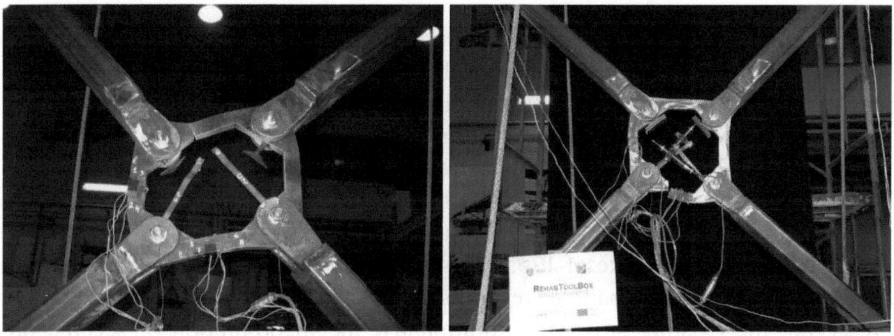

Fig. 12 Failure of the central dampers of the panel after cyclic tests—P1DHA2 on the *left* and P3DHA6 on the *right*

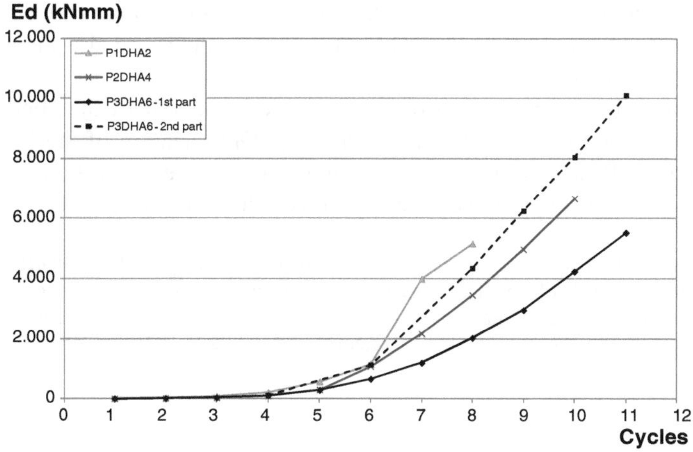

Fig. 13 Panels testing results. Energy dissipated in each cycle

For the three tested panels it was verified that when there was a change in the direction of the movement, the diagrams presented deformations without increasing of the force. This was associated to the existence of gaps in the joints between the different elements that form the Panels, namely, between the central damper and the diagonals and between the latter and the vertical and horizontal frames of the panel. For this reason, one of the future procedures to optimize the technology will be the reduction of the referred constructive gaps. Taking into account that the dissipated energy is function of the area enclosed by the force-displacement curve, it is estimated that when the gaps are reduced, the area increases allowing a better behaviour of the panels through the dissipated energy increase.

Conclusions

Stap, in collaboration with Instituto Superior Técnico (IST), has been developing a set of new complementary low intrusive devices and systems designed for structural rehabilitation, including seismic retrofitting, which will enable structural rehabilitation interventions to be achieved "surgically", with only tolerable disruption of the buildings. One of these systems corresponds to the Dissipation Panels.

Design of the Dissipation Panels is based on the role of the notable historic timber framed masonry walls that were used in the reconstruction of Lisbon after the catastrophic earthquake of 1755. This similarity allows for the interventions in the buildings to be compatible, reversible and low intrusive thus respecting and preserving the existing structure. Additionally, the incorporation of seismic protection systems in the conception of the Panels provides energy dissipation in view of reducing damage and limiting losses more effectively than the traditional approach.

A series of full scale cyclic tests to evaluate the performance of prototypes of the Dissipation Panels were carried out. The Panels herein presented take advantage of an incorporated hysteretic damper. Other types of dampers are also included in the scope of this research.

Experimental characterization of the dynamic behaviour of the conceived prototypes show good energy dissipation capacity. Optimization of the behaviour of the Dissipation Panels is in progress seeking to respond effectively to the constant search for economically competitive solutions with technical quality.

Acknowledgments This research was partially financially supported by PORLisboa\QREN \FEDER under RehabToolBox R&D Project—Project number 11975. This work was made in cooperation between Instituto Superior Técnico and STAP, SA. We acknowledge the support of all these institutions and the help of the laboratory staff in the installation of the experimental campaign.

References

1. Cóias V (2007) Reabilitação estrutural de edifícios antigos, 2nd edn. Argumentum, Lisbon
2. Candeias P (2008) Avaliação da vulnerabilidade sísmica de edifícios de alvenaria. Ph.D. thesis, University of Minho, School of Engineering, Portugal
3. Lopes M (2010) Construção pombalina: património histórico e estrutura sismo-resistente. In: Proceedings of the 8th national conference on seismology and earthquake engineering—SÍSMICA 2010, Aveiro, Portugal, 20–23 October
4. Guerreiro L, Craveiro A, Branco M (2006) The use of passive seismic protection in structural rehabilitation. Prog Struct Eng Mater 8(4):121-132. Wiley Interscience
5. Chopra AK (2001) Dynamics of structures. Prentice-Hall, New Jersey

Pombaline Buildings—Consequences of Some Rehabilitation Interventions

Jorge Mascarenhas, Lurdes Belgas and Fernando G. Branco

Abstract The Pombaline buildings of downtown Lisbon have undergone some forms of adulteration across time with a greater impact in recent years due to an increase in tourist activity. After the great earthquake of 1755, this area has been rebuilt according to an innovative, enlightened urban plan. The buildings have been built in a short period of time using a standardization and prefabrication system and also an innovative anti-seismic technique—the "Pombaline cage"—which was an unparalleled Portuguese contribution to construction technology and anti-seismic engineering. The recent conversion of historic buildings in hotels has led, in many cases, to total internal demolition while maintaining the facade in its original condition. This solution is not recommended in the case of historic centres because the maintenance of the internal structure and original materials of buildings is crucial to protect its authenticity. It may also undermine the performance of buildings throughout the quarter in the event of a major earthquake. This paper describes the original structure of a Pombaline building and the major alterations that these buildings have undergone as well as the implications on their overall performance in the event of an earthquake.

Keywords Pombaline downtown · Seismic vulnerability · Rehabilitation

J. Mascarenhas (✉) · L. Belgas
Polytechnic Institute of Tomar, Tomar, Portugal
e-mail: jmascarenhas@ipt.pt

L. Belgas
e-mail: lbelgas@ipt.pt

F.G. Branco
University of Lurio, Nampula, Mozambique
e-mail: fjbranco@dec.uc.pt

F.G. Branco
University of Coimbra, Coimbra, Portugal

H. Cruz et al. (eds.), *Historical Earthquake-Resistant Timber Framing in the Mediterranean Area*, Lecture Notes in Civil Engineering,
DOI 10.1007/978-3-319-39492-3_29

Introduction

The Pombaline buildings of downtown Lisbon have undergone some forms of adulteration across time with a greater impact in recent years due to an increase in tourist activity.

After the great earthquake of 1755, this area has been rebuilt according to an innovative, enlightened urban plan. The buildings have been built in a short period of time using a standardization and prefabrication system and also an innovative anti-seismic technique—the "Pombaline cage"—which was an unparalleled Portuguese contribution to construction technology and anti-seismic engineering.

The structural system of Pombaline buildings consists in a three-dimensional wood-lattice framework made up of stone masonry facade and gable walls and the floors.

The recent conversion of historic buildings in hotels has led, in many cases, to total internal demolition while maintaining the facade in its original condition. This solution is not recommended in the case of historic centres because the maintenance of the internal structure and original materials of buildings is crucial to protect its authenticity. It may also undermine the performance of buildings throughout the quarter in the event of a major earthquake.

There are also other situations in which the buildings undergo major changes that may impair their performance in case of earthquake, such as for example the construction of basements, the erection of additional floors and the increase of structural rigidity.

This paper is based on literature review and previous research work carried out by the authors. It describes the original structure of a Pombaline building and the major alterations that these buildings have undergone as well as the implications on their overall performance in the event of an earthquake.

The Importance of Preserving Pombaline Buildings

The Pombaline building system was created after the great 1755 earthquake. Developed in a chaotic manner, the ancient city of Lisbon was destroyed by the earthquake followed by a tsunami and a great fire in a sequence of destructive events that took place for several days.

Several technical solutions to rebuild the city have been considered but the one selected was to use the rubble debris of the earthquake. This solution would protect the city from the occasional floods originated by sea level variations and would allow a more efficient drainage.

Manuel da Maia, the main engineer of the kingdom, has been nominated to lead the rebuilding and he was in charge of defining the new technical criteria to be followed in the process. Captain Eugénio Santos and architect Carlos Mardel were the authors of the selected urbanistic proposal to rebuild the city.

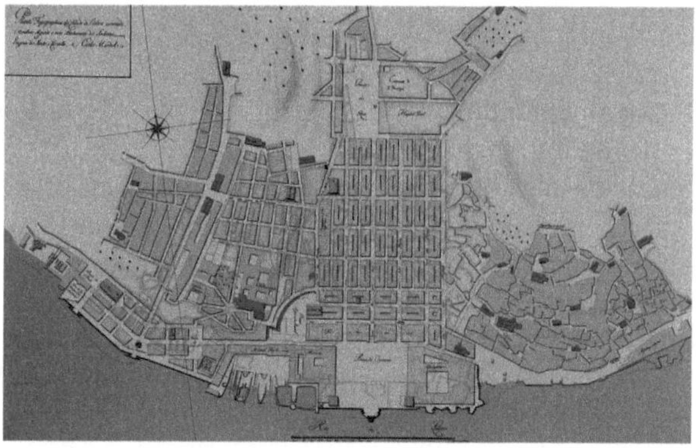

Fig. 1 General plan of the Pombaline urban mesh [3]

The plan was based on a standardized building process consisting of a regular and orthogonal set of main streets and modular typology for the new buildings and their main facades. The building aspect was defined hierarchically according to the importance of the street and the type of activity it would support. The plan presented innovative concepts in terms of workability and hygiene and contained some technical solutions to prevent future calamities, including improved safety precautions against fire [1].

The labyrinth of the medieval city was replaced by rectilinear streets inserted in an orthogonal mesh, thus regularizing the land located between the two Lisbon's main squares: Rossio and Terreiro do Paço (Fig. 1), following an enlightened, innovative plan [2].

Located in an unstable sedimentary terrain formed by a mix of sand and clay with variable thickness and built over a landfill resulting from the debris of pre-earthquake constructions, the buildings featured a pioneering anti-seismic structural system. This system was tested in a model produced over a dais and subjected to the vibrations produced by the march of a passing by military contingent to simulate the effect of an earthquake.

The construction was carried out under serious human and material resources constraints because it was urgent to quickly rehouse a great number of inhabitants and rebuild the city in a short time span. However, great care was devoted to the quality of construction and anti-seismic and fire safety. Carried out at city-scale, the Pombaline construction was only possible by resorting to standardization and industrialization of the building process.

The rebuild of the city was also used to anticipate the legislative reform that applied to construction at the time and new health and safety regulations have been introduced. Pombaline buildings are characterized by cost-effectiveness, strength and simplicity. With this kind of building a new building typology ("right" and "left" fractions) was implemented. All the aspects indicated are important factors, which justify the preservation of this historic heritage in its original design [4].

The Characterisation of Pombaline Buildings

Pombaline buildings are arranged in blocks of buildings. Each block is delimited by orthogonal streets creating rectangular blocks. According to the urban plan, all buildings had the same height, with a ground floor, three elevated floors and a dormer. This led to a group of buildings with identical dynamic characteristics, which provided a better seismic performance. The insertion of the Pombaline building in the block defines the room distribution inside the building (Fig. 2). Their relative position in the block leads to variation in the number of rooms with access to daylight, in the length of the facade of each building and the number of apartments in each floor. The imposed restrictions on the number of floors, the division of intermediate walls, the similarity in structural characteristics between the buildings in each block are factors that contribute to improve the seismic performance of the block. In every street, the building facades present a similar aspect, with small differences from street to street as a function of the relative importance of the street (Fig. 3).

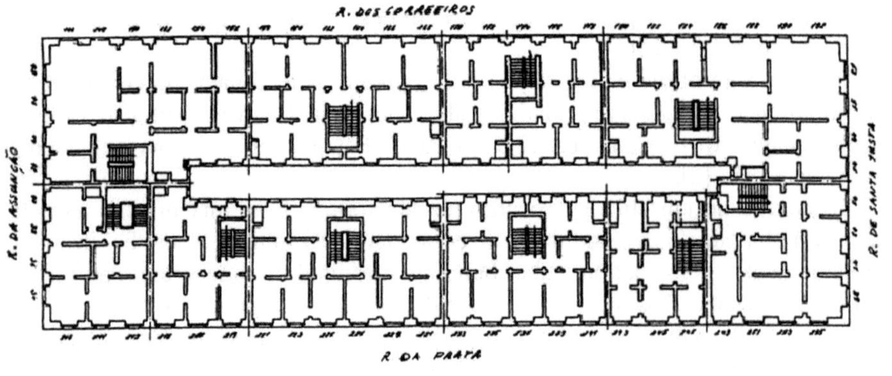

Fig. 2 Block plant [5]

Fig. 3 Example of a building facade [5]

The Pombaline building is also characterised by its interior architecture, with a number of typical characteristics common to all buildings: the cornices propagate along the whole block, without discontinuities; the block corners are coated in stone; the stonework surrounding the windows is aligned horizontally and vertically; the ground floor stonework has the same width of the upper levels, but the ground floor doors and windows have a higher vertical span than those of upper levels; balcony windows are usually present at the first level; on the upper levels, the windows usually are breast-high; above the cornice is a mansard roof; in secondary streets the roof has a triangular shape; there are dormer windows in the roof; the walls between two buildings are higher than the roof to prevent fire propagation.

The Pombaline building structure is easily identified by the existence of a characteristic wooden structure named "gaiola" ("cage") (Fig. 4a). This structure consists of a mutually perpendicular braced three-dimensional wooden frame. The buildings usually have an ample ground floor to allow the installation of trade or services.

The foundation of Pombaline buildings located in landfill and floods area are usually formed by arches which rest on a wooden crate system (Fig. 4b) directly supported by green pine stakes, thus creating a rigid system that improves the terrain load capability via consolidation and confinement Solid foundations

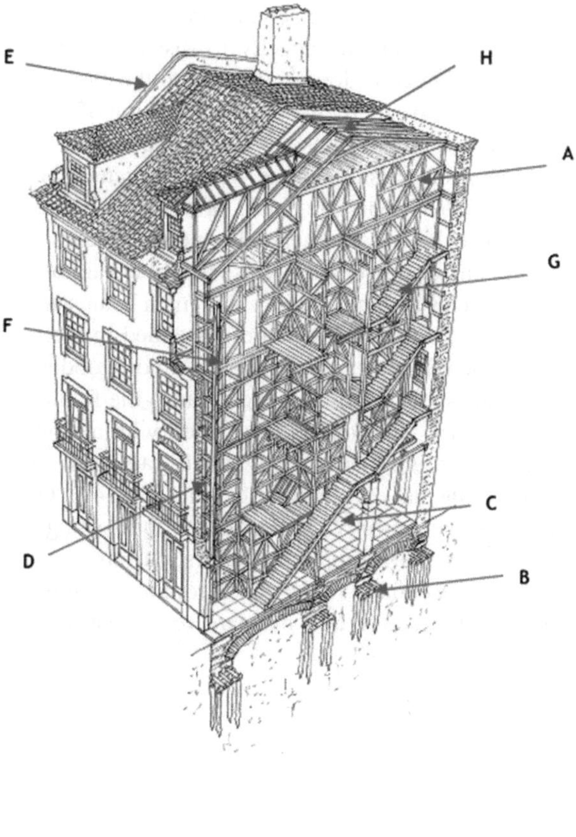

Fig. 4 Some constructive details characteristic of a Pombaline building

consisting of wooden railing serving header stakes can also be found. The ground floor structure differs from the structure of upper floors (Fig. 4c). The exterior walls are in stone masonry, with stone barrel vaults or vaults edge in brick masonry supported on walls, arches or pillars built in stone masonry. In order to regularise the surface and stabilise the arches of the domes, the top was filled with specimens from quake rubble. This whole system provided a greater rigidity to the structure at its base, in addition to functioning as an additional safety feature to prevent fire propagation in commercial areas or services [6].

Above the ground floor, there were three types of walls: mixed walls—made of plastered stone masonry and the wooden structure, the "cage"; Pombaline front walls—formed by a wooden trellis filled with ceramic plastered elements; partition walls—thinner, built in wooden lath and plastered.

The outer walls are built in stone masonry, plastered and connected to an inner structure in oak, the "cage", which provides the bracing (Fig. 4d). The link between these two elements was made through metal parts nailed to the wooden railing and leaded joints of the outer wall. These walls have an average thickness of 0.90 m. The walls separating buildings are made of plastered stone masonry with average thickness of 0.50 m and elevated above the roof elements for higher fire safety (Fig. 4e).

The front, located above the first floor, generally in both directions, plays a very important structural function in cage behavior, both for the absorbing part of the vertical load, and for the general bracing structure. These walls aimed at giving the building the required capacity to dissipate any energy transmitted by horizontal forces without considerable damage to the entire structure [7]. These characteristics are conferred partly by the good connection of the walls to the floor, roof and facade walls. These connections are very important as they form the constraints that limit the strains and tensions in the various structural elements that make up the building.

The higher floors are made of wooden beams coated on the upper and lower faces. The beams are generally perpendicular to the facade (Fig. 4f), made of rectangular section beams whose height is variable depending on the beam span, the wood species and the loading. The main beams are supported on support beams that run along the front wall or are embedded in the masonry walls. The upper coat of the wooden floor is generally made up of wooden boards. The lower coating of the floor consists of a lath perpendicular to the framework, nailed to the lower faces of the beam and subsequently skimmed with a lime and gypsum plaster.

The stairs are made of a wooden frame and supported at intermediate levels or on the walls that divide their sections (Fig. 4g). On the first floor, the stair walls are built in stone masonry. The roofs are tilted, shingles comprised of straw placed on a wooden structure comprising rafters, purlins, rows, a set of slats and rods (Fig. 4h). There are essentially two types of roofing: a simple, triangular, usually symmetrical, gable roof structure and mansard roofs that allow the top lever to have a higher ceiling height.

Modifications in Pombaline Buildings That May Lead to Different Structural Performance

Over time, a number of interventions in these buildings have been observed. These are usually associated with the introduction of new features and enhancements such as additional floor layers or adaptation to new uses, which completely alter the initial structural system and compromise the building's performance in the event of an earthquake. These interventions are of various types. They can consist in "embalming" the building around the interior which is demolished and only retains the facades. In this case, a new reinforced concrete structure is built, which changes the stiffness of the building at block level. The different stiffness between that building and adjacent buildings may lead to partial rout in the contact zone of the buildings if an earthquake occurs. Sometimes, rehabilitation interventions are carried out involving the renewal of individual apartments, leading to significant structural changes and discontinuities within the building.

Recently, due to increased tourism in the capital, some of these buildings have been converted into hotels. This has led to profound changes in the buildings to meet the requirements of the new use, in particular those associated with hygrothermal and acoustic comfort and fire safety. These changes can be specific, such as removing walls, widening spaces and changing the usage conditions of the floors, which is usually associated with an increase in acting loads and stresses on structural components. A more severe type of change is related with the height of the building such as increasing the number of floors without any prior study or stability project (Fig. 5a). The extra floors added lead to increased weight of the structure itself, increasing the efforts and displacements during the occurrence of earthquakes. Another type of burdensome interventions consists of removing structural elements at ground floor level to increase door frames or to create wider open-spaces (Fig. 5b). Built to meet the need to expand commercial areas, these openings reduce the area of vertical structural elements, which causes an abrupt reduction in stiffness at this level creating hollow floors and weakening the building in an area that will be most requested during the occurrence of an earthquake [8]. The removal of wooden elements, thus interrupting the three-dimensional cage structures (which are generally replaced by metal or concrete elements) and the removal of some columns and walls on the ground floor lead to serious damage in the original structure, increasing the vulnerability of buildings to seismic actions [9].

Several Pombaline buildings have undergone rehabilitation works to implement some facilities, such as bathrooms, water supply and drainage, electricity, telecommunications and gas, to meet the comfort and health requirements of twenty-first century users. However, sometimes such interventions do not respect the original structural elements (Fig. 5c) and cause irreversible damage and changes in performance as is the case of cutting the diagonal wood elements of the front walls to install pipelines. The installation of equipment, including lifts, in Pombaline buildings (Fig. 5d) with the introduction of a heavy metal structure imposes higher stiffness in a limited area of the building, thus influencing its seismic performance.

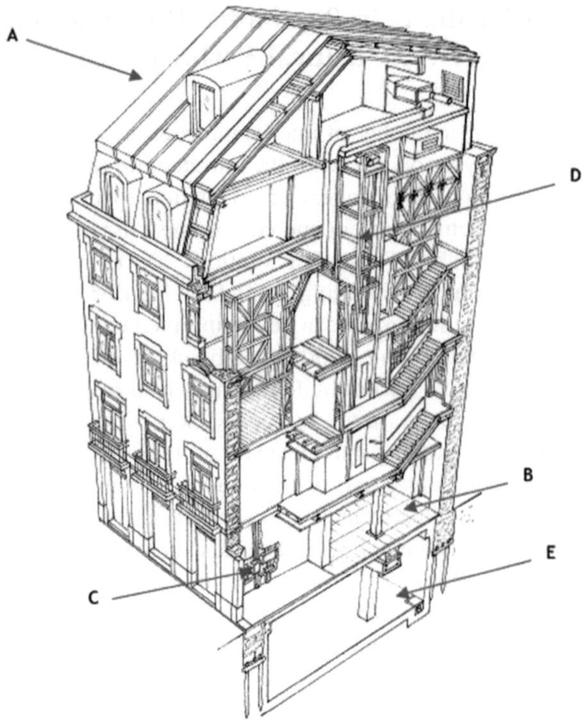

Fig. 5 Examples of the main changes usually introduced in Pombaline buildings

The construction of basements (Fig. 5e) prevents the normal circulation of water in the soil and changes the water level, thus creating conditions to increase the moisture content in the wooden elements, which may lead to rot and consequent section loss. As a consequence, higher deformation or differential settlements in buildings may arise. This differential settlement can cause cracking on the upper floors and sometimes masonry misalignment and cracking. When changing the original structure of Pombaline cage (St. Andrew's cross), the ability of these structures to support loads is changed with regard, not only to its bearing capacity, but also the capacity for absorbing horizontal inertia forces induced by earthquakes. Besides tampering with the bracing that these structures provide to the masonry facades, these changes will also affect the structure's capacity to dissipate energy. Such interventions are not desirable without prior safety and stability studies because they lead to reduced structural safety, in particular under seismic loads.

The overall performance of the building is affected by the spatial arrangement of the walls and the mass of structural elements [10]. Thus, the introduction of new elements such as metal or concrete should be avoided as it creates very sharp differential stiffness and may lead to increased eccentricity between the mass and rigidity, which can be observed in the building plan, by introducing additional torsional loads that can result in structural damage in the event of an earthquake.

Final Remarks

Pombaline buildings testify to the ingenuity and expertises demonstrated in very adverse conditions and are a landmark of earthquake engineering as the anti-seismic construction techniques were applied at a city scale for the first time. Therefore, they are a valuable asset that must be preserved, cared for and transmitted safely to future generations. The original Pombaline buildings have good anti-seismic performance taking into account the technical and material constraints at the time of build. However, some of these buildings have undergone repeated interventions with the purpose of adapting them to the new demands of comfort, features and uses. Some of these interventions were carried out without taking into account the original material and techniques characteristic of Pombaline buildings, severely affecting their original structure. Today, several Pombaline buildings present great seismic vulnerability due to interventions that reduced their ability to support horizontal actions. Under the guise of "rehabilitation", some of these buildings are now completely destroyed.

The preservation of the historical and heritage value of Pombaline buildings is crucial and therefore it is imperative to preserve the Pombaline cage in as many buildings as possible in future rehabilitation interventions.

References

1. Appleton J (2003) A Baixa pombalina: da Inovação ao Mito. In: Jornadas A Baixa Pombalina e a sua Importância para o Património Mundial, Câmara Municipal de Lisboa, Outubro
2. França J (1987) Augusto. Lisboa Pombalina e o Iluminismo. 3ª Edição, Editora Bertrand, Lisboa
3. Pena A (2008) Análise do comportamento sísmico de um edifício pombalino. Dissertação de Mestrado em Engenharia Civil, IST
4. Pombalina B (2005) Bases para uma intervenção de salvaguarda, Câmara Municipal de Lisboa, Coleccção de Estudos Urbanos, Lisboa XXI, Lisboa
5. Mascarenhas J, Sistemas de Construção V (2005) O Edifício de Rendimento da Baixa Pombalina de Lisboa. Processo evolutivo dos edifícios; inovações técnicas; sistemas construtivos. Livros Horizonte, Lisboa
6. Ramos L, Lourenço PB (2000) Análise das Técnicas de Construção Pombalina e Apreciação do Estado de Conservação Estrutural do Quarteirão do Martinho da Arcada, Engenharia Civil, U.M., Nº7, pp 35–46
7. Gomes RN (2011) Sistema Estrutural de Edifícios Antigos de Lisboa—Os Edifícios "Pombalinos" e os Edifícios "Gaioleiros". Dissertação de Mestrado em Engenharia Civil, IST
8. Teixeira MJ et al (2010) Avaliação experimental do Comportamento de Frontais Pombalinos sob a acção de cargas horizontais. Rev Assoc Portuguesa Anál Exp Tensões Mecânica Exp 18:47–53. ISSN 1646-7078
9. Lopes M (2010) Construção Pombalina: Património histórico e estrutura sismo-resistente. In: Actas da SISMICA 2010—8º Congresso de Sismologia e Engenharia Sísmica, Aveiro, Setembro
10. Appleton J (2003) Reabilitação de edifícios antigos, Patologias e tecnologias de intervenção. 1ª Ediçao. Amadora: Edições Orion

Experimental and Analytical Assessment of the Capacity of Traditional Single Notch Joints and Impact of Retrofitting by Self-tapping Screws

Karel Šobra, Richard de Rijk, Yasemin D. Aktaş, Coralie Avez, Isabela Burawska and Jorge M. Branco

Abstract The joints are the most crucial parts of a timber building and determine the overall structural behaviour, load-bearing capacity and failure mechanisms. Therefore, keeping the joints fully functional is of utmost importance to ensure a desired structural performance of timber buildings under various actions. Since replacement of damaged parts of an existing structure is expensive and in many cases very difficult to perform in situ, retrofitting to avoid failure becomes an increasingly widespread strategy. In this paper, the capacity and failure mechanisms of single notch joints before and after a simple retrofitting intervention by means of self-tapping screws were investigated. To this end, a series of tests were carried out during the 5th COST FP 1101 Training School, held in University of Minho, Portugal. The joints were first tested under compression, and the load-bearing capacity values obtained at the end of tests were compared to the capacity values

K. Šobra (✉)
Czech Technical University in Prague, Thákurova 7, 166 29 Prague 6, Czech Republic
e-mail: karel.sobra@fsv.cvut.cz

R. de Rijk
Eindhoven University of Technology, 5612 AZ Eindhoven, The Netherlands
e-mail: r.j.c.w.d.rijk@student.tue.nl

Y.D. Aktaş
University College London, Gower Street, London, WC1E 6BT, UK
e-mail: y.aktas@ucl.ac.uk

C. Avez
University of Mons, Rue du Joncquois 53, 7000 Mons, Belgium
e-mail: coralie.avez@umons.ac.be

I. Burawska
Warsaw University of Life Sciences, Nowoursynowska 166, 02-787 Warsaw, Poland
e-mail: izaburawska@o2.pl

J.M. Branco
ISISE, University of Minho, Campus de Azurém, 4800-058 Guimarães, Portugal
e-mail: jbranco@civil.uminho.pt

H. Cruz et al. (eds.), *Historical Earthquake-Resistant Timber Framing in the Mediterranean Area*, Lecture Notes in Civil Engineering,
DOI 10.1007/978-3-319-39492-3_30

calculated using theoretical models proposed in a variety of national codes. Then, tested joints were retrofitted using self-tapping screws. The retrofitting strategy aimed to prevent failure mechanism that was shown to dominate joints' behaviour in the unreinforced state, rather than to increase the load-bearing capacity or stiffness. The impact of retrofitting on the joints' performance was discussed and the success of the proposed intervention was further debated.

Keywords Traditional carpentry joints · Single notch joints · Retrofitting · Self-tapping screws

Introduction

Joints play a crucial role in determining the overall structural behaviour, load-bearing capacity and failure mechanisms under various loading conditions including seismic action. Among various joints that can be observed in timber buildings around the world, traditional carpentry joints take an important place. Single notch joints, known also as step, framed or cogging joints, are among the most commonly used traditional carpentry joints. In [1, 2], the authors tested double and single notch joints under compression and rotation to characterize behaviour and develop strengthening solutions. In some studies, the behaviour of single joints was deducted from results obtained from the testing of an entire building, such as in [3] where a traditional roof structure was tested to elaborate more on the impact of diversely retrofitted connections on the overall structural behaviour under monotonic and cyclic loading. In [4, 5], number of single notch joints were tested without and with mortise and tenon fitting, and conclusions were drawn with regards to the reinforcement/repair techniques as well as geometrical configuration of the joints (i.e., notch depth, skew angle, etc.). In another study [6], finite element analyses of single notch joints were carried out and unrealistic assumptions in design codes were pointed out. In [7, 8], the authors investigated the impact of common strengthening techniques, such as tension ties and stirrups, on single notch joints with varying geometrical features by means of a series of monotonic/cyclic tests.

In this study, a single notch joint was tested under compression and the obtained capacity value was compared against the capacity calculated using analytical models proposed by various national codes. Then the tested joint was retrofitted using self-tapping screws. Unlike conventional interventions, the reinforcement strategy that was followed here aimed to prevent failure mechanism that was shown to dominate joint behaviour in the unreinforced state, rather than to increase the load-bearing capacity or stiffness. It was analysed how retrofitting affects the failure mechanism and the capacity of the joint.

Testing of the Unreinforced Joints and Stress Values on Contact Surfaces

The tests described here were carried out during the 5th COST FP 1101 Training School that was held in University of Minho on 11–14 May 2015. The joints were first tested under compression (Fig. 1). For this aim, glued-laminated specimens with strength class GL 24 h were used (Table 1).

The unreinforced capacity under compression was measured as 42.5 kN (Fig. 2a). Assuming a linear stress distribution in the contact areas of the connection, the force measured in the actuator can be decomposed into two components perpendicular to the contact planes (Fig. 1) and this component can be divided over the contact area to obtain theoretical stresses (Fig. 2b). As these stresses are perpendicular to the contact planes of the connection they are acting under an angle to the grain direction of the elements.

The maximum stress level on an angle to the grain can be obtained by several theories. This has been done with the theories of Norris [9], Hankinson [10] and by decomposing $f_{c;0}$, $f_{c;90}$, f_v using the characteristic values for GL 24 h. The maximum stress levels in the contact surfaces, being at a 30° angle to the grain for the front-notch surface and 80° for the bottom-notch surface (Fig. 1), are shown for different theories (Table 2) in Fig. 3. However, this plot is very conservative since the characteristic values for GL 24 h represent a statistical 5 % lower limit of the materials strength

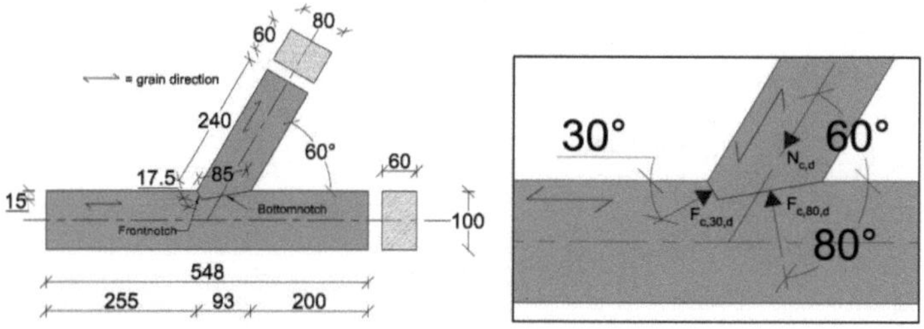

Fig. 1 Geometry of the tested Single notch joint

Table 1 Characteristic material properties of the test specimen (MPa)

$f_{t;0;k}$	$f_{t;90;k}$	$f_{c;0;k}$	$f_{c;90;k}$	$f_{v;k}$
16.5	0.4	24	2.7	2.7

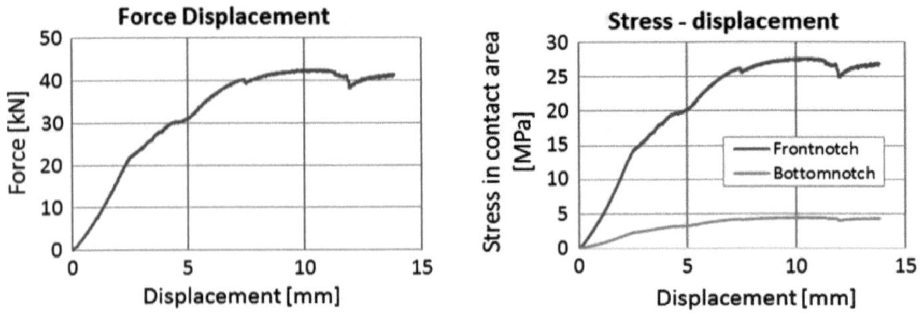

Fig. 2 (*Left*) Force displacement plot of the testing of single-notch joint, (*right*) Theoretical stress in the contact areas of the connection versus displacement of the actuator

Table 2 Stress values on contact surfaces according to a few theories

Hankinson	Norris	
$\frac{f_{c;0;k} * f_{c;90;k}}{f_{c;0;k} * \sin\alpha^2 + f_{c;90;k} * \cos\alpha^2}$	$\frac{f_{c;0;k}}{\cos\alpha^2 * \sqrt{1 + \left(\frac{f_{c;0;k}}{f_{v;k}}\right)^2 * \tan\alpha^2 + \left(\frac{f_{c;0;k}}{f_{c;90;k}}\right)^2 * \tan\alpha^4}}$	
$\mathbf{f_{0,c}}$ $\frac{f_{c;0;k}}{\cos\alpha^2}$	$\mathbf{f_{90,c}}$ $\frac{f_{c;90;k}}{\sin\alpha^2}$	$\mathbf{f_v}$ $\frac{f_{v;k}}{\sin\alpha * \cos\alpha}$

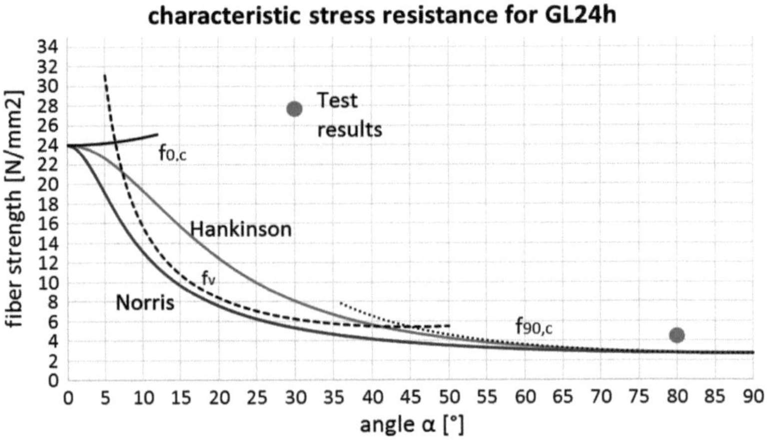

Fig. 3 Several theories giving material strengths when loading under an angle versus lab test results

Capacity Calculation According to Various Codes

As a uniform European standard on carpentry connections is not available, only national standards or national annexes containing guidelines for designing carpentry connections can be used for this purpose. In this study, the following three

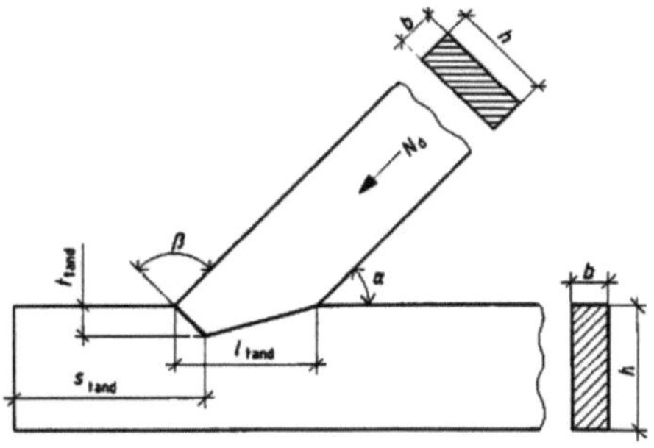

$$\sigma_{c;0;d} \leq f_{c;0;d} \quad \text{where} \quad \sigma_{c;0;d} = \frac{N_{c;d} \cdot \cos\alpha}{t_{tand} \cdot b}$$

$$\sigma_{c;90;d} \leq f_{c;90;d} \quad \text{where} \quad \sigma_{c;90;d} = \frac{N_{c;d} \cdot \sin\alpha}{l_{tand} \cdot b}$$

$$\sigma_{v;d} \leq f_{v;d} \quad \text{where} \quad \sigma_{v;d} = \frac{N_{c;d} \cdot \cos\alpha}{0.8 \cdot s_{tand} \cdot b}$$

Fig. 4 Dutch national annex to NEN-EN 1995-1-1 on carpentry joints

national standards with different theoretical bases for determining maximum stresses in the contact areas were used:

1. Dutch national annex to NEN-EN 1995-1-1 [11] (Based on decomposition of forces)

The Dutch national annex to Eurocode 5 gives guidelines for maximum stress levels based on decomposing the force in the diagonal into components which act on the depth and length of the notch. For comparing the test results with the guidelines in the code, the specimen geometry and characteristic material strengths of GL 24 h are used to determine the maximal force $N_{c;d}$ according to the formulas in the Dutch code as shown in Fig. 4.

The obtained results are given below:

$f_{c;0;k} \rightarrow f_{c;0;d} = \sigma_{c;0;d} =$ 24 N/mm^2	$\alpha = 60°$	$t_{tand} = 15$ mm	b = 60 mm	$N_{c;d} =$ 43.2 kN
$f_{c;90;k} \rightarrow f_{c;90;d} = \sigma_{c;90;d} =$ 2.7 N/mm^2	$\alpha = 60°$	$l_{tand} = 92.5$ mm	b = 60 mm	**$N_{c;d} =$ 17.3 kN**
$f_{v;k} \rightarrow f_{v;d} = \sigma_{v;d} =$ 2.7 N/mm^2	$\alpha = 60°$	$s_{tand} = 263.5$ mm	b = 60 mm	$N_{c;d} =$ 68.3 kN

The lowest of all values obtained for $N_{c;d}$ is 17.3 kN and it occurs for the $f_{c;90;k}$ parameter. This force is approximately 2.5 times lower than 42.5 kN measured in the lab test.

2. German national annex to DIN-EN 1995-1-1 [12] (Based on Norris's theory)

The German national annex to Eurocode 5 uses a derivation from Norris's theory to determine the maximum allowable stress under an angle to the grain. With the given formula, the maximum stresses in the front-notch and bottom-notch can be determined individually for different angles α of loading to the grain as shown in Fig. 5.

$$\frac{\sigma_{c,\alpha,d}}{f_{c,\alpha,d}} \leq 1$$

Where: $\sigma_{c,\alpha,d} = \frac{F_{c,\alpha,d}}{A}$

$$f_{c,\alpha,d} = \frac{f_{c,0,d}}{\sqrt{\left(\frac{f_{c,0,d}}{2 \cdot f_{c,90,d}} \cdot \sin^2\alpha\right)^2 + \left(\frac{f_{c,0,d}}{2 \cdot f_{v,d}} \sin\alpha \cdot \cos\alpha\right)^2 + \cos^4\alpha}}$$

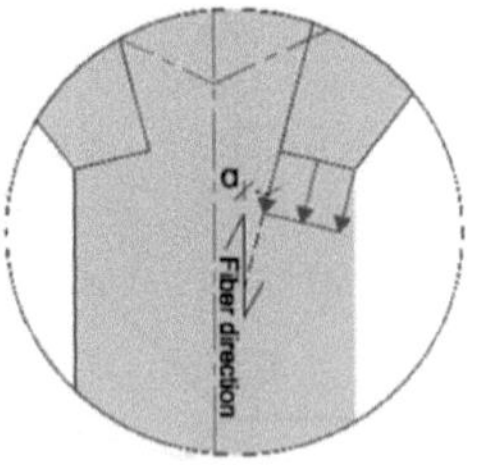

Fig. 5 German national annex to DIN-EN 1995-1-1 on carpentry joints

Front-notch	$\alpha = 30°$	$A = 17.5 * 60\ mm^2$	$f_{c;30;d} = \sigma_{c;30;d} =$ 10.23 MPa	$F_{c;30;d} =$ **10.7 kN**
Bottom-notch	$\alpha = 80°$	$A = 84.8 * 60\ mm^2$	$f_{c;80;d} = \sigma_{c;80;d} =$ 5.48 MPa	$F_{c;80;d} =$ 27.9 kN

Decomposing the force of 42.5 kN obtained at the end of testing into the same angular components of 30° and 80° (Fig. 1) give $F_{c;30;d} = 29$ kN and $F_{c;80;d} = 22.6$ kN, which means that the maximum allowable force in the front-notch is lower than the tested force and thereby governing (29 kN > 10.7 kN).

3. Swiss SIA 265 [13] (Based on Hankinson's theory)

The Swiss code uses a slightly altered version of Hankinson's equation to determine maximal stresses at an angle. These results are used to calculate the minimal dimension of the connection, but since these are already determined by the test specimens, again a maximal force in the diagonal (named F_{Ed} in the Swiss formula's, see Fig. 6) can be determined.

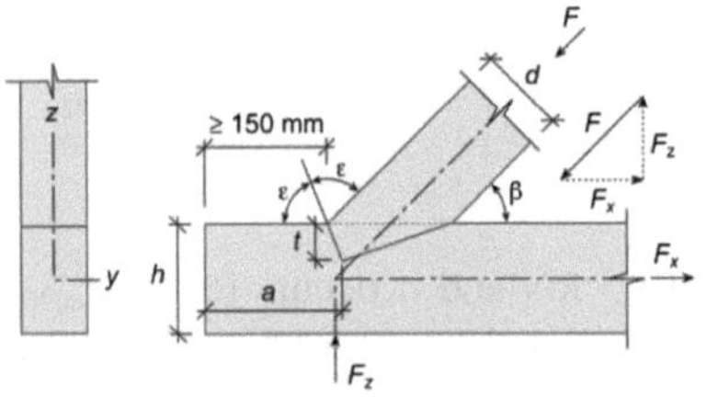

Shear-plane length: $a \geq \frac{F_{Ed} \cdot \cos\beta}{b \cdot k_{red} \cdot f_{v,d}}$ – $k_{red} = 0{,}6$ for sawn timber – $k_{red} = 0{,}8$ for glulam timber

Notch depth: $t \geq \frac{F_{Ed} \cdot \cos\beta}{b \cdot f_{c,\alpha,d}}$ – $\alpha = \frac{1}{2}\beta$ for Single Step Joints – $\alpha = \frac{3}{4}\beta$ for Double Step Joints

Strut height: $d \geq \frac{F_{Ed}}{b \cdot f_{c,\alpha,d}}$ – $\alpha = \beta$

Where

b is the width of the strut;

β is the angle between force F_{Ed} and the fiber direction.

$f_{c,\alpha,d}$ must be calculated as follows:

$$f_{c,\alpha,d} = \frac{0{,}8 \cdot f_{c,0,d} \cdot f_{c,90,d}}{0{,}8 \cdot f_{c,0,d} \cdot \sin^2\alpha + f_{c,90,d} \cdot \cos^2\alpha}$$

Fig. 6 Swiss SIA 265 on carpentry joints

The obtained results are given below:

Using a	Using t	Using d
$\beta = 60°$ b = 60 mm $k_{red} = 0.8$ $f_{v;d} = 2.7$ MPa a = 263.5 mm	$\beta = 60°$ b = 60 mm $f_{c;\alpha;d} = 7.59$ MPa t = 15 mm	b = 60 mm $f_{c;\alpha;d} = 7.59$ MPa d = 80 mm
$F_{Ed} = 68.3$ kN	$F_{Ed} =$ **13.7 kN**	$F_{Ed} =$ **36.4 kN**

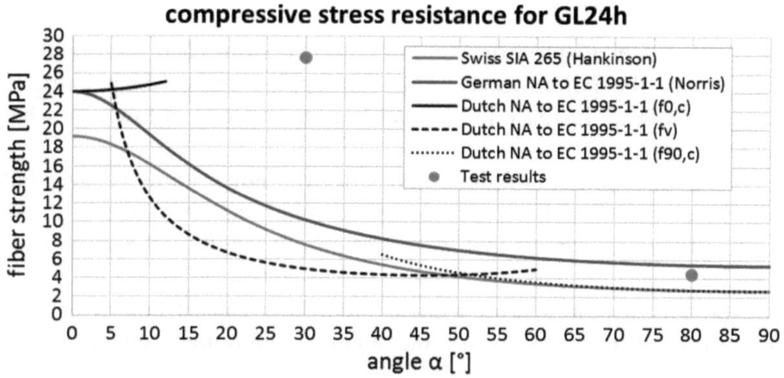

Fig. 7 Test results versus national standards

The formulas for t and d give smaller maximal forces than the applied force of 42.5 kN in the lab test, with t being governing since it gives the lowest allowable normal force in the diagonal.

The maximum allowable stress under an angle according to these codes can also be plotted for loading angles between 0° (parallel to the grain) and 90° (perpendicular to the grain) together with the results of the scaled lab test (Fig. 7).

Retrofitting Front Notch Joint Against Particular Failure Nodes

A variety of retrofitting/strengthening techniques are used to alter the failure mechanism or increase the load-bearing capacity or stiffness of traditional carpentry joints and structures built by these. Among the most commonly used techniques is the use of internal or external metal connectors such as stirrups, bolts, binding strips, clamps, glued-in rods and tension ties (e.g., [14–17]). In addition to these, self-tapping screws have been increasingly widespread in the last decade, as they

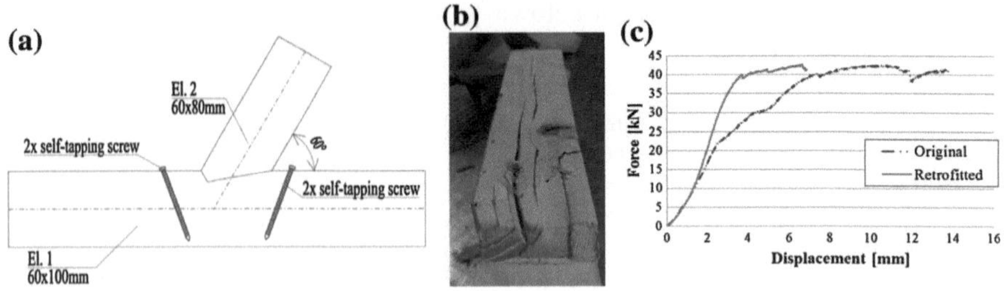

Fig. 8 Technique 1-reinforcement against shear failure of heel: **a** retrofitting with self-tapping screws, **b** obtained failure and **c** force displacement relationship from loading cell for both original and retrofitted joint

prevent connected elements from getting detached from each other and alter the load-path, hence the failure mode [17]. Self-tapping screws help joints become stiffer and result in an increase in the load-bearing capacity at low cost when used with care [18, 19].

One of the most common failure modes of front notch joints is shear failure in the heel. To avoid this failure, already tested joint was retrofitted at the heel by couple of self-tapping screws that were 100 mm in length and 8 mm in diameter (Fig. 8).

However, placing the screws without predrilling caused cracks on the upper surface of the joint (Fig. 8b). Even if the obtained failure mechanism looks like the joint experienced shear failure, it is the initial cracking due to the screws that caused delamination of wood where the crack follows annual ring. Despite these problems, the load bearing capacity of the joint remained more or less the same, but the average stiffness increased by approximately 50 % (Fig. 8c). Thanks to the geometry of the joint, where two elements are connected under angle of 60°, shear failure was not observed but the joint was damaged just by compression of contact surfaces.

Then, retrofitting with self-tapping screws was repeated with two different geometrical configurations. Firstly, the screws were used to retrofit only the contact area of horizontal element (Fig. 9a). The post-elastic bearing capacity obtained following this retrofitting work was almost equal to the original one, however not only the average but also the initial stiffness in the linear range was almost half of the initial state (Fig. 9c).

Finally, in order to increase both the stiffness of the joint as a whole and its bearing capacity and also to minimize influence of large compressive deformations, contact areas of both elements have been reinforced using self-tapping screw (Fig. 10a). The bearing capacity of unreinforced joint was reached, however a sudden failure was observed by cracking that follows the annual ring.

At force displacement relationships of both types of reinforcement of contact areas is possible to see that those interventions are effective in terms of overall capacity of the joint, although lower stiffness and higher deformation were observed what could cause secondary problems in a truss construction.

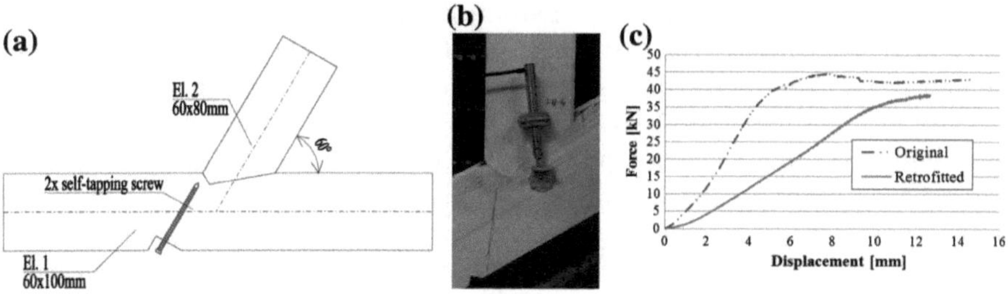

Fig. 9 Technique 2-reinforcement of horizontal element contact area: **a** retrofitting with self-tapping screws, **b** obtained failure and **c** force displacement relationship from loading cell for both original and retrofitted joint

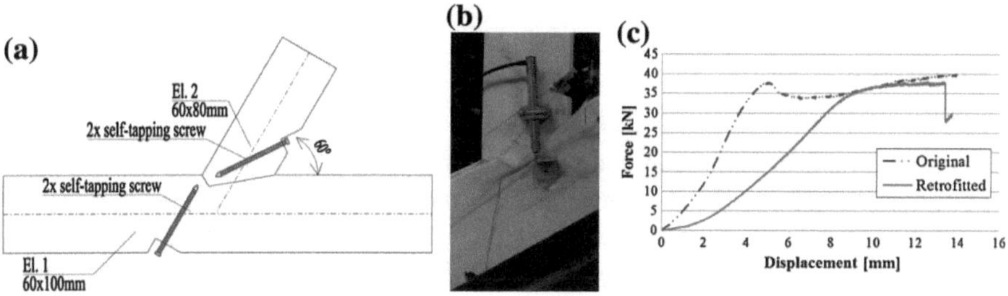

Fig. 10 Technique 3-reinforcement of contact areas of both elements: **a** retrofitting with self-tapping screws, **b** obtained failure and **c** force displacement relationship from loading cell for both original and retrofitted joint

Discussion

In this study, the load bearing capacity of a front notch joint of given geometry and dimensions under monotonically applied compression was measured between 38 and 45 kN. The specimen deformed plastically in both contact areas, and local shear failure (on one side only) was observed due to a knot very near to the notched area. Shear failure of the notched joint caused brittle failure of the connection, not allowing a large plastic yield deformation. Then the load bearing capacity was calculated using three different national codes, and it was seen that, among all codes that were used, the German National Annex to Eurocode 5 gives the least conservative values. This can be explained by the use of an amplification factor equal to 2, applied on the characteristic shear resistance of the material as well as the characteristic perpendicular-to-grain compressive resistance. These parameters, when doubled using the above-mentioned amplification factor, gives quite different results than the Dutch National Annex to Eurocode 5 and the Swiss SIA, which uses a factor of 0.8 to reduce characteristic shear capacity and parallel-to-grain compressive strength. However, despite this difference, all national codes seem to result in a significant underestimation of the joint capacity for low angle values.

Then already tested joints were retrofitted using self-tapping screws in three different configurations. The first technique (Fig. 9) was used for the aim of preventing shear failure. By means of this intervention, the original load-bearing capacity and the initial stiffness of the unreinforced joint was reached, however shear failure was not efficiently prevented. In the second and third techniques, the original load-bearing capacity was almost reached, while the initial and overall stiffness (stiffness of a joint as a whole) of the joint significantly dropped. This behaviour may be caused partly by weakening of retrofitted cross-section, which was caused by preparing the surface to drill the screws to the required position and under required angle (slots seen on bottom parts of skew element at Figs. 8 and 9 and also at vertical element at Fig. 9) and partly that tested specimens already reached their elastic bearing capacity.

Conclusion

The international or European codes do not provide means to calculate capacity of traditional carpentry joints. Only a few national annexes suggest analytical approaches to calculate the capacity of front notch joints. However, three of these national codes that were examined as part of this study, i.e., German and Dutch National Annexes to Eurocode 5 as well as Swiss SIA, seem to significantly underestimate the capacity, therefore the need for revisiting the underlying assumptions and fine-tuning these methods to successfully reproduce the capacity of front notch joints is apparent. Also, it was concluded that the efficiency of retrofitting front notch joints using self-tapping screws should be further investigated, as the tests showed that this intervention can significantly reduce the initial and overall stiffness of a joint as a whole.

Acknowledgments The authors are grateful to COST Action for funding the FP1101 and RILEM 245 Training School at University of Minho in Guimarães, Portugal, and to Dr. Jorge M. Branco and other members of Civil Engineering Department for making this study possible.

References

1. Parisi MA, Cordie C (2010) Mechanical behaviour of double-step timber joints. Constr Build Mater 24(8):1364–1371
2. Parisi MA, Cordie C, Piazza M (2008) Seismic strengthening of carpentry joints in traditional timber structures. In: AIP conference proceedings, CP1020, 2008 seismic engineering conference commemorating the 1908 Messina and Reggio Calabria earthquake
3. Parisi MA, Piazza M (2002) Seismic-behavior and retrofitting of joints in traditional timber roof structures. Soil Dyn Earthq Eng 22(9–12):1183–1191
4. Palma P, Cruz H (2007) Mechanical behaviour of traditional timber carpentry joints in service conditions—results of monotonic tests. In: ICOMOS IWC—XVI international symposium:

from material to structure—mechanical behaviour and failures of the timber structures. Florence, Venice, Vicenza
5. Palma P, Garcia H, Ferreira J, Appleton J, Cruz H (2012) Behaviour and repair of carpentry connections—rotational behaviour of the rafter and tie beam connection in timber roof structures. J Cult Heritage 13(3):S64–S73
6. Villar JR, Guaita M, Vidal P, Arriaga F (2007) Analysis of the stress state at the cogging joint in timber structures. Biosyst Eng 96(1):79–90
7. Branco J (2008) Influence of the joints stiffness in the monotonic and cyclic behaviour of traditional timber trusses—assessment of the efficacy of different strengthening techniques. Universidade do Minho, Guimarães
8. Branco JM, Piazza M, Cruz P (2011) Experimental evaluation of different strengthening techniques of traditional timber connections. Eng Struct 33(8):2259–2270
9. Norris CB (1962) Strength of orthotropic materials subjected to combined stresses. USBA, Wisconsin
10. Hankinson RL (1921) Investigation of crushing strength of spruce at varying angles to the grain. Air Serv Inf Circ. 3(259)
11. NEN (2013) NEN-EN 1995-1-1 Dutch NA 8.11 Ambachtelijke verbindingen. Nederlands Normalisatie-Instituut, Delft, pp 12–16
12. DIN (2010) DIN-EN-1995-1-1 German NA 12.1 Versätze. DIN Deutsches Institut für Normung, Berlin, pp 88–89
13. SIA (2012) Schweizer norm SIA 265—Holzbau. SIA, Zürich, pp 77–78
14. Branco J, Cruz P, Piazza M, Varum H (2006) Experimental analysis of original and strengthened traditional timber connections. In: 9th world conference on timber engineering, Portland
15. Branco JM, Piazza M, Cruz PJS (2011) Experimental evaluation of different strengthening techniques of traditional timber connections. Eng Struct 33:2259–2270
16. Palma P, Garcia H, Ferreira J, Appleton J, Cruz H (2012) Behaviour and repair of carpentry connections—rotational behaviour of the rafter and tie beam connection in timber roof structures. J Cult Heritage 13:S64–S73
17. Branco J, Descamps T (2015) Analysis and strengthening of carpentry joints. Constr Build Mater (in press)
18. Dietsch P, Brandner R (2015) Self-tapping screws and threaded rods as reinforcement for structural timber elements—a state-of-the-art report. Constr Build Mater (in press)
19. Lathuillière D, Bléron L, Descamps T, Bocquet J-F (2015) Reinforcement of dowel type connections. Constr Build Mater (in press)

Repairing of a Timber Truss Through Two Different Techniques Using Timber Elements and Screwed Metal Plates

Daniele Cibecchini, Niclas Bjorngrim, Hélder S. Sousa, Marco Jorge and Jorge M. Branco

Abstract Structural reinforcement of timber buildings may be needed due to different reasons such as change of use, deterioration due to lack of maintenance, exceptional damaging incidents or loading, after changes in regulatory specifications, or interventions to increase structural resistance. In this work, two different techniques were considered for repairing a timber truss that was previously assessed on laboratory (test facilities of University of Minho) and taken up to failure during a load-carrying test. A collar beam truss, with more than one hundred years, was tested considering a vertical point load on each main rafter. Failure of the timber truss was located in the sections of the rafters near the loading positions by bending. Repairing techniques, based on the use of timber elements for one of the rafters and on screwed metal plates for the other rafter, were evaluated and compared to the original unstrengthened condition. The efficiency of the combined repairing techniques was evaluated taking into consideration the structural performance of the collar truss, namely its displacement and ultimate load capacity. In this paper, the results of the experimental tests are discussed attending to the analytical calculation of the contribution of the repairing techniques. Also, the different failure scenarios, for original and strengthened truss, were analyzed and compared.

D. Cibecchini
GESAAF—Department of Agricultural, Food and Forestry Systems,
University of Florence, Florence, Italy
e-mail: daniele.cibecchini@unifi.it

N. Bjorngrim
Department of Wood Science and Technology, Division of Wood Technology,
Luleå University of Technology, Skellefteå, Sweden
e-mail: niclas.bjorngrim@ltu.se

H.S. Sousa (✉) · M. Jorge · J.M. Branco
ISISE—Department of Civil Engineering, University of Minho, Guimarães, Portugal
e-mail: sousa.hms@gmail.com

M. Jorge
e-mail: marco@civil.uminho.pt

J.M. Branco
e-mail: jbranco@civil.uminho.pt

H. Cruz et al. (eds.), *Historical Earthquake-Resistant Timber Framing in the Mediterranean Area*, Lecture Notes in Civil Engineering,
DOI 10.1007/978-3-319-39492-3_31

Keywords Timber trusses · Repair · Timber elements · Screwed metal plates

Introduction

Historical timber trusses are part of the architectural heritage on the Mediterranean Region and represent one of the most important type of ancient structures [1]. To conserve their worth and the original material, structural repairs and reinforcements may be needed due to different reasons such as change of use, deterioration, exceptional damaging events or loadings, after changes in regulatory specifications, or interventions to increase structural performance. Different structural typologies of timber trusses exist and the knowledge of the mechanical behaviour of these structures is needed to design appropriate strengthening and repair interventions. However, testing the overall behaviour of timber trusses is often not possible and few studies deal with such topic. Branco et al. [2] investigated the behaviour of two traditional king-post trusses on a full-scale load-carrying test under symmetric and asymmetric loading, to identify suitable reinforcement strategies. Parisi and Piazza [3] tested a full-scale roof compound truss (king-post truss superimposed above queen-post truss) of Silver fir (*Abies alba* Mill.), of mid-XIX century, to validate a numerical model developed for the analysis of timber structures in seismic conditions. Barbari et al. [4] tested a full-size prototype of traditional timber truss, to verify the mechanical behaviour of an original joint connection system between the top-chord and the tie beam.

In this work, a collar truss is analyzed considering the results of a full-scale load-carrying test. This type of roof structures was developed to provide more internal room height without increasing the overall height of the construction. The collar beam truss is constituted by a tie beam at the mid-height [5], with a different connection system to the inclined rafters than the tie beam of traditional king-post trusses [6–8]. This beam resists to the rafter separation, which carry the bending moment, and minimizes their span. The most common and dangerous damages for this type of truss structure, may occur at joint connection of the rafters with the tie beam and at the rafters ends (e.g. for rot due to water infiltration). Also, due to its configuration, they are very susceptible to horizontal loading such as in the case of seismic action, because they transmit a significant thrust to the supporting walls (often masonry walls). In these cases, repairing interventions are needed to avoid the collapse of the structure. Currently, the use of steel and fiber reinforce polymers (FRP), made of glass or carbon fibers, are the most common solutions for the reinforcement of timber elements in existing structures [9–13]. The choice of the appropriate strengthening and repairing method depends on the work situation and on the safety level required.

The aim of this study is to compare two different intervention techniques that can be used onsite, in a short time period, taking into consideration the structural performance of the truss during both load-carrying tests (before and after intervention).

Materials and Methods

The Collar Beam Truss. The timber truss analyzed was salvaged during the restoration of the Chimico Laboratory's roof in Coimbra, Portugal, built at the end of 18th century. Four pine timber trusses with 14.2 m span, constituted the roof structure. Two of them had later additions and strengthening interventions. The collar beam trusses were composed with an iron tie at the base and a timber tie beam at mid-height, Fig. 1. The iron tie allowed to prevent problems of transmitting horizontal loads that occur as a result of the vertical loading which could be increased in an earthquake. Due to large damage and degradation, the trusses were brought down and assessed and considered unsuitable for the new use of the building. Two of them, were dismantled and moved to the Laboratory of the Civil Engineering Department of University of Minho in Guimarães, and reassembled to perform mechanical tests and research studies. The present work considers the analysis of one of these collar beam truss.

Load-carrying Tests. A full-scale load-carrying test was carried out by loading both rafters between the strut (south brace and north brace, Fig. 1) and the joint connection with the tie beam (Fig. 2). Linear variable displacement transducers (LVDT) were used to measure the displacements of the timber elements and to evaluate the entire behaviour of the truss under loading condition. The vertical load was applied with displacement control (displacement rate of 0.05 mm/s) by two hydraulic jacks fixed to the floor slab of the laboratory. A wood wedge was placed between the rafter and the loading heads, in order to have a better load distribution to minimize local indentations. Steel frames were used to restrain out-of-plane movement of the truss. In addition, the horizontal movement of the rafters were restrained, aiming at replicating the on-site conditions (Fig. 1) where a metal tie rod was used to restrain horizontal movement and to minimize the thrust transmitted to the masonry walls. Taking into account these boundary conditions, the test setup was made in order to simulate pinned support conditions. Load was applied until visible failure was obtained in both sides of the truss. Then, two repair interventions were made and the same testing procedure was performed.

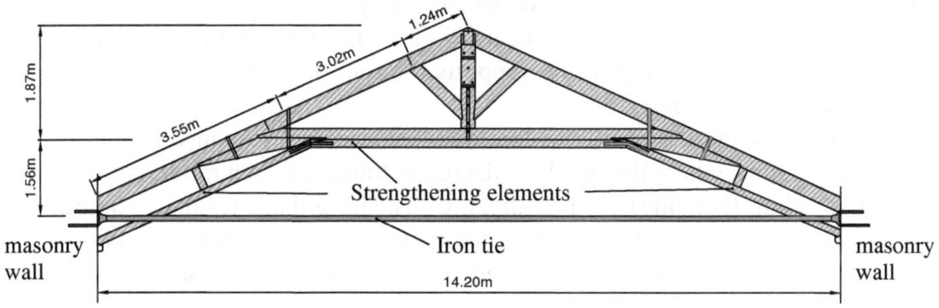

Fig. 1 Roof structure of the Chimico laboratory in Coimbra

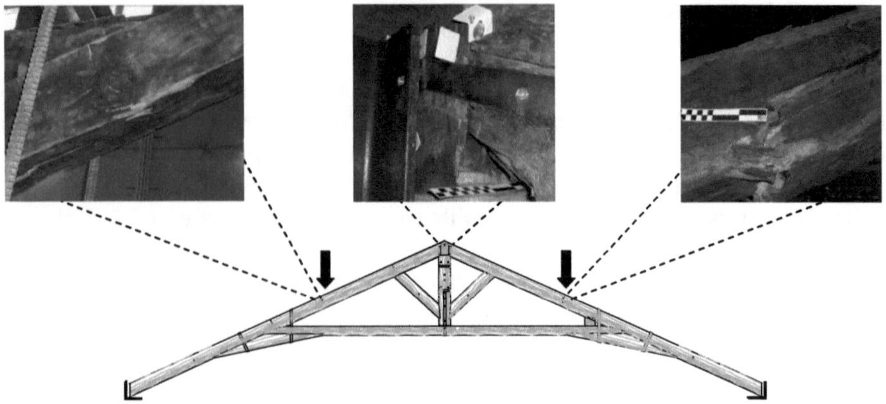

Fig. 2 Schematic of test setup and details of the failure of the rafters and of the post-rafter connection

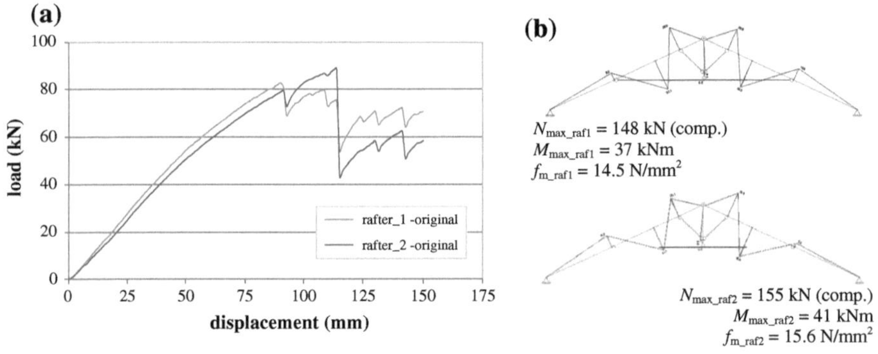

Fig. 3 Full scale test on the truss in original conditions: **a** test results measured at the loading points on the rafters; **b** simplified model and bending strength of the failed sections

Failure of the truss was found in both rafters near the loading position and on the connection between one rafter and the post (Fig. 2). The failure of the rafters coincided with sections where significant knots were found on the tensioned side of the element. Slight crushing of the supports (bottom of the rafters) was also found (due to decayed segments).

Figure 3 presents the results of the load-displacement response of the truss on the loading points, where the first load decrease, found for each rafter, correspond to the initial failure (bending) on the rafters. Considering those results, a simple truss model was calibrated to determine the ultimate stress on each failed section. The cross-section dimensions were obtained through the geometrical survey of each element complemented with the information of non-destructive tests (namely pin penetration tests and drilling resistance tests) [14]. The connection between elements were simulated by hinges and the stiffness of each element was calibrated

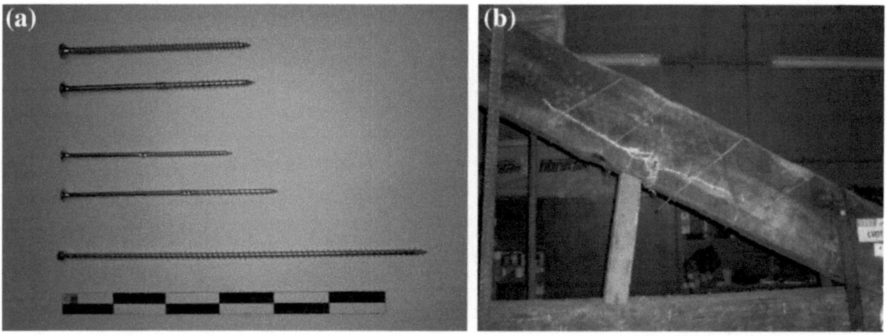

Fig. 4 **a** Metallic screws used for the repair of the timber rafters; **b** exemplification of the placing of a strut and of the inserting of inclined screws to close the existing cracks

considering the displacement of the tie-beam and of the rafters on time of failure. The stresses found for each initial failure of the rafters are presented in Fig. 3b. It is noted that the stress conditions for the failure of the second rafter (rafter_2) considered a decrease of approximately 10 % of the cross-section height of the rafter that had the first failure (rafter 1). This reduction was obtained by the analysis of the local failures of rafter 1 and calculating the height that allowed to maintain the same ultimate bending strength of that section along the test.

Repair Interventions. The initial step made within the repair interventions was to reestablish the state of deformation of the rafters prior to failure. To that aim, hydraulic jacks were used and a strut was placed between the tie-beam and the rafter while screws (VGZ 7 × 340) were inserted in the rafters, in a 45° angle with respect to the observed crack (Fig. 4b). The objective of these screws was only to allow to close the crack and recover the initial configuration of the truss, and they should not be considered as influencing the strength of the element. After having inserted the screws the strut was removed and the repair intervention was continued. Two different techniques were considered for the intervention on each side of the truss.

Repair with Timber Elements. In the first case, the cross-section was reinforced with timber elements placed on the lateral sides of the cross-section by means of screws. On the edges of the new timber elements taken from remaining elements of the same original roof, screws (HBS 6 × 200) were inserted to connect the new timber element to the existing rafter, while perpendicular screws (HBS 6 × 160) were inserted along the segment where failure occurred. In this case, the contribution of the timber elements to the strength capacity of the cross-section mainly concerns the increase of its geometry (increase of width) using elements with the similar mechanical properties. Regarding the cross-section at the failure segment, for this type of repair, the contribution of the existing element should not considered when determining the cross-section load bearing capacity. The inertia and area of the cross-section is taken only by consideration of the added timber elements. In this repair configuration (Fig. 5a), since each element had half of the

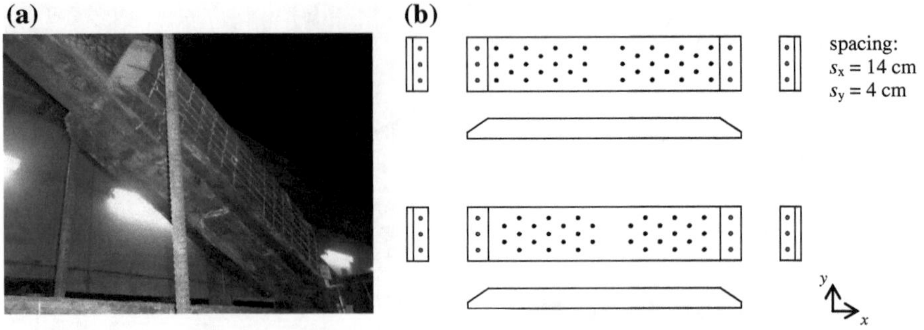

Fig. 5 Repair intervention on a rafter using timber elements: **a** repair intervention onsite; **b** schematic of the location of the screw connectors

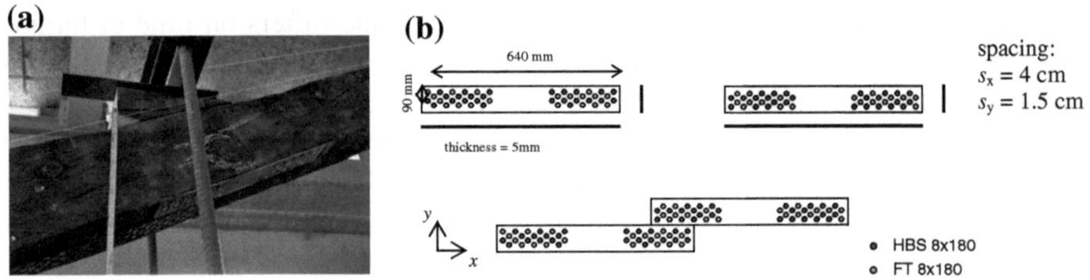

Fig. 6 Repair intervention on a rafter using steel plates: **a** repair intervention onsite; **b** schematic of the location of the screw connectors

width of the original cross-section and the same height, the area and inertia of the repaired cross-section is equal to the original one. Therefore, the bending strength of the repaired section is estimated to be similar to the original section, since the newly added elements were taken from a timber element with similar mechanical properties of the original rafter.

The main function of the screws used in this system, is to ensure that the stress field is uniformly distributed along the rafter element and the new added elements. Due to the orientation of the screws regarding the load solicitation, these are mainly solicited in shear due to the axial stress present in the element. The verification of the load bearing capacity of the screws are made taking into account that the group of screws located either before or after the failed section must be sufficient to attend to the existing normal stress.

As one of the objectives of this repair was to reestablish the load bearing capacity of the original truss minimizing the use of screws, the residual cross-section of the rafter was calculated considering the residual load that was still sustained after the initial failure. This is only admissible due to the existence of data of the full-scale test. In this experimental campaign, a value for the residual area of 80 % of the original area was obtained according to the simplified numerical model. Due to the presence of knots located near the critical section, the residual cross-section was reduced to 65 % of the original section. In onsite conditions, where the load level capacity after initial failure is not possible to be quantified, the residual cross-section should conservatively not be considered.

The maximum normal stress (combination of compression stress and bending moment) observed during the first test was 166.3 kN. Considering the contribution of the residual cross-section, each side of the connection system should be able to sustain 58.2 kN. Attending to the configuration presented in Fig. 5b, the bearing capacity of the connections, calculated through Eurocode 5 [15] and also attending to the manufacturers the European Technical Approval [16], was equal to $F_{V,Rk}$ = 58.4 kN, therefore being suitable for the this load level. Disposition of screws attended to [15] recommendations and to the manufacturers' guidelines, aiming at maximizing the effective number of screws.

Repair with Metal Plates. The second intervention technique considered the use of steel plates placed on the bottom face of the cross-section, which corresponded to the tension side where failure was initiated. Two types of screws were used for this connection namely screws HBS 8 × 180 and fully threaded screws (8 × 180). The screws were inserted perpendicularly to the elements' bottom face fixing the metal plates to the timber element. Due to the configuration of the cracks, two metal plates (5 mm thickness) were considered. In order to ensure a suitable safety level, the load carrying capacity of both connection systems was preliminary calculated considering the ultimate load, according to [15, 16] (Fig. 6). The distribution of bending stress found in the mechanically jointed section considering the failure load level previously observed is presented in Fig. 7.

According to [15], considering a repair solution using only normal screws (with lower characteristic shear strength compared to the chosen fully threaded screws) leads to a value of load per fastener of 4.15 kN. For the proposed repair configuration a $F_{V,Rk}$ = 4.18 kN is obtained for each effective screw (using only HBS 8 × 180 screws). Attending to the stress level on the shear plane, a total of 18 screws would be necessary to efficiently fix each side of each steel plate. Disposition of screws attended to [15] recommendations and to the manufacturers' guidelines, aiming at maximizing the effective number of screws. Taking into account the use of fully threaded screws and to attend the number of effective screws a total of 22 screws were considered for each side.

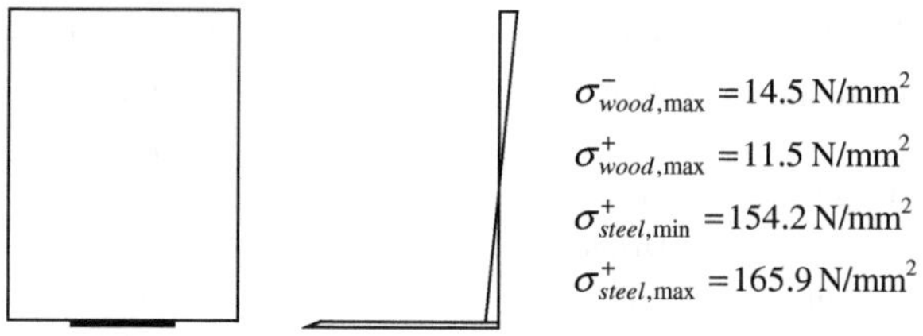

Fig. 7 Bending stress distribution on the mechanically jointed section timber with steel plates

Results and Discussion

Efficiency of the repair techniques. Figure 8 shows the load-displacement graphs, before and after the repair interventions, measured at the loading points on the rafters. It is possible to note that the ultimate load achieved with the original truss was not reached with the repaired one. In fact, the maximum load (at failure) of the repaired rafters was, using both repairing techniques, approximately 80 % of that of the original rafters (Table 1). In addition, as expected, a different stiffness was observed due to loss of rigidity of the connections and also due to local failures along the elements. In this case, for the same load level a larger displacement was obtained in the repaired truss evidencing a loss of rigidity. The screws used in both repair interventions presented no failures, but the stiffness of the metal plates were not sufficient to avoid local deformation.

The repaired truss did not achieve the same load level of the original truss due to the fact that different failure conditions were found (further detailed in the following topic). However, if the truss had to be repaired onsite, the obtained results could be considered suitable for serviceability states. Moreover, in the past, timber elements were overdesigned (very large cross-sections), therefore, for this case, a cost effective repair solution could consider a lower level of load than the original ultimate load.

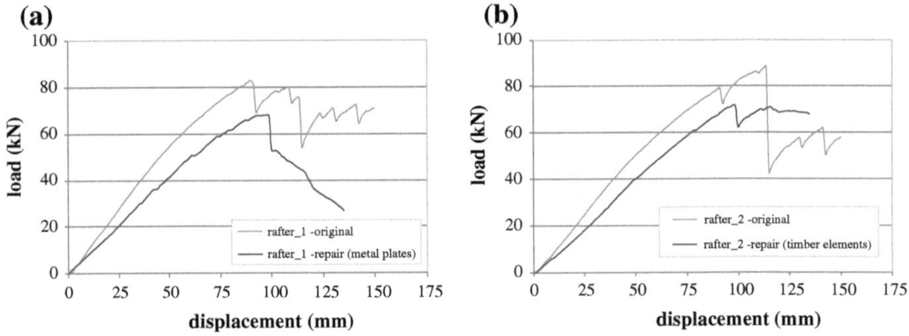

Fig. 8 Comparison between original and repaired truss with: **a** metal plates; **b** timber elements

Table 1 Ultimate load results and efficiency of the repair techniques

Description	Property	Unit	Rafter 1 (metal plates)	Rafter 2 (timber elements)
Original	$Load_{1max}$	[kN]	82.8	88.7
	$Displacement_1$	[mm]	90.1	114
Repaired	$Load_{2max}$	[kN]	68.1	71.4
	$Displacement_2$	[mm]	97.4	97.4
Efficiency	$Load_{2max}/load_{1max}$	[%]	82.2	80.6
	$Displacement^{a}_{load2max}$	[%]	144	126

[a]Ratio between the respective displacements of original and repaired rafters for the $load_{2max}$ value

Comparison between the two repair methods. Different failure scenarios occurred between the two rafters and, therefore, between the two repair methods. Figure 9 depicts the load-displacement graph. The first drop of the load corresponded to failure of the rafter 1 (around 72 kN), where the metal plates were placed. For the repaired side with timber elements, the load after the first drop slightly rises and then maintains the same value (around 70 kN) until the end of the test; while for the rafter 1 the displacement increases without any upturn of the load. This behaviour was due to the fact that other failures occurred in different places of the truss destabilizing the structure and not allowing the same load distribution as seen in the original truss.

Once the repaired timber cross-section strength of rafter 1 was overcome, the metal plates prevented the collapse of the truss shifting the stress distribution in other points of the structure. As a result, additional failures took place in rafter 1, namely on the top surface of the previous failed section (near the connection with the tie-beam), and after the strut (Fig. 9b). Moreover, the truss started to have a significant out-of-plane behaviour, as the rafters began to incline around their axis, especially rafter 1. Therefore, the use of metal plates was not able to reestablish the same load bearing capacity of the original truss (due to its low stiffness). On the other side, the added timber elements allowed a better performance and are recommended in case of fire as they are less sensible that exposed metal plates. However, it was not possible to fully determine the efficiency of the connection system in the repaired sections because failures were observed in other sections.

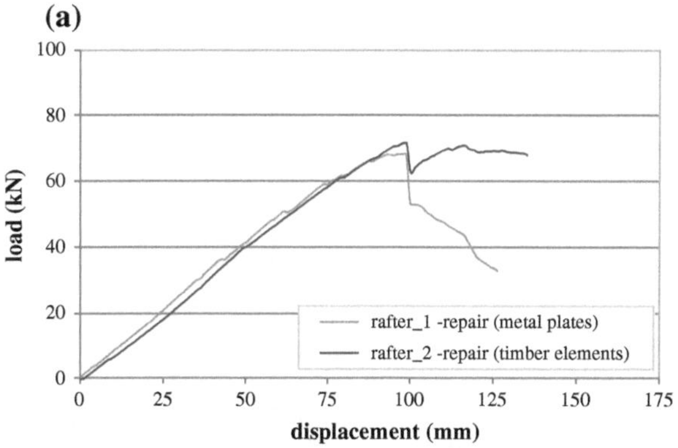

Fig. 9 Rafters in repaired conditions: **a** test results; **b** failures on the top side of rafter 1

Conclusions

An old collar timber truss was taken to failure in a full-scale load carrying test and the ultimate load and strength were recorded. In order to repair and reestablish the load bearing capacity of the original truss, two techniques were made with the use of screwed timber elements and screwed metal plates. Both tested repair techniques showed equivalent efficiency (around 80 %) in terms of load-carrying capacity. However, this result was influenced by other sections (not repaired), that were damaged in the first test. Also the utilized metal plates were not sufficiently rigid to avoid the increase in displacement and out-of-plane behaviour. Although the achieved level of load with both techniques was suitable for serviceability states, the use of timber elements was more compatible with wood and with the structural performance of the tested collar truss. When choosing which technique to use, ease of attachment and aesthetics need to be considered. Using timber elements is uniform with the materials of the old structure. Metal plates are fast to attach to the area needing reinforcement but the use of metal plates may interfere with the aesthetics of historical buildings and also the fire resistance. Finally, a careful assessment of the failure and of the effective residual timber section is important for choosing the most appropriate repair technique.

Acknowledgments The first and second authors would like to express their gratitude for the grant provided by training school: Assessment and reinforcement of timber elements and structures (Guimarães, Portugal) within the scope of FP1101 and RILEM TC 245. The present work was made within a homework task regarding that training school. The authors acknowledge the support of the Structural Lab from University of Minho (test facilities). The assistance of Eng. Ricardo Braz from Rotho Blaas in the definition of the repair solutions and material provided are acknowledged. The authors acknowledge also the support and assistance of Augusto de Oliveira Ferreira e Companhia Lda (AOF). This work was partly financed by FEDER funds through the Competitivity Factors Operational Programme—COMPETE and by national funds through FCT—Foundation for Science and Technology within the scope of the project POCI-01-0145-FEDER-007633

References

1. Tampone G (2001) Acquaintance of the ancient timber structures. 3rd international seminar of historical constructions. Guimarães, Portugal, pp 117–144
2. Branco JM, Piazza M, Cruz PJS (2010) Structural analysis of two king-post timber trusses: non-destructive evaluation and load-carrying tests. Constr Build Mater 24:371–383. doi:10.1016/j.conbuildmat.2009.08.025
3. Parisi MA, Piazza M (2002) Seismic behavior and retrofitting of joints in traditional timber roof structures. Soil Dyn Earthq Eng 22:1183–1191. doi:10.1016/S0267-7261(02)00146-X
4. Barbari M, Cavalli A, Fiorineschi L, Monti M, Togni M (2014) Innovative connection in wooden trusses. Constr Build Mater 66:654–663. doi:10.1016/j.conbuildmat.2014.06.022
5. Lourenço PB, Sousa HS, Brites RD, Neves LC (2013) In situ measured cross section geometry of old timber structures and its influence on structural safety. Mater Struct 46:1193–1208. doi:10.1617/s11527-012-9964-5

6. Branco J, Cruz P, Piazza M, Varum H (2006) Behaviour of traditional Portuguese timber roof structures. In: Proceedings of world conference on timber engineering, Portland
7. Lewandoski J (2006) Kingpost trusses. In: Rower Kenneth (ed) Historic American roof trusses. Timber Framers Guild, Becket, Massachusetts
8. Sobra K, Branco JM, Aranha CA (2014) Application of the component method to traditional dovetail joints of timber trusses. Mater Constr Sust 2:187–197
9. Akbiyik A, Lamanna AJ, Hale WM (2007) Feasibility investigation of the shear repair of timber stringers with horizontal splits. Constr Build Mater 21:991–1000. doi:10.1016/j.conbuildmat.2006.03.004
10. Schober KU, Rautenstrauch K (2007) Post-strengthening of timber structures with CFRP's. Mater Struct 40:27–35. doi:10.1617/s11527-006-9128-6
11. Richter K, Cruz H (2008) Bonding of timber. COST action E34—working group 1: bonding on site. Core Document, pp 7–98
12. Borri A, Corradi M (2011) Strengthening of timber beams with high strength steel cords. Compos B Eng 42:1480–1491. doi:10.1016/j.compositesb.2011.04.051
13. Nowak TP, Jasieńko J, Czepiżak D (2013) Experimental tests and numerical analysis of historic bent timber elements reinforced with CFRP strips. Constr Build Mater 40:197–206. doi:10.1016/j.conbuildmat.2012.09.106
14. Gomes ID, Kondis F, Sousa HS, Branco JM, Lourenço PB (2015) Assessment and diagnosis of two collar timber trusses by means of visual grading and non-destructive tests. HEaRT2015, Lisbon, Portugal
15. CEN (2004) EN 1995-1-1:2009—Eurocode 5—design of timber structures—part 1–1: general—common rules and rules for buildings. National specifications, national comments and national supplements concerning ÖNORM EN 1995-1-1. European Committee for Standardization (CEN), Brussels (EN 1995-1-1:2009)
16. ETA-11/0030 (2012) European technical approval issued by ETA-Danmark A/S for *Self-tapping screws for use in timber structures* approval issued first 11/08/2012; approval holder: Rotho Blaas s.r.l, Italy

Stiffness of Prosthetic Repairs for Historic Timber Beams

Coralie Avez and Thierry Descamps

Abstract Glued connections are often used as reinforcements or repairs of decayed parts in historic timber frames. Those timber frames are most of the time statically indeterminate structures, meaning that the ratios of stiffness between different members and joints of the structure are related to distribution of stresses within the structure. Hence, a poor repair, i.e. a too stiff or at the contrary a too flexible intervention may change dangerously the distribution of stresses within the structure, induce cracks and damage the whole structure. However, there are still no guidelines on how to evaluate the stiffness of glued connections and therefore, on how to predict the impact of these interventions on the force distribution in the timber frame. This paper focuses on glued-in rods, and uses finite element models to predict and compare the stiffness of glued-in rods in different configurations of repairing connections. This study shows sensitively different axial stiffness between those configurations: they may indeed vary in the ratio of one to two for the same strength. This highlights the importance of predicting the stiffness of glued-in rod connections to be able to make recommendations for the configuration which best respects the original structure.

Keywords Timber · Glued-in rods · Repairs · Historic buildings · Stiffness of connections · FEM

Introduction

Stiffness of connections in a wood structure is highly important because, since most timber frames are statically indeterminate structures, they may significantly influence the distribution of forces into the structure [1, 2]. Moreover, when dealing with

C. Avez (✉) · T. Descamps
University of Mons, Rue du Joncquois 53, 7000 Mons, Belgium
e-mail: coralie.avez@umons.ac.be

T. Descamps
e-mail: thierry.descamps@umons.ac.be

H. Cruz et al. (eds.), *Historical Earthquake-Resistant Timber Framing in the Mediterranean Area*, Lecture Notes in Civil Engineering,
DOI 10.1007/978-3-319-39492-3_32

old timber structures, connections are always semi-rigid [3]. When doing repairs or reinforcements, the closest attention should be paid to the prediction of stiffness of the proposed interventions. Otherwise, they can lead to unexpected and damaging changes in the distribution of forces between the members of the frame.

Glued connections [4], and especially glued-in rods, are widely used in timber construction as reinforcements or repairs [5, 6]. If numerous guidelines exist to evaluate their axial load capacity [7–14], there is still no prescription on how their stiffness should be predicted. Moreover, a glued-in rod stiffness depends on many parameters, such as the type of rod, type of glue, thickness of bondline, diameter of rod, anchorage length,…

The aim of this research is to compare the stiffness of different solutions for an intervention made of glued-in rods, with varying parameters (i.e. number of rods, anchorage length, rod diameter and type of rod), to evaluate to which extent those parameters impact the stiffness of each glued-in rod and thus, of the connection. This paper is limited to connections only loaded with axial tension, with the aim, in the long-term, of studying connections loaded with bending moment, shear and axial forces. The intervention thus studied here is a typical repair for timber trusses: replacement of the decayed part of a tie-beam (straight connection, in tension). These parts of a truss are indeed often decayed since they are connected to masonry walls, and therefore subject to high moisture content. Different configurations of glued-in rods connections exhibiting the same strength as the initial beam are proposed and studied, using a FE model based on cohesive surfaces [15, 16] in order to predict the axial stiffness.

Context and Hypotheses

The connection studied in this paper is a typical end of tie-beam repair (i.e. replacement of the decayed end of a tie-beam). We assume that glued-in rods are used to connect the initial beam with the new end, as illustrated on Fig. 1 (straight connection). The tie-beam is assumed to be loaded only in tension. Problem data and hypotheses are recapped in Table 1.

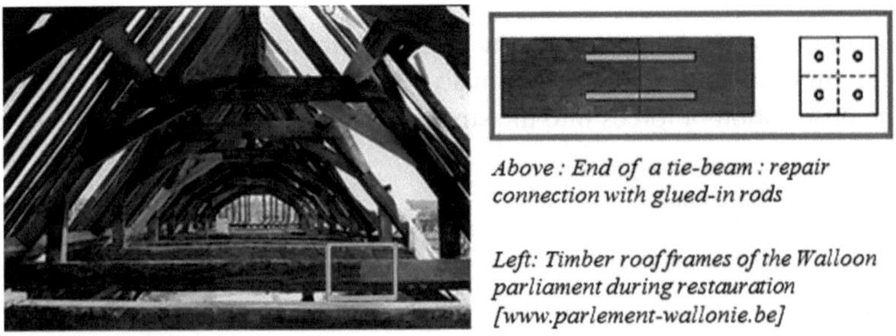

Fig. 1 Case study and example

Table 1 Data and hypotheses

Data for FE model		Assumptions	
Loading type	Tension only	Minimum edge distance	2.5 times the rod diameter [7, 8, 17]
Species	Oak: $f_{m,k} = 30$ MPa $f_{t,0,k} = 18$ MPa	Minimum spacing between rods	5 times the rod diameter (conservative value [17, 18])
Cross-section	20 × 20 (cm)	Bondline thickness	1 mm
Beam length	8 (m)	Glue properties	Pliogrip 7779's [19]

FE Model Description

To study multiple glued-in rods connections, single glued-in rods are first modelled. Thanks to the symmetry of single glued-in rods set in a wooden block, the FE model is axisymmetric.

Single glued-in rod strength or stiffness is then combined to account for the multiple glued-in rods connection. It is assumed that there is no interaction or reduction of efficiency between multiple glued-in rods as the minimum spacing between rods is respected.

Material properties. The *wood* is modelled as a homogeneous elastic orthotropic material, whose properties are recapped in Table 2.

As the *rods* can be in steel, FRP or even hardwood, different types of rods are considered in this paper. Their main properties are the Young modulus and the tensile strength (usually found in manufacturer's technical information or ETA); they are listed in Table 3. The rod material is modelled as an isotropic elastic material.

Adhesives used for glued-in rods can be polyurethanes (2C), epoxies (2C), PRF, … [10]. Two-component polyurethane adhesives are widely used in wood construction and especially for glued-in rods since they exhibit good performances (stiffness, strength) with a smaller cost than epoxies. Polyurethane 2C develops a

Table 2 Wood properties [21]

E_1 (MPa)	E_2 (MPa)	E_3 (MPa)	ν_{12}	ν_{23}	ν_{13}	G_{12} (MPa)	G_{23} (MPa)	G_{13} (MPa)
10,000	640	640	0.3	0.3	0.3	600	25	25

Table 3 Properties of different types of rods [22, 23]

Threaded steel	E = 210,000 MPa	f_y = 640 MPa
CFRP	E = 127,500 MPa	f_y = 1860 MPa
Aramid RP	E = 85,000 MPa	f_y = 1410 MPa
GFRP	E = 35,000 MPa	f_y = 470 MPa
Nylon	E = 2930 MPa	f_y = 80 MPa

good bond with wood [20], and cures (i.e. develops a good cohesion/strength in the thickness of the bondline) at room-temperature thanks to the chemical reaction between the two components. Hence, nearly no pressure is needed, it can be used in thick joint (0.5 mm to maximum 5 mm [19]) and develops little shrinkage (leading to no cracks or voids in the bondline).

Here, the glue modelled is Pliogrip®7779, a 2C polyurethane structural adhesive [19]. It is chosen since results from rods glued with this particular adhesive were used [15, 16] to calibrate and validate the FE model addressed in this paper.

The bondline is modelled using a cohesive surface; its properties are shown in Table 3.

Cohesive surface. Cohesive surfaces were developed in fracture mechanics to predict and study the propagation of a fracture, taking into account the energy required to widen an existing crack. Serrano proposed to use them to model the bondline of glued-in rods [9, 24].

A "cohesive surface" models the glue through a non-thickness element (i.e. an interaction) taking into account all successive behaviours happening during the adhesive failure (i.e. the traction-separation laws in tension and in shear, see Fig. 2).

Most of the necessary properties to model the glue through cohesive surfaces (i.e. E and σ, see Table 4) can be found in technical data [19]; or can be calculated using the information from these data (G from E and Poisson's coefficient ν). The critical fracture energy G_f and the shear strength τ_f can either be found in the literature [9, 24–26] (as first approximations) or calibrated on experimental results. In this model, they have been preliminary calibrated and validated previously [15, 16] with experimental results from work by Tannert et al. [22].

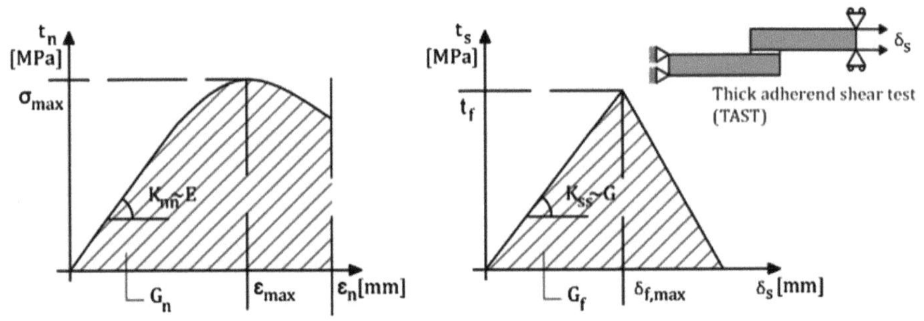

Fig. 2 Adhesive behaviour (in tension and shear)

Table 4 Glue properties

E (MPa)	G (MPa)	$\sigma_{tension}$ (MPa)	τ_f (MPa)	G_f (N/mm)
1184	414	29	5	2

Configurations of Connection

Connection geometry. Different potential configurations for the repair connection are proposed and studied: knowing the minimum edge spacing and rod spacing (see Table 1), three configurations (i.e. number of rods and diameter of these rods) are possible, as shown on Fig. 3 and Table 5.

Glued-in rods geometry. On the assumption that the connection tensile capacity has to be at least equal to the initial tie-beam strength, the minimum strength of each rod is assessed.

Timber frames are assumed to be made of oak. The target axial strength for the connection, i.e. the tie-beam strength tensile capacity is 553.85 kN (namely $f_{t,0,d} * A_{tie\text{-}beam}$).

Therefore, the minimum strength (i.e. the targeted strength) of each rod of the different configurations is determined (see Table 6).

The only parameter left is the anchorage length: anchorage lengths of glued-in rods considered in this study are set as the minimum length that the glued-in rod (knowing other parameters described above) must have to exhibit the target strength. This anchorage length can be found on Fig. 4, which shows the tensile

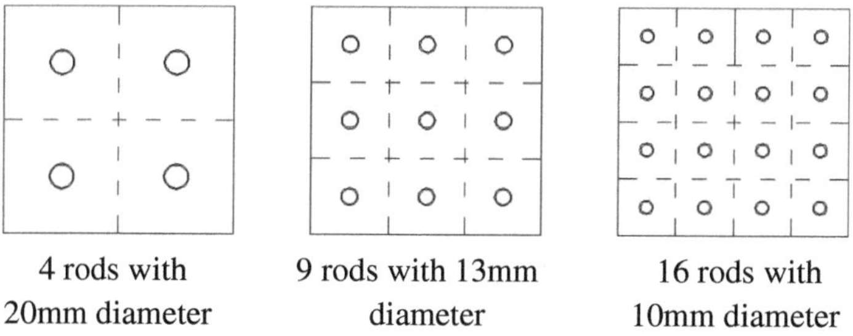

Fig. 3 Configurations of proposed glued-in rods connection

Table 5 Geometry of proposed connections (mm)

	4 rods	9 rods	16 rods
Edge spacing	50	33.3	25
Rod spacing	100	66.7	50
Wood cross-section per rod	100 × 100	66.7 × 66.7	50 × 50

Table 6 Targeted strength of each rod of the connection (kN)

	4 rods	9 rods	16 rods
Strength (kN)	138.47	61.54	34.62

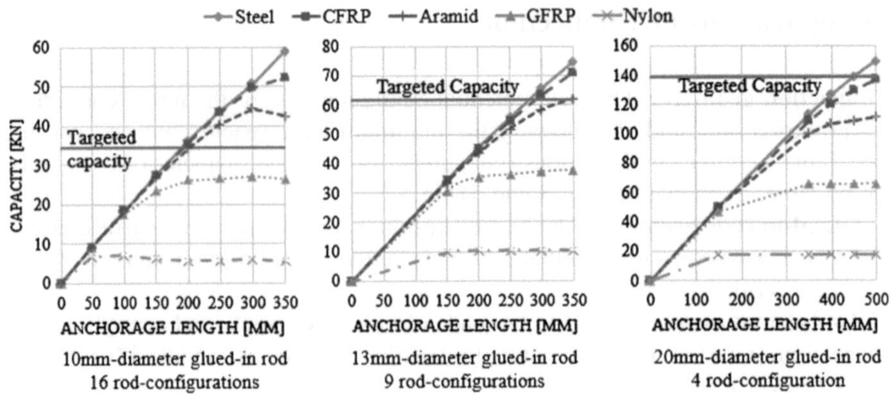

Fig. 4 Capacity of single glued-in rods depending on the anchorage length (for different configurations)

capacity of different types of glued-in rods depending on their anchorage lengths. This capacity has been obtained numerically by using the cohesive-based FE model.

Figure 4 shows that the smaller the stiffness of the rod (i.e. the Young modulus of the material), the less efficient the glued-in rod is:

- For the same value of anchorage length, glued-in rods made of steel (E = 210,000 MPa) outperform glued-in CFRP rods (E = 127,500 MPa), which outperform Aramid RP rods (E = 85,000 MPa), and so on.
- For every type of glued-in rods, a ‘stagnation’ of performance is visible on the graph, i.e. the capacity of the glued-in rod does not increase with the anchorage length beyond a certain point. But for more flexible rods, this stagnation of performance occurs more rapidly than for stiff glued-in rods, i.e. for smaller anchorage lengths. For instance, no stagnation of performance is visible (for anchorage lengths up to 350 mm) for steel rods, while GFRP capacity does not increase beyond 200 mm.
  For nylon rods (E = 2930 MPa) and GFRP rods (E = 35,000 MPa), the target capacity is never reached, even for high anchorage length!

The anchorage length that will thus be studied and whose stiffness will be compared (i.e. the minimum length of the glued-in rod to get the targeted strength), depending on the type of rods, are thus found on Fig. 4. For instance, for glued-in rods with a diameter of 10 mm, the glued-in rods studied here are (see Table 7):

- Steel rod with 200 mm anchorage length
- CFRP rod with 200 mm anchorage length
- Aramid rod with 250 mm anchorage length

Table 7 Anchorage lengths studied for each configuration (mm)

	4 rods (mm)	9 rods (mm)	16 rods (mm)
Threaded steel	450	300	200
CFRP	/	300	200
Aramid RP	/	350	250

Same types of rods with higher anchorage lengths obviously reach the targeted strength but this increase of length makes no sense (increase of the cost for unnecessary gain of strength).

Geometries of glued-in rods (related to the rod material) that reach the target strength with a diameter equal to 13 mm (target of 61.54 kN, see Table 6) and 20 mm (target of 138.64 kN) are determined on the same principle (see Fig. 4) and summarized in Table 7.

One may notice that the capacities plotted in Fig. 4 correspond to pull-out strength by failure of the glue but do not take into account the tensile strength of the rod (to highlight the influence of the anchorage length on the capacity). Hence, it has to be specified that:

- For 10 mm-diameter nylon glued-in rods, tensile strength is limited to 6.28 kN ($=f_y * \pi * r_r^2$).
- For 10 mm-diameter GFRP glued-in rods, tensile strength is limited to 36.91 kN.

Single Glued-in Rod Stiffness

Stiffness of glued-in rods listed above (with different rod diameter, type of rod and anchorage length) are predicted using the FE model addressed before in this paper. Since the model is linear, the stiffness $K_{1/2rod}$ is simply the slope of the load-displacement curve, measured at a deformation of 0.1 % (therefore, at a displacement of 0.2 mm for 200 mm-long glued-in rods for instance). The reference point for the displacement is shown on Fig. 5; the obtained stiffness $K_{1/2rod}$ therefore accounts for half a glued-in rod (i.e. set in a piece of wood only on one side).

Comparison of stiffness for rods with a diameter equal to 10 mm is shown on Fig. 5. Glued-in rods exhibit different stiffness $K_{1/2rod}$ that ranges from 36.822 to 54.646 kN/mm (ratio of stiffness ~1/1.5, see Table 8).

Regarding glued-in rods with a 13 mm-diameter (9 rods-configuration) and steel glued-in rod with 20 mm-diameter (4 rods-configuration), the stiffness are summarized in Table 8.

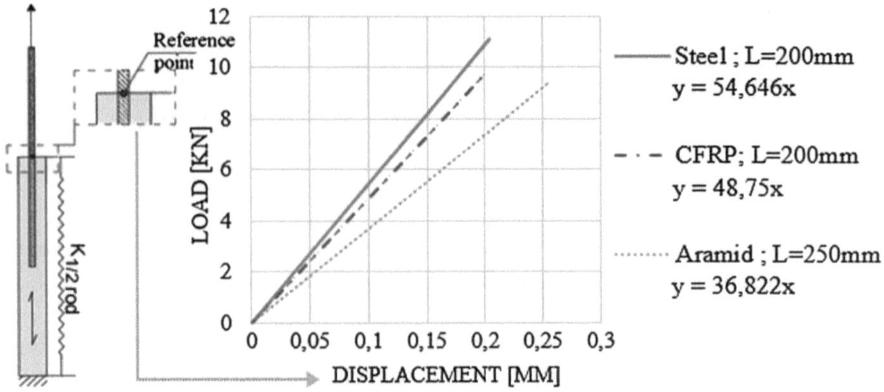

Fig. 5 Load-displacement curves for a single glued-in rod with 10 mm diameter ($\varepsilon \approx 0.1\,\%$)

Table 8 Stiffness of different single glued-in rods (kN/mm)

	20 mm	13 mm	10 mm
Steel	97.27	63.951	54.646
CFRP	/	57	48.75
Aramid	/	45.549	36.822

Multiple Glued-in Rods Connection Stiffness

Knowing the stiffness of each *half* glued-in rod, we can evaluate the stiffness of the whole connection consisting of *n* rods by using an equivalent spring model. This principle is illustrated in Fig. 6. From the stiffness point of view, each glued-in rod is made of two equivalent springs connected end-to-end (each one is actually a half of the glued-in rod, see Fig. 6, Scheme 2). The whole glued-in rod connection is then made of those glued-in rod springs connected side-by-side (Fig. 6, Scheme 3), therefore their stiffness must be summed together.

Hence, the stiffness of the whole multiple glued-in rod connection is given by:

$$K_{connection} = n * K_{rod} = n * \frac{K_{1/2rod}}{2}. \tag{1}$$

The stiffness of the connection for the configurations studied are shown on Fig. 7. Depending on the configuration, stiffness of the multiple glued-in rod connection ranges between 197 and 437 kN/mm, and may thus vary in a ratio from more than 1 to 2.2! The stiffest connection is the configuration made of 16 glued-in steel rods (rod diameter equal to 10 mm) exhibiting an anchorage length of 200 mm.

When working on historic timber frames, it is thus important to evaluate both the original stiffness (of elements and connections) and the stiffness of the proposed

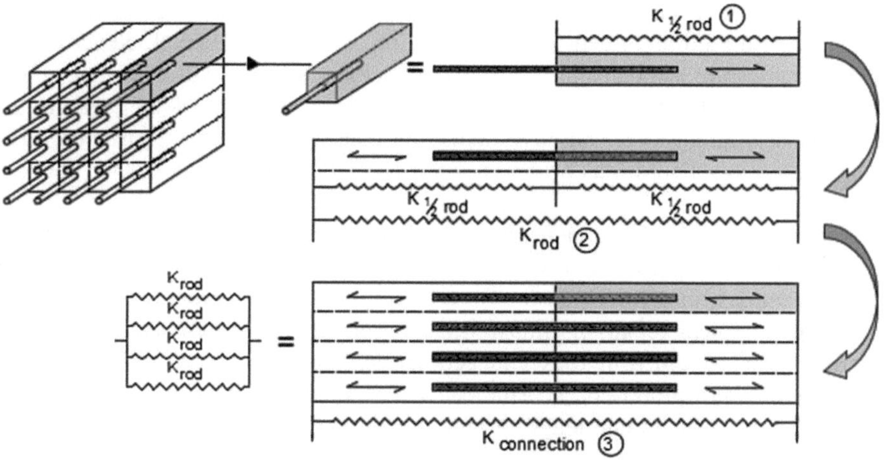

Fig. 6 Stiffness of the connection: parallel circuit composed of 16 glued-in rods with an individual stiffness K_{rod}

Fig. 7 Stiffness $K_{nconnection}$ of multiple glued-in rod connections (kN/mm)

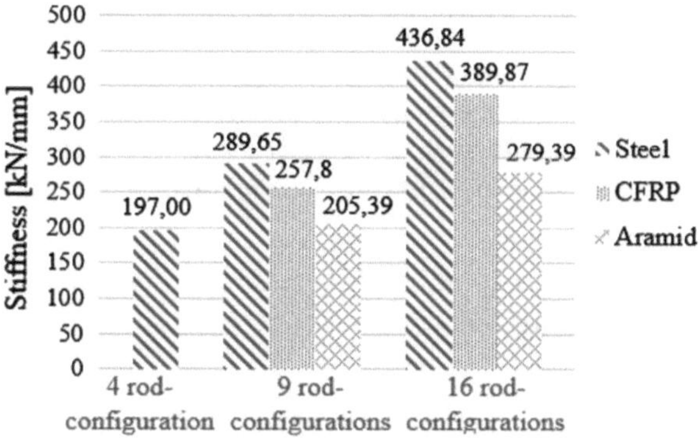

intervention, in order not to change significantly distribution of forces within the structure. This re-distribution should though be studied on real cases (i.e. real timber frames), since each structure would not be affected to the same extent by the same modification of connections' (or members') stiffness.

Conclusion

Glued connections are widely used in historic timber roof trusses to repair or reinforce decayed/weakened parts. Among glued interventions, glued-in rods are quite common (i.e. rods glued into a piece of wood), mainly to link new pieces of wood replacing decayed ends to original members (end of tie-beam repairs).

However, there are still no guidelines on how the stiffness of a glued-in rod, and therefore of a multiple glued-in rod connection, could be predicted. Nevertheless, timber frames are statically indeterminate structures. Hence, stiffness of interventions in a wood structure is highly important since a too stiff (or on the contrary too flexible) new connection can have damaging consequences on the structure. The closest attention should thus be given to the prediction of stiffness of the proposed interventions.

This paper used a finite element model based on cohesive surfaces and developed previously to predict and compare the axial stiffness of different configurations of interventions, therefore focusing on connections. The intervention thus studied here was a typical repair of timber trusses: end of a tie-beam repair. Different configurations of glued-in rods connections exhibiting the same strength as the initial beam were proposed and studied. The studied configurations had different number of rods, anchorage length, rod diameter and type of rod.

Results show that two connections of glued-in rods having the same strength can exhibit stiffness which vary in a ratio from 1 to 2.2!

Further studies will use this prediction of stiffness to make recommendations (based on real cases) regarding the connection that best fits the initial structure, i.e. whose stiffness best respects the initial static scheme of the timber frame. The Finite Element Model dedicated to glued-in rod should also be validated on experimental results. Then, prediction of stiffness of glued-in rods connections using FEM should be extended to the study of other types of interventions on historic timber trusses, loaded not only with axial forces but also shear and bending moments, such as repairs of moment-carrying connections.

References

1. Descamps T, Lambion J, Laplume D (2006) Timber structure: rotational stiffness of carpentry joints. In: 9th world conference on timber engineering, Portland, USA
2. Descamps T, Leoskool L, Van Parys L et al (2014) Influence of the joint stiffness in old timber hyperstatic frames. In: World conference on timber engineering, Quebec, Canada
3. Branco J, Descamps T (2015) Analysis and strengthening of carpentry joints. In: Construction and building materials
4. Descamps T, Avez C, Carpentrier O et al (2015) Historic timber roofs modelling: prosthesis and resin repairs. In: SHATIS-3rd international conference on structural health assessment of timber structures, Wroclaw, Poland
5. Moore D (1998) Rotafix resiwood systems: estimating for timber resin repairs
6. Steiger R, Serrano E, Stepinac M et al (2015) Strengthening of timber structures with glued-in rods. In: Construction and building materials
7. Tlustochowicz G, Serrano E, Steiger R (2011) State-of-the-art review on timber connections with glued-in steel rods. In: Materials and structures, vol 44, pp 997–1020
8. Stepinac M (2012) State-of-the-Art review on timber connections with glued-in steel rods. In: International Council for Research and Innovation in Building and Construction—Meeting forty six, Vancouver, Canada
9. Serrano E (2001) Glued-in rods for timber structures—A 3D model and finite element parameter studies. Int J Adhes Adhes 21:115–127

10. Johanssen C-J, Bengtsson C (2002) GIROD-glued-in rods for timber structures: final report. SMT4-CT97-2199
11. Rossignon A, Espion B (2008) Experimental assessment of the pull-out strength of single rods bonded in glulam parallel to the grain. In: Holz Roh Werkst 66, pp 419–432
12. Steiger R, Gehri E, Widmann R (2009) Pull-out strength of axially loaded steel rods bonded in glulam parallel to the grain. In: Materials and structures, vol 40, pp 69–78
13. Feligioni L, Lavisci P, De Ciechi M et al (2003) Influence of glue rheology and joint thickness on the strength of bonded-in rods. In: Holz als Roh-und Werkstoff, vol 61, pp 281–287
14. Natterer J, Sandoz JL, Rey M (2004) Traité de Génie Civil de l'Ecole Polytechnique Fédérale de Lausanne- vol 13-Construction en bois: Matériau, Technologie et Dimensionnement, Lausanne: Presses Polytechniques et universitaires romandes, pp 243–248
15. Avez C, Descamps T, Van Parys L (2015) Development of a cohesive surface-based model predicting glued-in FRP rods behaviour. In: AB2015-3rd international conference on structural adhesive bonding, Porto, Portugal
16. Avez C, Descamps T, Van Parys L (2015) Development of a cohesive crack model for glued timber connections. In: Matinée des Chercheurs UMons (MDC2015), Mons, Belgique
17. German National Application Document DIN V EN 1995-1-1/NA 1:2044-12
18. Gonzales E, Tannert T, Avez C (2015) Timber joints with multiple glued-in steel rods. In: AB2015-3rd international conference on structural adhesive bonding, Porto, Portugal
19. Pliogrip structural adhesives, Technical Datasheet, Document 2432 V2 F2, 2012, Ashland Inc
20. Cognard P (2006) Collage du bois-Colles. Adhésifs. Liants-bm7636 in «Techniques de l'ingénieur»
21. European standard-EN 338-Timber grades
22. Tannert T, Faghani P, Zhu H et al (2014) Timber joints with glued-in FRP rods. In: World conference on timber engineering, Quebec, Canada
23. Datoussaïd S (2014) Calcul des Structures et problèmes non linéaires. Université de Mons, Mons
24. Serrano E (2004) A numerical study of the shear-strength-predicting capabilities of test specimens for wood-adhesive bonds. Int J Adhes Adhes 24:23–25
25. Campilho R, Banea MD, Da Silva LFM (2013) Modelling adhesive joints with cohesive zone models: effect of the cohesive law shape of the adhesive layer. Int J Adhes Adhes 48–56
26. Cognard P (2002) Collage des matériaux-Mécanismes. Classification des colles.bm7615 in «Techniques de l'Ingénieur»

Consolidation of a Collar Truss with a Steel Cable System

Giacomo Crivellari and Jorge M. Branco

Abstract Post-tensioning techniques for strengthening existing timber elements have been used throughout history with excellent results. Former applications on wooden trusses have already shown some of the advantages of such a consolidating system, but very few tests were performed on the topic. During an experimental campaign, a full-scale historical timber truss was deeply analysed and tested in its original (unreinforced) condition, reaching a considerable level of damage. It was later repaired with a steel cable system, using prestresses to recover the structure bearing capacity, and tested again. The results show that the repair allowed a recovery of at least the 80 % of the ultimate load-carrying capacity. The complete failure was not achieved, but it is likely to be presumed that the structure could have regained the totality of its strength. The data acquired in both tests were later compared in order to understand in detail the behavior of the truss.

Keywords Timber truss · Consolidation · Post-tensioning

Introduction

Timber structures have been built since ages and are worldwide spread. Between all the historical realizations, timber trusses are an outstanding combination of aesthetic, functionality and structural optimization of each load-carrying member. Many different layouts were generated according to the particular conditions, but nowadays all of them share the necessity to be preserved and, in the worse cases, to be consolidated, for guaranteeing their conservation.

G. Crivellari (✉) · J.M. Branco
ISISE—Department of Civil Engineering, University of Minho, Azurém, 4800-058 Guimarães, Portugal
e-mail: crivellari.giacomo@gmail.com

J.M. Branco
e-mail: jbranco@civil.uminho.pt

H. Cruz et al. (eds.), *Historical Earthquake-Resistant Timber Framing in the Mediterranean Area*, Lecture Notes in Civil Engineering,
DOI 10.1007/978-3-319-39492-3_33

Many retrofitting techniques have been developed, and lots of them have been systematically reported in dedicated manuals. However, while some applications were largely used and taught, others have been slowly left aside, with the risk of forgetting them and losing part of the technical and constructive heritage. For instance, the retrofitting of historical timber elements with *prestressed system* has been performed several times in the past, but still there are many unknowns that discourage new applications and limit its development. The post-tensioning of steel elements, as a specific way of applying the prestresses, has been proposed since the beginning of the 19th century [1], and it has been applied both for new construction and for restoration. Some inspiring cases, realized on historical timber trusses by some Italian conservators, have already shown the potential of this strengthening method, as the roof structure of theatre of Sarteano [2], or the one of Savona [3].

In order to compensate the lack of knowledge on post-tensioning methods for retrofitting timber structures, an experimental campaign has been carried out by the authors. A full-scale collar truss has been saved during the dismantling of an historical factory, and it has undertaken several tests to be fully characterized and prepared for the application of a prestressed steel cable system. The twofold purpose of the campaign focused on the realization of the retrofitting and on the comprehension of its behavior while interacting with the timber structure.

Preliminary Studies

Geometry assessment. While still unmounted, the truss has been carefully studied with different approaches. At first, all the composing elements have been renamed following the pattern chosen during the dismantling process (Fig. 1). Furthermore, every piece has been discretized in segments 40 cm long, and all sections have been analysed with different *non destructive tests*. The first step was the geometrical survey of the structure, and the relative differences in the gross cross section ends, have been accurately measured. A part from some local reductions in the cross sections, due to the nature of the material or to the carpentry joints, the timber elements presented a regular shape, with mean values of 195 × 230 mm^2. The detailed measures are reported in Table 1.

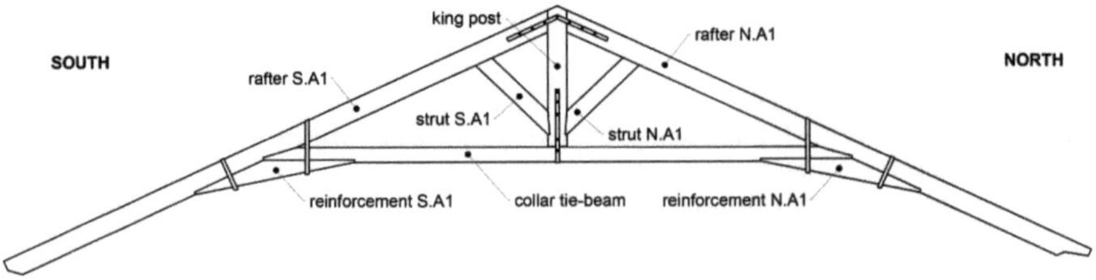

Fig. 1 Scheme of the collar truss analysed

Table 1 Measures of the elements composing the collar-beam truss

Dimension (mm)	Rafters	Tie beam	Struts	King post	Reinforcements
Base	8000	7650	1220	1950	2230 (max)
Height	235	205	230	270	280 (max)
Length	195	190	205	205	200

Fig. 2 Decays affecting the truss in the South support (*left*) and in the North one (*right*)

Decay visual mapping. A second step consisted in the identification of the damages present on the truss and their mapping. It was observed that fungi and xylophagus insects affected the wood (Fig. 2). The most deteriorated part was the area closest to the supports, as typically happens in wooden structures embedded in walls. The graphical restitution of the decay was overlapped with the data offered by the consecutive tests, to obtain a crosschecking of the results.

Non Destructive Tests. To have a qualitative idea of the decay distribution and depth, a spring-loaded penetrometer was used on every surface, testing every section in twenty-four different spots. Considering that tests depth was maximum 40 mm, the results were just related to the surface state. A further test was done by using a Resistograph®, checking every segment in five different locations, both on the bigger and smaller dimensions. This technique allows to spot inner voids or deep decaying phenomena, but in this particular case, no major defects were found in the central part of the elements. An ultimate test, the ultra-sound method, was used to assess the decay different levels between various parts of the same elements, but the fissures present in the timber affected considerably the results.

Bending test. Following the requirements of the EN 408:2003 [4], bending tests were performed on spare pieces belonging to a similar truss. During the dismantling of the original roof structure, some other elements were saved, that have been then tested and which results have been compared to the truss under analysis. One rafter and two tie beams became the specimens for the bending test, on which a total of 18 experiments were done, considering both the local and the global behavior of the elements. The outcomes were the estimation of the *Modulus of Elasticity* (MoE) and of the density, which have been later compared (Table 2) with the national Portuguese codes [5] and another study, specific on the wood specie used in this case, the *Pinus pinaster* [6]. The differences in the values should be correlated with

Table 2 Characteristics for the *Pinus pinaster*, according to different references

	Bending strength f_m (MPa)	MoE (GPa)	Density (kg/m^3)
Bending tests	49.0–63.0	10.9–13.6	556–593
LNEC [5]	18–35	12–14	580–610
Other [6]	49.9–62.3	12.3–14.0	560–589

the historical and partially decayed character of the analysed structure, which has been assessed being older than 100 years.

The density evaluation has been done using a weight, calculating the ratio with the *apparent volumes* of the specimens. Considering the absence of major inner voids, and the limited level of decay of the elements, is possible to claim that the density results were acceptably close to the real ones.

Assemblage and Test Set-up

The truss has been then remounted, assuming the configuration of its original location. Thanks to the expertise of an experienced carpenter, it was possible to assemble the elements with few, minor modifications, using all the elements in the correct positions, and interlocking them with the original joints. The metal bands that were tightening together the most delicate connections had to be replaced by new ones. The steel plates have been realized according to the original design, trying to replicate the same effects on the structure. The old wrought iron nails have also been changes in favor of steel screws, keeping into account the resisting sections of both old and new elements, with no alteration in the stiffness of the connections. To avoid a certain localized rupture in the support, the most decayed parts of the rafters were removed, cutting them respectively about 20 cm on the North side, and 40 cm in the southern one.

For the set-up of the laboratory facilities, few preliminary test were previously performed, in the framework of an international workshop [7]. In fact, a part from the truss here studied, another similar one was successfully reassembled and tested, allowing a proper preparation for the experimental configuration. The testing layout finally chosen is shown in Fig. 3.

As it possible to appreciate from the scheme, the experiments have been carried out with specific boundary conditions reflecting some particular assumptions that have been previously made. Firstly, the supports did not allow any major horizontal displacement, resembling a double-hinged configuration. Two wooden supports were interposed between the rafter and the steel elements, as sacrificial elements that could have permitted some small adjustments. Although the resemblance of the original configuration could have suggested a different supporting system (as masonry walls deform, and so allow horizontal displacement), to guarantee safe and controlled experiments this set-up have been preferred. Moreover, as the aim of the tests was to focus on in plane behavior of the timber structure, some lateral

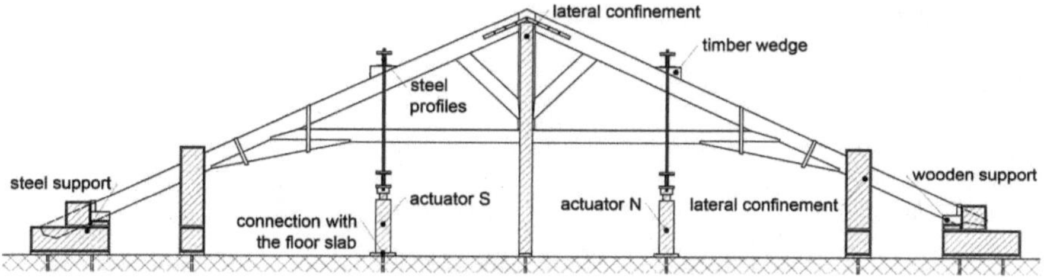

Fig. 3 Scheme of the testing set-up

confinements have been placed, coupling them on both sides of the truss, in order to keep it vertical while loading it. Three couples of confinements have been place, one in the middle, and two bracing the rafters.

The loading system consisted in two hydraulic actuators, firmly anchored to the concrete floor slab of the laboratory, that were connected to the timber structure through a system of steel sections. The load was imposed in the upper part of the truss rafters, exactly in the middle point between the strut connection and the tie beam joint. The actuators were pulling down the steel over-structure, and consequently, loading the truss. The devices were displacement-controlled, allowing to set a constant loading speed, and were capable of developing a maximum force of 500 kN, within a displacement range of 150 mm. Although such high values of force were never reached, it was instead necessary to use the totality of the displacement range.

First Test

The timber truss, completely assembled and placed in the testing set-up, has been equipped with a full set of *LVDTs*, which were distributed all around the structure, and located in the most meaningful positions. The loading speed was set to 0.05 mm/s and the structure was initially stressed with a preload of 20 kN. With this initial force, the truss adjusted all the gaps and irregularities deriving from the reassembling. After this initial input, the set-up was retightened and the first test was performed. It took in total 50 min, with a maximum displacement equal to 150 mm in the actuators. The actuators induced in the truss, a maximum force of 269 kN, divided in 142.5 kN for the North jack, and 126.5 kN in the southern one.

The structure was heavily damaged, with visible fissures in the rafters, located just below the loading points, and in correspondence with the strut–rafter connection (Fig. 4). Many minor cracks appeared all around the truss, and the tie beam joints were significantly opened during the test. The wooden support, separating the rafters from the steel bases, started to crush under the compression stresses. The overall structure showed a tendency of deforming in the out-of-plane direction, with a torsion around the tie beam that affected mainly the upper part of the truss. The deformation

Fig. 4 Fissures in the South rafter in the bottom part (*left*), and in the strut–rafter joint (*right*)

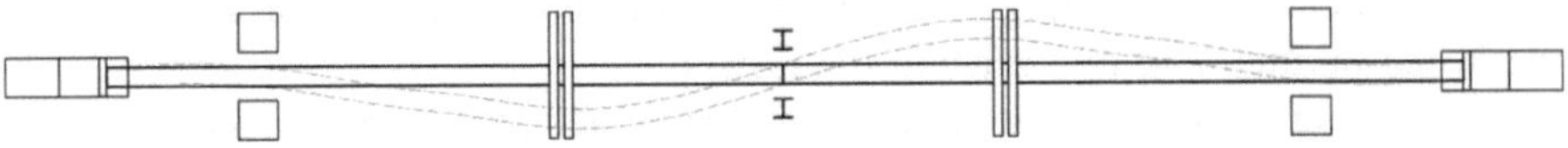

Fig. 5 Out-of-plane deformation of the truss during the test

is shown in Fig. 5. After the test, this latter damage was almost completely reabsorbed by the structure.

Retrofitting System

After the un-loading of the truss, it was possible to start the repairing operations. Firstly, the cracks over the struts were closed using 500 mm, fully threaded, inox screws. Secondly, the rafters were lifted with the help of a crane, and straightened again. At the same time, the post-tensioned retrofitting system was conceived. It was designed to counteract the bending stresses that was harming the most the structure, and in the meantime transform the imposed load in stresses that could be better borne by the timber elements (see Fig. 6). Some references were found in the state of the art by former realization of other authors [8], but the specific conditions imposed an original design. The basic requirements of the intervention were the *light* approach to the existing elements, a rapid execution with limited budget, and the feasibility of such a technique with real case boundaries (i.e. impossibility to operate from the top for the presence of a roof cover). A fundamental criterion was also the control of the applied prestress. More restriction were imposed by the many obstacles of the set-up and testing system. It was in fact impossible to operate anywhere below the line of the tie beam, considering the presence of all the laboratory devices.

The cable system was twice symmetrical, both to the vertical and longitudinal axis. In fact, all the retrofitting was done with two cables, made with galvanized

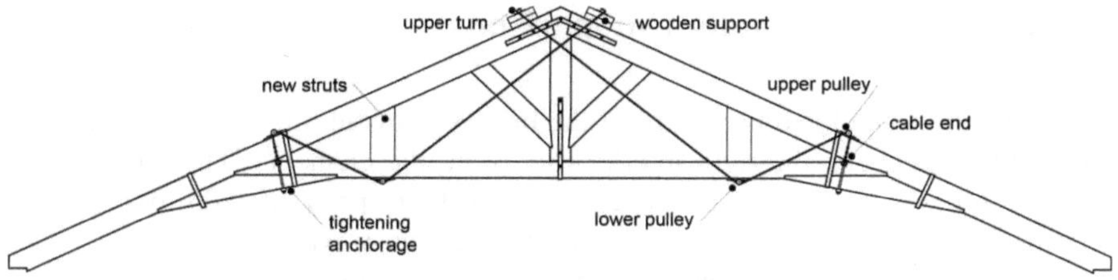

Fig. 6 Retrofitting intervention

Fig. 7 Details of the retrofitting. *Upper turn* (left), *lower pulley* (middle) and *anchoring* (right)

steel, 14 mm diameter, 6 × 19 WSC, with ultimate strength of 139 kN. Each line was starting from the *tightening anchorage*, and after some directional changes at the *upper* and *lower pulleys*, it was directed on the other side of the truss passing around the *upper turn* (see Fig. 7). This imply that the retrofitting system was simply laid on the truss, and very few screws were put to fix the elements, making the intervention almost totally reversible. Between the new additions and the historical structure, some wooden sacrificial elements were interposed. All the metal pieces were realized following this specific design, but an accurate choice of commercial sections allowed a drastic reduction of costs and production time.

The prestresses were induced in the structure through the *anchoring* pieces. Each cable end was tied to a steel ring, connected to a metal threaded rod. These, were firmly kept in position by steel nuts, pushing against a massive metal bar in contact with the truss. While tightening the nuts, the rods were pulling the cables, inducing the necessary tension. With a dynamometric key was possible to evaluate the torsion applied to the system, and with it calculating the axial force in the cables. The post-tensioning was done progressively and as uniformly between the four application points. In total, a prestress of 29 kN was applied in the retrofitting system.

Moreover, new struts were placed just below the loading point, to redistribute the stresses through the whole timber structure. As would be possible to appreciate in the results section, the cable system induced multiples benefic actions on the truss. A part from the main purpose, the counteraction of the loading stresses, the cables helped to tight both the rafters—king post connection, and the tie beam—rafter ones.

Second Test and Results Comparison

Following the same loading pattern of the first experiment, the truss was preloaded with 20 kN, to check the right interlocking of all the elements. Then the real test started, and was imposed again a constant loading speed of 0.05 mm/s, taking 43 min before being stopped. Due to safety reasons, it was not feasible to reach again the rupture of the structure. In fact, the out-of-plane deformation was too significant to assume that the truss would not have failed under a torsional effect, with consequent problems in the testing devices, laboratory facilities and for the personnel involved. However, thanks to the data recording system, good results were collected and analysed.

During the second test a maximum load of 220 kN was reached, divided between 108 kN on the South actuator and 112 kN in the North one. The estimation of the goodness of the repairing in terms of load-carrying capacity shows that the structure regained the 82 % of its strength, which is a remarkable achievement if compared to the residual state after the first test. It was also noticeable the change in the structural behavior, change that was highlighted from the outputs form the LVDTs set-up (see Fig. 8). Are here presented some significant ones, in order to explain these modifications. The information gathered during the first test are represented through dashed lines, and the second test is drawn with continuous ones. On the horizontal axis are shown the deformation registered by the transducers, expressed in millimeters, while on the vertical direction is shown the total load applied in that moment, in kN.

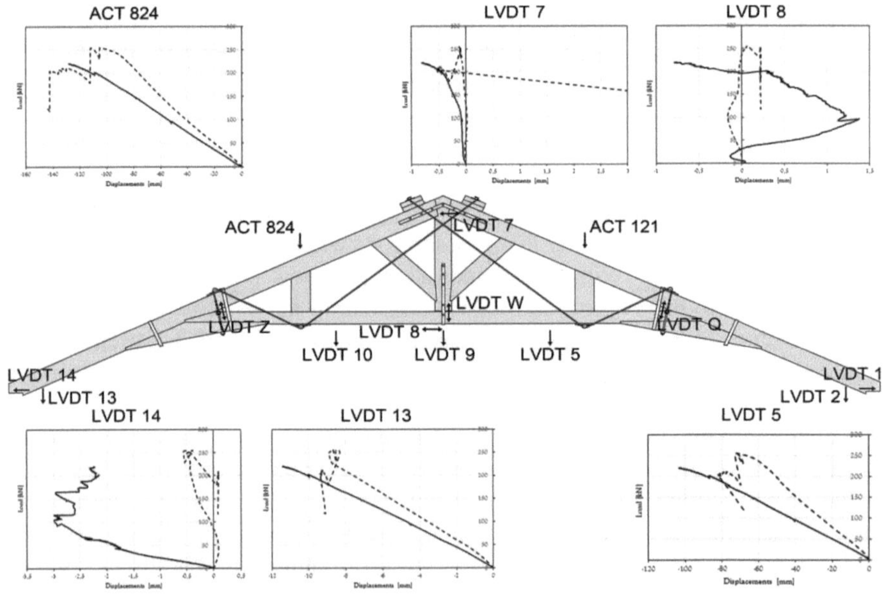

Fig. 8 LVDTs disposition and result comparisons between first (*dashed*) and second (*continuous*)

The data presented have been selected between all the outcomes deriving from both the first and the second test. These information have been qualitatively interpreted to better understand the behavior of the truss during the two experiments. Although the structure was slightly asymmetrical, both in geometry and in the displacement response, the differences were so small to become negligible for the overall comprehension of the tests. The asymmetry is understandable from the graph *LVDT 8*, which shown how the middle point in the tie beam was shifting from one side of the other of the central vertical confinement. The graph *ACT 824* refers to the inner controller of the actuator. During the first test, an abrupt rupture happened, generating the drop after the linear behavior. In the second test, the failure was not achieved, but the graphs shows how the upper rafter was moving more under the same loading. At the same time, the lower part of the structure was also moving faster, as displayed in graph *LVDT 5*. This represents a uniform movement of the whole upper part of the truss, implying an improvement of the stiffness in this area. From *LVDT 13* and *14* is understandable the change in the reactions at the support level. The horizontal thrust was much bigger after the retrofitting, when compared to the first test, and contemporaneously the rafter was rotating more. These modifications imply a less ductile behavior of the upper part of the truss, which consequently loaded more the supports. Eventually, the graph *LVDT 7* is showing the effectiveness of the steel cable system in the confinement of the carpentry joints. The sensor has been placed in the upper part of the connection, and while during the first test there was almost no change in its dimension, in the second experiment is was considerably under the compression effect, generated by the post-tensioning.

Conclusions

A historical timber truss has been reassembled, analysed and then tested in laboratory conditions. With a total load of 268 kN, the failure of the rafters was reached, plus some more damages in the other elements. The test has been recorded by a full set of LVDTs placed in the crucial positions. The truss has been later retrofitted with a system of post-tensioned steel cables, and retested under the new conditions. The maximum force borne was equal to 220 kN, which represent a recovery of the 82 % of the original load-carrying capacity. The comparison between the displacements measured in both tests reveals a change in the structural behavior, as the activation of alternative resisting mechanisms, showing that the residual capacity of the truss has been used to resist the load.

Acknowledgments The present work was made within a Master thesis of the Erasmus Mundus Advanced Masters course in Structural Analysis of Monuments and Historical Constructions. The authors acknowledge also the support and assistance of Augusto de Oliveira Ferreira e Companhia Lda (AOF).

References

1. Gasparini DA, da Porto F (2003) Prestressing of 19th century wood and iron truss bridges in the U.S. In: Huerta S (ed) Proceedings of the first international congress on construction history. Madrid, pp 978–986
2. Tampone G (1996) Il restauro delle strutture di legno. Hoepli, Milano, p 321
3. Paolini L, Messina C, Marradi AM (1989) Il restauro della copertura del teatro Chiabrera a Savona, in "Legno e restauro", a cura di G. Tampone. Messaggerie toscane, Firenze
4. EN 408:2003 (2003) Timber structures—Structural timber and glued laminated timber—Determination of some physical and mechanical properties. European Committee for Standardization
5. LNEC (1997) Madeira para construção – Ficha M2: pinho bravo para estruturas, Lisbon
6. Carballo J, Hermoso E, Fernandez-Golfin JI (2009) Mechanical properties of structural maritime pine sawn timber from Galicia (Pinus pinaster Ait. ssp. Atlantica). Insituto Nacional de Investigacion y Tecnologia Agraria Y Alimentaria, Madrid
7. Branco JM et al (2015) FP1101 and RILEM TC245 RTE Training school on assessment and reinforcement of timber elements and structures. HEaRT, Lisbon (in press)
8. Jurina L, Bassoli A (2011) Il recupero della copertura lignea della Chiesa di San Carpoforo a Milano, in Ingegneri, n. 5, y. 3. Maggioli editore, Milano, pp 10–11

A Timber Roof Structure for the Milreu Rural House Rehabilitation

Jorge N. Bastos

Abstract Under the European Union (EU) support programs, the Portuguese heritage agency carried out several actions to develop archaeological sites with adequate infrastructures to be integrated into leisure routes. The archaeological site of the Milreu Roman ruins is located in the southernmost province of Portugal—the Algarve, near the small village of Estoi. This village is in the central Algarve region, 10 km north of the city of Faro and the coastline. Archaeological excavations conducted in the late XIX-th Century, by the archaeologist Estácio da Veiga showed the existence of both a roman Villa and a temple dedicated to the cult of water. Among other constructions on this site, a medieval rural house was built some centuries ago over the existing Roman ruins. In the 1950s, the German Archaeological Institute also excavated this location discovering unique Roman artifacts and archaeological structures (mosaic floors, masonry walls, water tanks, roman baths). In the mid-1960s, the Ministry of Public Works—Historical Monuments and Buildings Division (M.O.P.—D.G.E.M.N.) carried out a safeguard program to protect the XVI-th Century Rural House by installing a reinforced concrete (RC) roof system over the existing masonry walls. Several decades after the intervention, extensive deformations occurred followed by severe wall cracking. The proposed rehabilitation strategy for this rural building used traditional building techniques to provide an adequate rehabilitation standard. The RC roof was dismantled and removed, and it was replaced by a traditional timber roof, which resulted in a large decrease in the dead load (DL) mass, and reduced the seismic risk of this historical building.

Keywords Timber roof · Rehabilitation · Heritage · High seismicity · Aesthetics

J.N. Bastos (✉)
School of Architecture, University of Lisbon, Polo Universitário do Alto da Ajuda, Rua Sá Nogueira, PT-1349-063 Lisbon, Portugal
e-mail: jnbastos@fa.ulisboa.pt
URL: http://faculdadedearquitectura.ulisboa.pt/

H. Cruz et al. (eds.), *Historical Earthquake-Resistant Timber Framing in the Mediterranean Area*, Lecture Notes in Civil Engineering,
DOI 10.1007/978-3-319-39492-3_34

The Roman Ruins of Milreu—The Historical Site

The Algarve region is located in the southernmost part of Portugal. In the beginning of the Christian era, it used to be a Roman province with important settlements along the coast line, with a major road from Silves (in the West) to Seville, in south of Spain. Later on, in the early VIII-th Century, it was under Arab rule and it was integrated into the Al-Andaluz kingdoms of South of Spain. In 1249, the Christian armies conquered Tavira and Faro, and the Arabs departed from this region which was integrated in the Portuguese Kingdom. Currently, is an important leisure destination not only for Portuguese inhabitants as well as for other EU citizens.

Under the EU cultural programs financial support was obtained in the 1990s to intervene in the archaeological sites and to install adequate infrastructures, creating a cultural circuit for the visitors. The archaeological site of Milreu, almost 10 km north of the city of Faro, is located in the Algarve central region along the main road that linked Silves to Seville. It is well-known for its Roman ruins: (a) a Roman Villa with farm buildings; and, (b) a Roman Temple dedicated to the cult of water, [1–3], see Fig. 1.

In the late XIX-th century it was extensively excavated by the Portuguese archaeologist Dr. Estácio da Veiga, the acquired data was recorded and published and important unveiled artifacts stored in museums. Afterwards, the different building's ruins were covered up again for protection against vandalism. In the early 1950s, the German Archaeological Institute under the supervision of Dr. Theodor Hauschild also conducted important excavation's campaigns revealing important artifacts (buildings, mosaics, Roman structures) allowing a more detailed understanding of this important site organization [1–3].

A singular building sitting on the top of the existing roman ruins layers—the XVI-th Century Rural House, experienced successive expansions until the XIX-th century. This Rural House is one of the few examples of ancient rural constructions in Algarve and, it was also described in the middle 1970s by the Portuguese Nobel

Fig. 1 The archaeological site of Milreu, Algarve. **a** Plan and building's location. **b** The IV-th century A.D. water temple

Fig. 2 The Roman ruins of Milreu and the XVI-th century rural house. **a** The rural house and the water temple. **b** The Roman mosaics—the baths, the dauphins' carpet and a "smiling" squid

prize winner José Saramago in his book "Journey in Portugal", as a singular abandoned building were the shepherd and their flocks slept, [4], see Fig. 2a.

The Rural House

Under EU programs to improve vacation regions, the Milreu archaeological site began to be renovated in the 1990s with: (a) the conservation of the existing mosaics and other structures; (b) the installation of a well-detailed pedestrian circuit; (c) the XVI-th century Rural House rehabilitation; and, (d) the construction of a new visitor's interpretation center with multimedia support. The Rural House rehabilitation process was very important and particular care was taken to define the methodology and the design solutions to be followed. The building plan has a N-S longitudinal axis orientation and the roof is made with a singular design solution—three inclined surfaces [2, 5], see Fig. 3.

One of the longitudinal exterior walls (west side) is tapered inwards reducing the north end wall length by 2.30 m. The external building dimensions in plan are approximately: (a) total length, $L = 25.40$ m; (b) longer width (south end wall), $B_1 = 11.90$ m; (c) shorter width (north end wall), $B_2 = 9.60$ m; (d) external contour wall height, $H = 4.0–7.0$ m; and, (e) average wall thickness, $t = 0.60$ m. The walls' height is variable because of the different archaeological excavations around the building [2, 3], see Fig. 4.

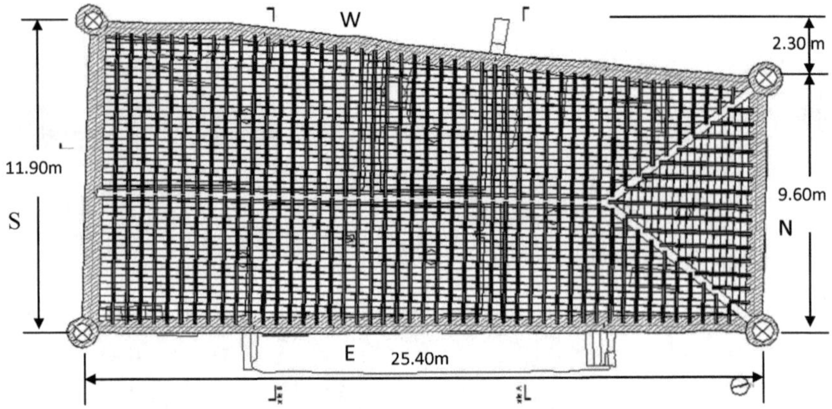

Fig. 3 Milreu rural house—existing RC roof before rehabilitation

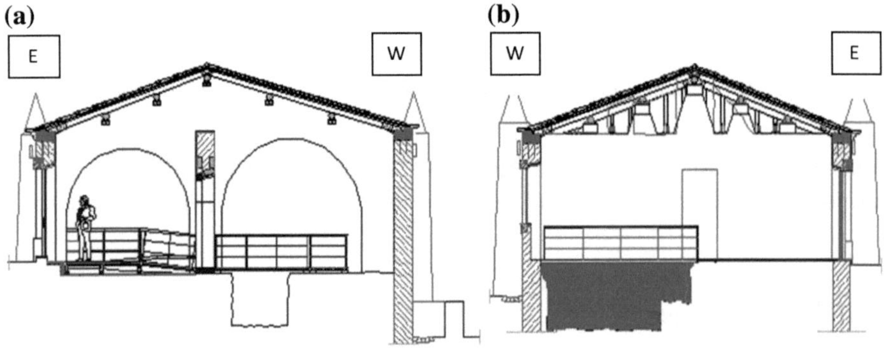

Fig. 4 Different transverse walls planes in the building. **a** South zone. **b** North zone

The roof height above the top surface of the exterior walls is constant, $H_r = 2.10$ m, which yields different slope ratios considering two extreme transverse cross sections: (a) south end wall (extended dimension)—shallower roof slopes (1: ~2.83); and, (b) north end wall (narrower section)—steeper roof slopes (1: ~2.3). This variation in the roof system transverse dimensions, while maintaining the level for the top crowning longitudinal beam leads into a warped roof surface, when the building becomes narrower in plan, see Figs. 3 and 4. The interior transverse walls have an approximate thickness identical to the exterior ones of 0.60 m. They are oriented along the E-W direction with different spacing along the longitudinal interior wall length [2], see Fig. 5.

The Rural House used ancient-type building materials: (1) masonry walls made of small stones set with lime mortar; (2) several layers of lime mortar wall finishing painted with white wash lime paint; (3) a wood roof structure combined with an external layer of reed bars; (4) overlapping clay roof tiles for exterior cover and rain

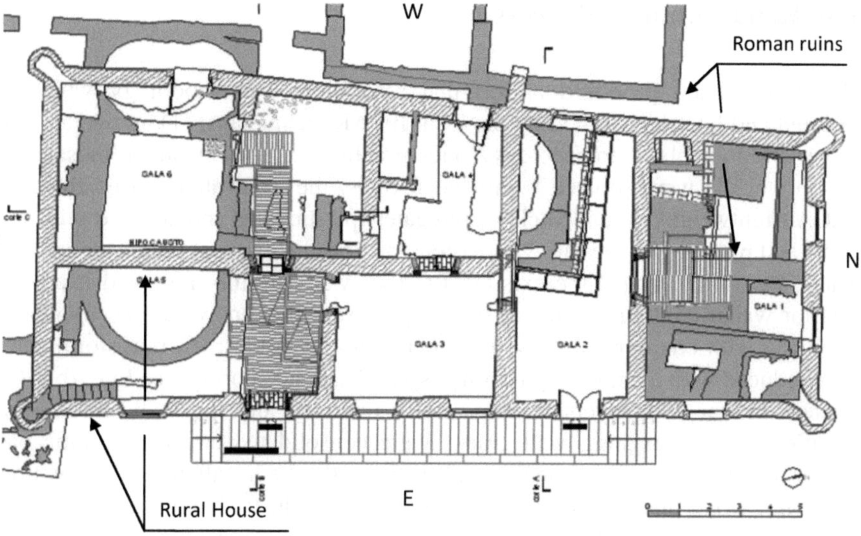

Fig. 5 The rural house and the underlying Roman ruins (general plan survey)

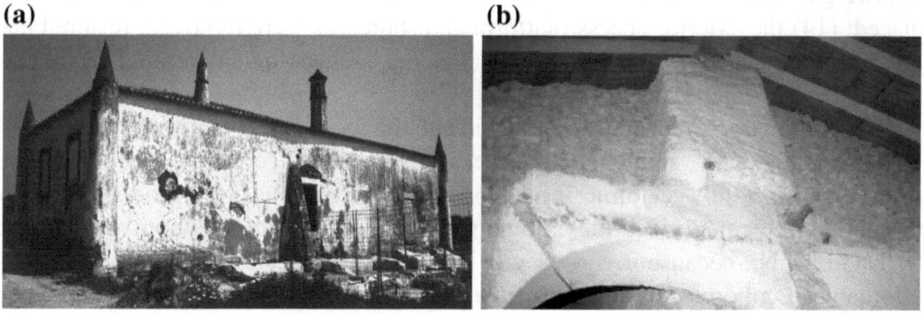

Fig. 6 The rural house before intervention with the heavy roof solution. **a** 1970s—abandoned building (west elev.). **b** Precast concrete roof with clay hollow blocks

protection; (5) the reduced height interior masonry walls used short masonry pylons that reached the ceiling skin saving material; (6) in each corner of the building plan, a conical hollow turret, with narrow vertical openings (slits) allowed the existence of an interior defensive observation post to warn possible of incoming hostile attacks, see Figs. 5 and 6.

The different spacing used in the interior transverse walls with a thickness of 0.60 m is related to the building spatial organization and the existing functions. Most of these interior walls have a reduced maximum height similar to the exterior walls. This solution is very practical because it allows transverse roof ventilation which is deemed necessary during the hot Algarve summers.

The Rehabilitation Process

The design team proposed an original rehabilitation methodology which was adjusted and later on adopted by the local IPPAR technicians: (1) an overall global survey (topographic, archaeology, photographic); (2) the Roman mosaics and masonry walls were protected with geotextiles, the excavated spaces filled with LECA (lightweight expanded clay aggregates) plastic bags and covered with wood floors and planks, [5], see Fig. 7a; (3) construction site external perimeter and lower planes were drained; (4) the existing RC roof was carefully removed leaving the exterior walls' contour beam; (5) the deeper excavated areas had the foundations stabilized with contour masonry walls made of clay bricks; (6) rain storm actions were reduced with a temporary roofing assemblage and shoring, see Fig. 7b; (7) the existent mortar wall layers were carefully removed and the wall structure data recorded; (8) the existing wall cracks were consolidated and a new lime based mortar finishing was placed on the wall surfaces; (9) the installation of a new timber roof, with wood planks overlapping to protect from dust; (10) traditional clay roof tiles were used; (11) all the carpentry work (doors and windows wood frames) and finishing (lighting, pavement tiles) followed traditional techniques; (12) an organized visitors tour and access to different rooms was made with the installation of lightweight steel platforms and observation decks; (13) information signs were placed; (14) the exterior access platform and staircases were removed, restored and put back in the same location; and, (15) the archaeological artifacts and new findings were restored.

The building was thoroughly surveyed and the roof system that was placed in the 1970s, made with small height precast prestressed concrete (PPC) beams (h = 0.12 m), hollow ceramic infill blocks and a concrete layer produced locally topping both elements, see Figs. 3 and 6b. This construction solution was adopted by D.G.E.M.N., because it was easy to build, fast to assemble and, above all, very affordable. Economic reasons and time constraints to stop the interior building's degradation and the Roman structure's exposure to rain and other natural elements

(a) (b)

Fig. 7 The construction phase—LECA plastic packed bags and roof shoring. **a** Roman mosaics and ruins protection. **b** Temporary roof cover and shoring (SW view)

led to adopt this "temporary" solution, while more definite studies were developed. Nearly two decades later, the PPC roof was carefully removed and replaced by a traditional timber roof structure with local materials—clay tiles, wood assemblages and members. The existing masonry walls were fractured and unstable. Any excessive vibrations with electrical hammers could trigger a local collapse mode in the walls or, in the limit, an extensive wall collapse scenario.

Although the construction deadlines were successively extended, the initial rehabilitation cost estimate was under the limit of 250,000 euro, for a gross area of nearly 300 m^2.

Seismic Performance Improvements

Before the rehabilitation process, the existing roof dead load (DL), q(DL), created by a precast prestressed concrete (PPC) slab, with shallow rectangular support beams topping the contour masonry walls and traditional clay tiles for the exterior skin, was 4.50 kN/m^2. The static load path introduced by this PPC roof system can be considered different from the original building solution. The roof loads are mainly supported by three longitudinal N-S axis: two exterior façade walls and one interior central wall.

This 1970s structural concrete roof solution supported on the longitudinal masonry walls had secondary effects. The building's longitudinal mid-axis had the masonry walls raised as far as the roof apex with clay hollow blocks. After the PPC beams and hollow clay blocks were placed, the interior longitudinal reinforced concrete (RC) top beam was cast together with the precast roof slab elements and the lower RC longitudinal beam topping the exterior wall contour. This shallow roof with the transverse precast beams developing a horizontal thrust at the top of the 4.00 m tall masonry walls, which generated rotations and cracks over the wall surfaces. In the joint regions located where the longitudinal and transverse walls met, detachment, cracking and material crushing could be observed. The masonry walls' finishing layers were removed and replaced by a new lime mortar layer, the joints and cracks filled in and repaired, after some "in situ" tests were carried out in order to evaluate real local limitations—materials, workmanship [5, 6], see Fig. 8.

The design proposal for the building's rehabilitation, introduces a timber roof made with maritime pine, which restores the assumed original solution and enhances the architectural value of this historic construction. The exterior covering material is made by traditional clay tiles (0.8 kN/m^2), a support layer made by overlapping wood boards (0.24 kN/m^2) fixed over rectangular section wood purlins (0.06 m $\times$ 0.12 m) closely spaced at 0.40 m o.c., (0.15 kN/m^2).

Under these purlins are located circular wood beams (D = 0.20–0.25 m), running longitudinally and supported at one third their span along the transverse walls, see Fig. 4, which corresponds to an equivalent D.L. of 0.25–0.40 kN/m^2. The timber roof solution with the same clay roof tiles as the previous heavier PPC roof, has a D.L. average value q(DL) = 1.50 kN/m^2. There is a large decrease on the roof

Fig. 8 "In situ" lime mortar tests on masonry walls

D.L. which decreases from 4.50 to 1.50 kN/m^2. The much lower value combined with the mobilization of both longitudinal as well as transverse walls drastically decreases the D.L. value on the top of the masonry walls, see Figs. 4 and 5.

This "smooth" static load path distributes the dead load evenly through the masonry walls top regions, instead of concentrating the loads along the three major longitudinal support walls.

The seismic design improvement was obtained with this design solution by: (1) replacing the masonry walls surface mortar layers by a new continuous lime mortar finishing; (2) the existing joints and cracks in the walls and connection regions were infilled and consolidated; (3) the PPC roof system removal greatly reduced the roof mass from 4.50 kN/m^2 to nearly 1.50 kN/m^2; (4) the timber roof solution provided a more evenly distributed load path throughout all walls; and, (5) the proposed architectural solution became more adequate and provided an added cultural building value to a previous abandoned construction.

Before the timber roof was installed, the masonry walls were consolidated and a better confinement level was provided with a new continuous lime mortar surface layer. The total roof DL was greatly reduced by 2/3 from 121.5 to 40.5 tons, which improves the building seismic response by reducing the earthquake actions at roof level.

Conclusions and Recommendations

This study's purpose is to present a unique Roman archaeological site in a major vacation region in the South of Portugal. Under a special EU–Portuguese Government supported project an extensive renovation program was implemented. Several major problems had to be solved to bring this historic location and building rehabilitation work to excellent results, see Figs. 9 and 10.

Fig. 9 The XVI-th century rural house after rehabilitation—east elevation

Fig. 10 The timber roof—interior ceiling view and transverse wall supports (north room)

The design solution required an extensive study of other local traditional constructions. However, the most important achievement was the choice made by using a maritime pine timber structure with a reduced DL value of 40 tons. That was only one third the DL value of the existing structural concrete roof system. The equivalent seismic load action of this mass at roof level was greatly reduced.

Other construction requirements had to be met such as: (1) the replacement of the external mortar layers on the masonry walls by a new lime mortar finishing; (2) the consolidation of joints and cracks to improve the structural response and stability; and, (3) better aesthetics obtained with the maritime pine timber design which greatly improved this historical building image, see Fig. 10.

The rehabilitation process was extensive and costly in part due to the lack of experience with local contractors on using traditional materials: timber members and assemblages; lime mortar wall coverings; iron works for doors and windows; and, limestone masonry.

A great deal of patience and in situ training was required to achieve the desired final results. Probably, the most impressive results were obtained with the timber roof system being adopted and the carefully chosen lighting system.

At a total cost of approx. Euro 250,000—for a 300 m^2 building, the final result was excellent both for the owner and the visitors. This experiment can be used for other public rehabilitation works in this Portuguese region.

References

1. Parreira R (2004) Notas para um plano de salvaguarda e valorização das Ruinas de Milreu (Estoi, Faro), (in Portuguese), Revista Património – Estudos no 6. IPPAR, Lisboa, pp 150–156
2. Ramalho M (2004) Casa Rural de Milreu: percurso histórico de um edifício (in Portuguese). Revista Património – Estudos no 6. IPPAR, Lisboa, pp 162–167
3. Teichner F (2004) Breve descrição dos vestígios arqueológicos identificados sob a Casa Rural de Milreu (Estoi, Faro), (in Portuguese). Revista Património – Estudos no 6. IPPAR, Lisboa, pp 157–161
4. Saramago J (1990) Viagem a Portugal (illustrated with images) (in Portuguese). Editorial CAMINHO, Lisboa
5. Bastos J (1994) A Recuperação da Casa Rural de Milreu, no Algarve (Diagnóstico e Projecto) (in Portuguese). Instituto Português do Património Arquitectonico e Arqueológico (IPPAR), Direcção Regional do Algarve, Faro
6. Bastos J (2004) A recuperação da Casa Rural de Milreu: uma visão do projectista (in Portuguese). Revista Património – Estudos, no 6. IPPAR, Lisboa, pp 172–177

Design and Maintenance of a Long-Span Timber Roof

Jorge N. Bastos

Abstract The historical Citadel of Almeida is located in the northeast part of Portugal, near the Spanish border and close to the town of Ciudad Rodrigo. This geographical location makes this region the natural entrance gate of invading armies to access the heart of Portugal and trying to reach the capital town of Lisbon. Therefore, the Portuguese border region has several military fortresses and, probably, the most unique one is twelve-point granite star of Almeida. Under a EU program to re-vitalize historical villages, the horse-riding school building (the *manège*) restoration project received the largest financing grant, in Phase 1, reaching 2.25 m Euro. Due to the lack of business investments, the hinterland Portuguese regions have shown during the past three decades a continuous population decrease and a migratory flux into the coastal regions, or even to other countries. The main purpose of this EU program is to help attracting new residents to these interior regions and to maintain the current residents in their locals with adequate business activities. The rehabilitation of several historical landmarks helps to provide a local sense of identity and improved living conditions by installing better infrastructures (electrical power, water supply and sewage treatment plants, telecommunication systems). Within these general program objectives, the leisure activities are one of the best "attractors" for tourism. The rehabilitation of the Almeida *manège* is integrated into the combination of three major guidelines: (a) historical heritage; (b) tourism and economy; and, (c) leisure activities. The designer rehabilitation criteria was: (1) to use a traditional timber solution, e.g., a scissor's type truss; (2) to use improved joint connections, e.g., split rings; (3) to carefully detail steel supports and connections; and, (4) to innovate as required by the architectural functional program. After more than sixteen years under continuous use, the *manège* ensemble built in the Citadel of Almeida with extensive use of timber material can still be considered both an aesthetically sound, cultural

J.N. Bastos (✉)
School of Architecture, University of Lisbon, Polo Universitário do Alto da Ajuda, Rua Sá Nogueira, PT-1349-063 Lisbon, Portugal
e-mail: jnbastos@fa.ulisboa.pt
URL: http://faculdadedearquitectura.ulisboa.pt/

H. Cruz et al. (eds.), *Historical Earthquake-Resistant Timber Framing in the Mediterranean Area*, Lecture Notes in Civil Engineering,
DOI 10.1007/978-3-319-39492-3_35

enriching, and economically feasible rehabilitation solution to be enjoyed by all kinds of visitors

Keywords Timber roof · Rehabilitation · Heritage · Low seismicity · Aesthetics

Introduction

In the 1990s, a design proposal for the rehabilitation of a horse-riding school (the "*manège*") located in the historical Citadel of Almeida was submitted and approved by the City of Almeida and the Portuguese Heritage Agency—IPPAR. A new heritage restoration policy—the "Aldeias Históricas" (Landmark Villages) was implemented by the Portuguese Government, through European Union funding, to develop the hinterland regions and attract new populations. In the first phase, the horse-riding school building had the largest budget for the restoration works—2.25 m Euro. It became a major tourist attraction and it was combined with the rehabilitation of different fortress sectors, in this unique landmark Portuguese village. The construction constraints and the degree of isolation that Almeida, increased the challenge of this restoration process, see Fig. 1.

In the first visit to the construction site an abandoned building could be observed, Figs. 1 and 2, where only large granite blocks stacked on the top of each other existed. Several design questions about the most adequate intervention solution were made by the Owner: (a) to intervene or not; (b) if an intervention

Fig. 1 The Citadel of Almeida (1990s) and the *Manège*, Beira Alta Province, Portugal

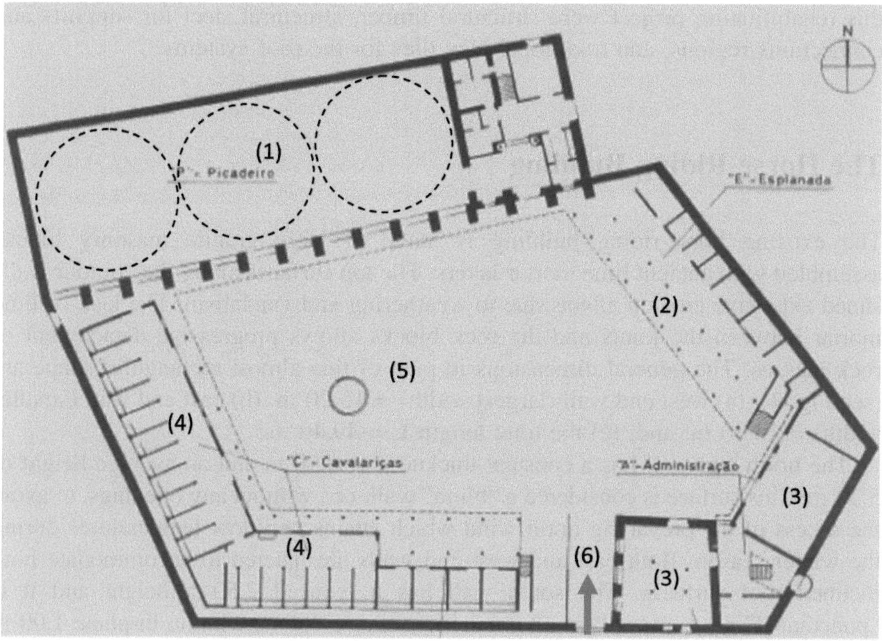

Fig. 2 The *Manège* ensemble (design proposal)—(*1*) the horse-riding building; (*2*) esplanade; (*3*) offices; (*4*) horse boxes; (*5*) courtyard; and, (*6*) entrance gate

solution was chosen which materials must be adopted; (c) the structural solution is completely visible and what would be the most adequate truss solution; (d) all the parameters being considered, the adopted solution would be under an economic reach; (e) the access conditions would impose any limits on the size/weight of the construction materials and structural members.

The *manège* ensemble is located in one of the Fortress ramparts where the artillery supplies were stored. The main building (1)—the horse-riding space located in the northern zone was identified with this function because of the proportions of the interior space which are 1:3, usually known as a three circles space, to be adequate for equestrian training, see Fig. 2, and an interior horizontal "line" carved on the granite blocks surface, 1.50 m from the ground surface bordering the whole contour. This line would serve to help fix a slightly inclined wood panel to avoid the horse rider to be crushed against the wall during training sessions. Adjacent to the training ground and in the same building was the guard's dwellings, remodeled to have toilet spaces and access areas to the exercise area and, on the first floor, an observation deck/lounge.

The initial design proposal for this complex was organized around the large central courtyard (5), with a water tank: (a) a long two story esplanade (2), with storage spaces underneath on the ground floor; (b) the offices area (3), in the SE zone; and, (d) the horse boxes (4), in the SW region. The main materials used for this rehabilitation project were structural timber, structural steel for supports and connections regions, and traditional clay tiles for the roof systems.

The Horse-Riding Building

The existing horse-riding building is made of large granite masonry blocks assembled with ancient lime mortar layers. The top surfaces along the contour walls shoed extensive erosion effects due to weathering and vandalism. The lack of lime mortar between the joints and the rock blocks allows progressive detachment of rock masses. The general dimensions in plan of this almost rectangular shape are (see Fig. 3): (a) west end wall (largest width) = 15.20 m; (b) east end wall (smaller width) = 13.80 m; and, (c) the total length L = 49.40 m.

The north end wall has a constant thickness of 0.90 m and an average height of 5.20 m. This surface is considered a "blind" wall, i.e., without any openings, to avoid the access of the prevailing north wind which attains very low temperatures during the winter season. Both east and west end walls are tapered to accommodate both inclined roof surfaces. The south wall has a reduced 2.50 m height and it is "punctuated" by fourteen "giant" granite masonry columns with an in-plane 1.00 m

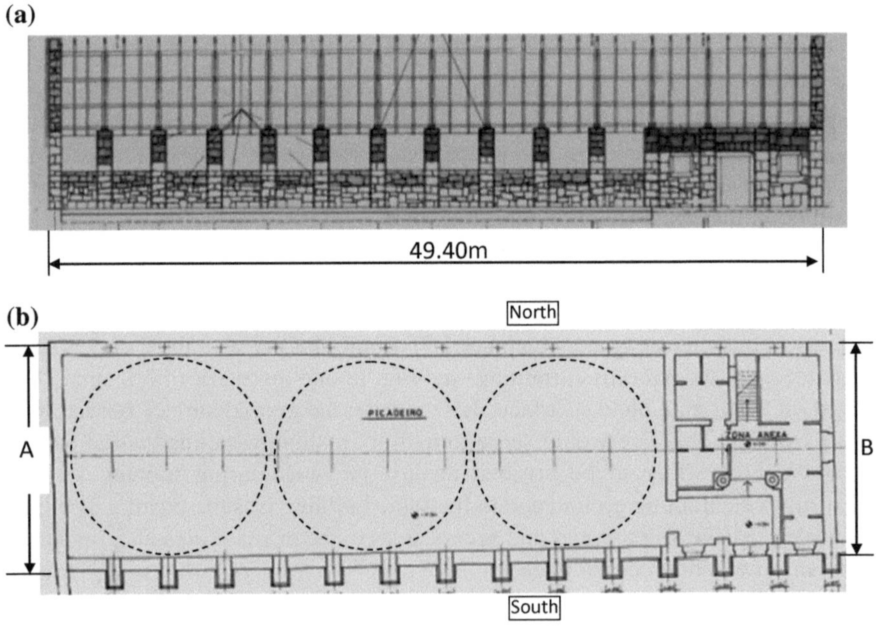

Fig. 3 The horse-riding school—building characteristics. **a** Longitudinal south elevation. **b** General plan (A = 15.20 m; B = 13.80 m)

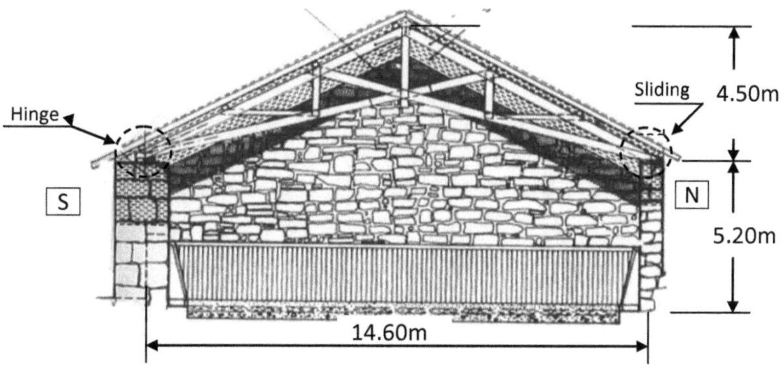

Fig. 4 The timber scissor's truss with largest span—*manège* area (west zone)

by 1.80 m cross section and a total height of nearly 5.00 m. This wall with a 0.70 m thickness encloses the training ground space along eleven successive spans. Near the last northern three spans, the height increases to nearly 5.00 m. The central span has a wide access door and the two side spans one raised window in each span, see Fig. 3.

The interior training ground space has the characteristic three circle structure with the interior plan dimensions of 37.60 m length by a variable decreasing width of 13.60–12.50 m. This variable transverse width created a major design problem: if the building top crowning line must be kept horizontal, the roof surfaces have to become warped surfaces; on the other hand, to have plane roof surfaces the top roof crowning line needs to be inclined. The first option was chosen because the 3.50 m purlins, the top secondary beams, the roof timber skin and the clay roof tiles could easily adapt to the warping effect. Eleven timber trusses had to be designed separately because the span varied by 0.10 m from one truss to the other. The first truss (west side) has a 14.60 m span between supports, whereas the last one (east side) has a reduced value of 13.50 m, see Fig. 4.

The Design Proposal for a Low Seismicity Zone

The wood material adopted for the truss design was first grade southern yellow pine, from Oregon State (USA), because very long members with 10–12 m were available from the supply company. The specified design stress, $f_{wd} = 10.0$ MPa (100 kgf/cm^2). The external contour members, i.e., the superior inclined "leg" and the inferior raised "line" are made of twin rectangular 0.12 m × 0.30 m cross sections, separated by a 0.12 m gap between them. This gap allows the insertion of the smaller interior secondary members which are subjected to much lower axial forces. The wood truss members were connected with 0.10 m (4 in.) split rings centrally pierced by 16 mm diameter zinc coated threaded rebars, fixed with washers and nuts on both ends, [1]. This assembling technique, extensively used in

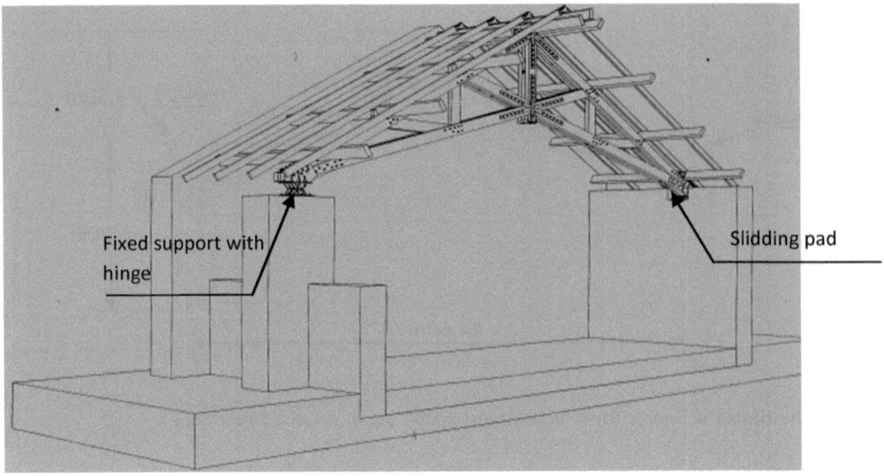

Fig. 5 The scissor's type timber truss—3-D rendering

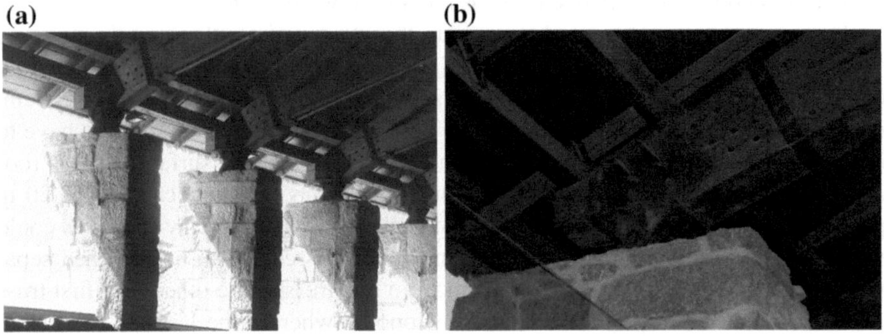

Fig. 6 South façade—fixed supports with hinges details. **a** Giant column with fixed hinged supports. **b** Timber truss and steel support detail

the US during WWII to build large span industrial plants, proved to be both feasible and economically sound. The adopted timber truss solution was of the scissor's type because the available *manège* space is smaller than the usual dimensions for a standard three-circle model. The other ancient Lisbon spaces with similar functions have 17.0 m × 51.0 m ("Museu dos Coches") or 18.0 m × 54.0 m ("Colégio dos Nobres"). Therefore, by lifting the lower truss line, the smaller Almeida horse-riding school would get an increased spatial open space. Both a 1:100 scale model and a 3-D rendering were developed to study this scissor's type timber truss, see Fig. 5.

This timber scissor's truss design concept required specially detailed structural steel supports to accommodate the different loading actions combinations. A fixed support with a hinge to allow small in-plane rotation movements was placed on the top surface plane of the "giant" columns, see Fig. 6, and the other movable support can absorb horizontal movements (ex. temperature, creep, humidity variations) with a 0.05 m (2 in.) thick Neoprene rubber pad. The expected horizontal seismic actions are considered to be extremely low, as the seismic coefficient, β, is less than 5 %, due to the actual rock foundations and the building location in a Portuguese region with the lowest specified ground motion acceleration value.

The other two load actions being considered were due to wind forces and snow loads as legally required by Portuguese law and design code, R.S.A.-1983 [2]. The wind action is relatively reduced by the roof slope angle, $\alpha \approx 27°$, which is very favorable to obtain reduced values with the wind pressure and suction coefficients. This roof angle corresponds to a 1:2 (H:L) slope, which is very easy to specify in the construction site. The required snow loads are a function of the local altitude, h = 750–800 m. For this altitude value, the obtained snow loads, s, [horizontal projection (h.p.)] are respectively, $s_1 = 2.1$ kN/m^2 (210 kgf/m^2) and $s_2 = 1.5$ kN/m^2. If this value is compared with the Spanish code snow loads, for the neighboring town of Ciudad Rodrigo, which has the same altitude value, a much smaller value is obtained. The M.V.101-1962 [3] design snow load value is uniform and equal to s = 0.72 kN/m^2 (72 kgf/m^2), which is nearly 1/3–1/2 the Portuguese code values.

The local climate conditions are characterized by a dry continental climate, i.e., extreme peak temperature values between the summer and the winter seasons, with reduced rainfall values. The two locations near the border are at the same altitude and, the Spanish design code probably reflects more accurately the actual snowfall conditions, see Fig. 7a, b.

The extensive number of technical drawings and accurate detailing of joints and connections proved to be a correct decision for this rehabilitation work located more than 420 km from the Lisbon design office. The timber members preparation and pre-assemblage plant yard was located in Coimbra which is conveniently at mid-distance between Lisbon and Almeida, and large size pre-assembled members were carried by a truck that could go through the access Fortress gates.

The Rehabilitation Process and Maintenance

The ensemble rehabilitation was initiated in early January 1994 (design phase) and the completed in April 1999 (construction phase). During the year of 1994, the contractor had some difficulty to find suppliers for very long timber members with adequate sections, as well as, for the split ring connectors. As soon as the construction works started several other problems emerged for the design solutions for the other ensemble buildings. These facts resulted from misleading decisions taken by the Heritage Agency (IPPAR) which assumed that a fast track process (design-build) could be used with local designers in a difficult rehabilitation process.

When the local designer proposed a reinforced concrete (RC) 15 m-span truss for the horse-riding building the IPPAR agency assumed that it was unsuitable.

Fig. 7 The Almeida *manège* (2014). **a** The building main south elevation and the courtyard. **b** The building north, "blind" façade (*left side*) and west (*right side*) elevation

During the construction phase, the *manège* constructions other than the horse-riding building were being "designed" on a daily basis during construction site visits, and the contractor submitted new invoices based on change-of-the-order actions, the IPPAR agency prompted the rehabilitation process to be halted. An invitation followed to design team to develop a global solution for the remaining part of the *manège* ensemble. Consequently, all the other parts were re-designed using not only timber structures (ex. the esplanade, the horse boxes), as well as, RC frames (ex. the offices, the guard's house) and, sometimes, a hybrid combination of both materials when it was deemed necessary, Figs. 8 and 9.

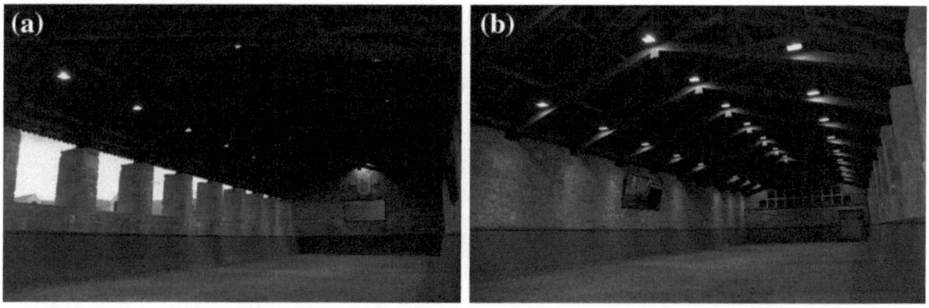

Fig. 8 The interior space and the large span timber scissor's trusses (2014). **a** The west end wall. **b** The east end wall and observation area

Fig. 9 The other re-designed spaces with timber and RC materials (2014). **a** Esplanade and administrative offices. **b** Guard's house—RC frame structure

Fig. 10 The Almeida *manège* near the inauguration day (1999)

After nearly twenty years under continuous use, the exposed timber structures around the courtyard (ex. esplanade) were painted with a protective coating to extend the useful material's lifespan. The scissor's timber trusses are expected to undergo an extensive surveying and maintenance program.

Conclusions and Recommendations

A structural timber roof system was proposed, after comparing with other different material design options (steel, reinforced concrete and glulam), because of aesthetics, economy, feasibility and an adequate heritage material, for the Fortress of Almeida *manège* ensemble, see Fig. 10.

Due to the relatively small size of this horse-riding indoor space (approx. 13.50 m × 40.50 m), a scissor's truss was adopted to increase the available height and improve the spatial conditions. The timber members were made of southern yellow pine from 12.0 m long pieces were cut to the proper size members. As the building plan was slightly tapered along its width, with a 1.40 m difference in the span length from both ends, fourteen different plane trusses had to be designed individually, for the entire building length. Steel custom designed supports were used: the fixed hinged support topped the "giant" granite masonry columns and the other support, on the top of the exterior wall contour had a sliding neoprene rubber pad to allow for horizontal movements (temperature, humidity, creep). Design seismic loads were relatively small as compared with the required larger design code values, particularly, for the snow loads actions.

This building rehabilitation process was one of the most successful projects sponsored by the Portuguese Heritage Office in the Landmark Villages ("Aldeias Históricas") EU program due to the excellent craftsmanship and the use of structural timber. In this solution the structure is completely exposed to the public view and detailing plays a major role. Low maintenance and a simple technical assistance to the contractor's team also deserves to be mentioned. Seldom major timber structures are designed in current Portuguese in heritage rehabilitation works and the main objective of this study is to demonstrate its large potential as an adequate heritage rehabilitation material.

References

1. Mateus TJE (1961) Bases para o dimensionamento de estruturas de madeira, (in Portuguese). Laboratório Nacional de Engenharia Civil (L.N.E.C.), Lisbon
2. R.S.A. (1983) Regulamento de Segurança e Acções para Edifícios e Pontes (in Portuguese). Imprensa Nacional – Casa da Moeda (I.N.C.M.), Lisbon
3. M.V. 101 (1962) Normas Españolas – Acciones en la Edificación, (in Spanish). Ministério de la Vivienda, Madrid

Part VI
Timber Structures

Assessment of the Mechanical Characteristics of Timber Floors Through Simple Dynamic Identification

José Coelho, João M. Guedes, Bruno Silva and Tiago Ilharco

Abstract The non-destructive in situ characterization of timber floors is difficult and may involve expensive procedures, such as loading tests. Alternatively, it is possible to estimate some mechanical properties by assessing the characteristics of timber beams through sonic or penetration tests (e.g. pilodyn and resistograph); however, the existing correlations are still not very reliable. Thus, this project aims assessing the characteristics of timber floors using dynamic identification procedures, a simple and low cost technique when compared to loading tests. In particular, and through the identification of the main frequencies, this research aims estimating the mechanical behaviour of timber floors, including the characteristics of the beams to walls connections.

Keywords Timber floors · Dynamic identification · Mechanical assessment

J. Coelho · J.M. Guedes · B. Silva (✉)
Faculdade de Engenharia da Universidade do Porto (FEUP), Rua Dr. Roberto Frias, s/n, 4200-465 Porto, Portugal
e-mail: bsilva@fe.up.pt

J. Coelho
e-mail: ec10163@fe.up.pt
URL: http://www.fe.up.pt/

J.M. Guedes
e-mail: jguedes@fe.up.pt

T. Ilharco
NCREP—Consultoria em Reabilitação do Edificado e Património, Lda, Praça Coronel Pacheco, 2, 4050-453 Porto, Portugal
e-mail: tiago.ilharco@ncrep.pt
URL: http://ncrep.pt/

H. Cruz et al. (eds.), *Historical Earthquake-Resistant Timber Framing in the Mediterranean Area*, Lecture Notes in Civil Engineering,
DOI 10.1007/978-3-319-39492-3_36

Introduction

The rehabilitation of old buildings is becoming a priority issue in Portugal. It's a process that demands the assessment of the state of conservation and the understanding of the mechanical behaviour of the materials and structural elements that constitute these buildings. Among these elements, timber floors are particularly important, having a relevant participation in the behaviour of the entire structure. The floors are not only responsible for creating functional spaces, but also for establishing connections between walls, promoting a better box behaviour.

The non-destructive in situ characterization of timber floors is difficult and may involve expensive procedures, such as loading tests [1]. Alternatively, it's possible to estimate some mechanical properties by assessing the characteristics of timber beams through sonic or penetration tests (e.g. pilodyn and resistograph); however, the existing correlations are still not very reliable [2]. This study aims at assessing the characteristics of the old timber floors through dynamic identification, i.e. by measuring the dynamic behaviour of the floors. With a small device, such as a seismograph or an accelerometer, this work seeks measuring the vibration frequencies of timber floors and analysing the results to estimate the in situ characteristics of these elements. Using a structural analysis software, the frequencies obtained experimentally are compared to those obtained through the numerical simulation of the tested floor, allowing estimating some of its mechanical characteristics. In particular, this exploratory research foresees the estimation of the behaviour of timber floors by verifying the influence of the various elements that constitute them, namely their characteristics and links. The benefits of this technique are, among others, its fully non-destructive characteristics, the easy transportation of the equipment, the fast processing of the results and the low cost of the whole process. The present research is based on results obtained through the application of this technique to a set of buildings located at Porto.

Timber Floors Structural Systems

A set of 3 old buildings located at Porto, Portugal, that was undergoing renovation works were selected for this study. This allowed assessing to real cases, namely to the in situ dimensions and characteristics of floors structural elements. The buildings under analysis, as most of the old buildings in Porto and Porto region are made of stone masonry walls at the contour and timber walls at the interior (although exceptions exist—stone masonry walls inside and timber walls outside) and timber floors and roofs [3]. The floors' structural system is typically made of timber beams with a span of not more than 5 m and struts in between. The beams are materialized by tree trunks, usually Chestnut or Pine, either fully round, partially cut or, when part of a floor built around the beginning of the XX century, fully cut into rectangular section elements. To guarantee the horizontality of the floor, wood billets are placed between the beams and the floorboards. The boards are nailed to the beams, as well as the ceiling of the lower floor (a structure made of wooden thin laths placed perpendicularly to the beams supporting a wood and plaster covering). Figure 1 shows some of the elements that constitute these timber floors.

Dynamic Identification

The procedure used in the dynamic identification is quite simple. A seismograph measures the vertical time history accelerations of the floors and/or of the floor components. Then the accelerations are transformed into frequencies using a FFT algorithm, and the results are compared to those obtained using a numerical model of the structure. The numerical simulations were performed with the software *Autodesk Robot Structural Analysis*. It was used a seismograph from the Laboratory of Earthquake and Structural Engineering of the Faculty of Engineering of Porto University (*GeoSIG-GSR18*); the FFT analysis was performed using *GeoDAS* software. Before each test, it was assessed the floor's geometric/structural characteristics: span, cross-section and spacing of beams, cross section and spacing of struts and thickness of floor boards. Some of these characteristics were only

Fig. 1 Timber floors: **a** with wooden floorboards; **b** with plaster ceiling view from *above*; **c** and **d** plaster ceiling view from *below*

possible to assess because the buildings were under renovation works that involved strengthening or any other type of intervention at the floors' level. The numerical models were built taking into account the inspection and diagnosis reports made by *NCREP—Consultoriaem Reabilitação do Edificado e Património, Lda.*, the design office that was in charge of the intervention projects involving the case studies selected in this campaign.

Testing Campaign

Three sets of tests are detailed in this paper. The first set was performed on a 4.3 m span floor at level 1 above ground floor of a building located at *Torrinha* street. A timber-based partition wall located below the floor, transversal to the timber beams, was later demolished to allow a new architectural layout. Although the wall was not originally placed to support the beams, the demolition increased the floor vibration level, demanding the reinforcement of the floor structure with 3 new glulam timber beams. The floor was tested before and during the intervention, i.e. at its original state and, afterwards, with the new beams and the original plaster ceiling (that was not removed), but without floorboards (Fig. 2a).

The second set of tests was performed on two different floors (at level 1 and level 2 above ground floor) of a building located at *Rosário* street, Fig. 2b. Also in this case, the floors were tested in the before and during the intervention phases. In the intervention phase, both tested floors had no floorboards and only level 1 floor kept the plaster ceiling. Moreover, the floor at level 1 had a 2.8 m span and only one of the extremities of each beam was embedded in the masonry walls; the other extremity was simply supported on the wall taking advantage of the reduction of the wall thickness from level 0 to level 1. The floor of level 2, also with a 2.8 m span, had both extremities of the beams embedded in the masonry walls and was tested with and without struts.

The third set of tests was performed on a 4.3 m span floor at level 1 above ground floor of a building located at *Monte dos Judeus* street. The floor was entirely

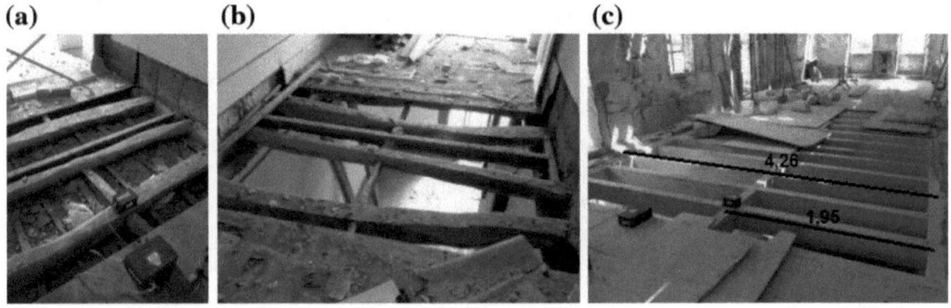

Fig. 2 Pictures of the three case studies: **a** Torrinha, **b** Rosário and **c** Monte dos Judeus

new and made of glulam timber beams and struts, Fig. 2c. When the test was performed, there was no ceiling or floor covering; the structure was made of border beams connected to the masonry walls, transversal beams linked to the border beams and struts between transversal beams at mid-span. The elements were connected to each other through metallic connectors.

Geometrical Assessment

The geometrical layout of each floor structure was assessed. It were considered 5–7 transversal alignments defining intersection points to each beam axis that were fully characterized. For each point it was measured the relative position to a reference longitudinal axis and estimated the dimension of its cross-section. Table 1 and Fig. 3a exemplify the procedure and present the results for *Torrinha* street building floor.

Table 1 Cross-section "S" and distance "X" in cm to the red reference line in Fig. 3 of the beams' intersection points to the transversal black lines (Torrinha floor)

		1	2	3	4	5
V1	S	ø18	ø18	ø17	ø15	ø15
	X	66	60	50	46	42
VN1	S	12 × 17	12 × 17	12 × 17	12 × 17	12 × 17
	X	87	85	79	77	77
V2	S	ø15	ø17	ø19	ø21	ø22
	X	135	124	117	114	96
V3	S	ø18	ø17	ø16	ø16	ø14
	X	193	187	174	173	159
V4	S	ø12	ø13	ø16	ø16	ø19
	X	211	209	203	201	199
VN2	S	12 × 17	12 × 17	12 × 17	12 × 17	12 × 17
	X	248	241	230	224	223
V5	S	ø20	ø19	ø19	ø16	ø16
	X	302	306	295	282	280
V6	S	ø15	ø15	ø17	ø19	ø20
	X	365	370	356	351	343
VN3	S	12 × 17	12 × 17	12 × 17	12 × 17	12 × 17
	X	391	395	395	398	401
V7	S	ø16	ø15	ø13	ø13	ø12
	X	426	427	428	434	444

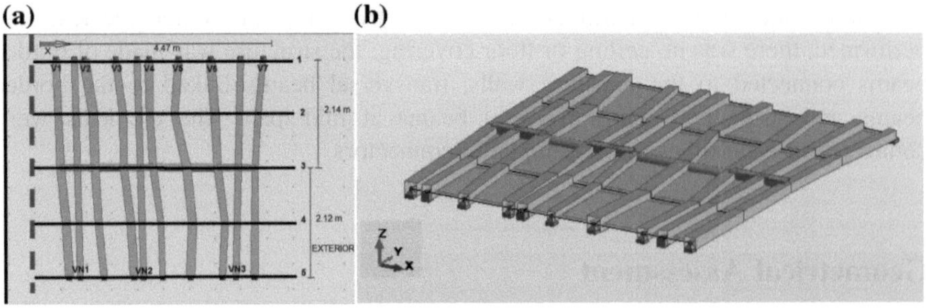

Fig. 3 Torrinha floor: **a** reference lines and measures (see Table 1) and **b** numerical model

Numerical Simulation

Using still *Torrinha* street as reference, Fig. 3b shows the numerical model that was constructed to simulate the floor structure without the floorboard, but still with the ceiling and already with the new beams (reinforcing elements). Before fitting the numerical to the experimental results, it were performed a series of numerical tests to estimate the sensitivity of the results to certain timber floors characteristics. Taking this into account, Table 2 displays the results of the numerical simulation of a beam considering different strength classes, showing that a change from C16 to C50, which doubles the Young modulus, gives an increase of just 16 % on the first frequency, i.e. the influence of the timber strength class in the natural frequencies of the floor is expected to be low, since the increase of the Young modulus is associated to an increase of the density.

Afterwards, the same beam, but considering a D30 class ($E = 11$ GPa and $\rho = 5.2$ KN/m^3), was numerically tested with different built in conditions at the borders, i.e. the level of "prevented rotation" (100 % means fixed rotation). The results displayed in Table 3 show that a much more important effect is expected from the variation of the border conditions, i.e. a higher difference on the dynamic behaviour is expected from the border conditions than from the material characteristics.

Considering this, Table 4 presents the properties that were calibrated in the numerical model of *Torrinha* street to fit the experimental frequencies. Both the Young modulus and the density were considered fixed/known and equal to those of

Table 2 First vibration frequencies of a 10×20 cm^2 $\times$ 5 m beam with 80 % built in at the borders

Strength class	E (GPa)	ρ (KN/m^3)	f (Hz)
C16	8.0	3.1	41.6
C20	9.5 (18.8 %)	3.3	42.9 (3.1 %)
C30	12.0 (50.0 %)	3.8	46.0 (10.6 %)
C50	16.0 (100.0 %)	4.6	48.3 (16.1 %)

Table 3 First vibration frequencies of a 10 × 20 cm^2 × 5 m beam with different border conditions

Built in degree (%)	Built in at one extremities f (Hz)	Built in at both extremities f (Hz)
0	16.2	16.2
30	19.5 (20.4 %)	22.5 (38.8 %)
60	22.0 (35.8 %)	28.0 (72.8 %)
100	24.7 (52.5 %)	35.0 (116.0 %)

Table 4 Properties considered (Torrinhastreet) (see Fig. 3)

Model	Elements	Support conditions	Built in degree (%)	E (GPa)	Density (KN/m^3)
Torrinha floor after reinforcement	Old beams (V)	Hinged	–	11	5.2
	New beams (VN)	Built in	40	9	3.1
	Old struts	Built in the beams		11	5.2
	New struts			9	3.1
	Ceiling	–	–	1	27.0

D30 class, and the calibration consisted on fitting the built in (embedding) conditions of the beams in the stone walls, giving 40 %.

Numerical *Versus* Experimental Results

Table 5 shows the results obtained experimentally, namely the first, or the two or three first detected main natural frequencies of the floors *versus* the numerical frequencies obtained after calibrating the numerical model. The seismograph was positioned in different beams and in different positions in the beams (1/4 and 1/2 the span) to catch the highest number of possible vibration frequencies. For *Torrinha* street, the 3 frequencies correspond to the 3 first vibration modes of the whole floor, i.e. the same set of values was obtained when positioning the seismograph on the 6 beams separately, showing that the beams work together, even when there is no floorboard, provided that there are still the struts and/or the ceiling structure. Notice that the plaster ceiling homogenizes the vibration behaviour of the floors, but decreases their global vibration frequencies as result of its important mass, much higher than the floorboards mass.

As for *Rosário* street, the results show that the 3 tested beams responded with the same first vibration frequency and a second mode with similar frequencies, meaning that the ceiling structure at level 1 also homogenised the behaviour of the floor. As for level 2, the struts clearly stiffened the floor, since the first mode of each of the three beams, referred to f1, f2 and f3 gave higher frequencies before removing the

Table 5 Experimental versus numerical frequencies for the case studies

Case study			Frequencies (Hz)	
			Model	Experimental
Torrinha	Original state	f1	17.1	15.1
	Reinforced	f1	10.4	9.3
		f2	12.7	12.7
		f3	18.7	17.6
Rosário	Level 1	f1	20.6	21.0
		f2	24.3	23.4–25.4
	Level 2 (with struts)	f1	36.3	34.7
		f2	37.3	37.1
		f3	–	28.3
	Level 2 (without struts)	f1	32.4	32.2
		f2	36.9	36.8
		f3	24.7	24.2
Monte dos Judeus		f1	53.8	24.9

struts. This result confirms that the struts may improve the floors behaviour by imposing a more uniform behaviour, "forcing" the beams to work together; the struts make the floors behaviour less dependent from each beam singular characteristics, even if in *Rosário* case the link between struts and beams seemed to be not very strong.

Concerning the dynamic identification, the fitting of the numerical frequencies to the experimental results of *Rosário* and *Torrinha* case studies allowed estimating that only 40 % of the full built in conditions should be considered for the beams extremities embedded in the stone masonry walls.

A part from the global behaviour, some of the experimental results of *Rosário* street exhibited a quite unusual result: a thicker old beam (ϕ19 cm) responded with a first vibration frequency (24.1 Hz) quite lower than a thinner beam (ϕ14 cm—36.1 Hz). The supporting conditions were identical and, therefore, the border conditions could not explain the difference. The only observable dissimilarity was in the axis shape that for the thicker beam was quite curved (along the horizontal plane), in opposition to the thinner beam that had a straight axis. Thus, a numerical test was performed simulating two beam shaving the same mechanical characteristics and supporting conditions of the calibrated model for the thicker beam: one with a straight axis and the other with the real curved axis. The dynamic analysis gave frequencies of 24.7 Hz for the curvilinear beam (vibration mode perpendicular to the beam plane, i.e. vertical) and 40.4 Hz to the rectilinear beam. Notice that the frequency of the second vibration mode of the curvilinear beam (vibration mode perpendicular to the first mode, i.e. horizontal) was 76.6 Hz, while for the rectilinear beam was still 40.4 Hz. This result shows the importance of the axis shape to the behaviour of the floor.

Finally, the *Monte dos Judeus* case results showed a high discrepancy between the numerical and the experimental results that could not be explained by the built in degree at the extremities that, for the present case, should be very high (estimated to be higher than 80 %, in order to simulate the beams' stiff link to the contour ring beam strongly connected to the walls); the model frequencies for 80 % were much higher, more than double, than the experimental ones. By analysing the in situ conditions, the conclusion was that the influence should come from a series of stones (mass) that was placed on beams far from the area where the beams were tested. Numerical simulations considering that extra load show that, due to the good/stiff connections between beams and struts, the structure (still without planking) works and responds as one and the extra mass, although far, has a strong influence on the results, reducing deeply the vibration frequencies and stressing the major importance of the struts to the floors global behaviour when well connected to the beams.

Conclusions

The aim of this study is to evaluate the behaviour of timber floor structures by using dynamic identification, i.e. the experimental measurement of the natural vibration frequencies of the floors and of the floors structural elements. In particular, it was possible to measure the embedding characteristics of the beams in the stone masonry walls, estimating, in the case of the old floor structures of the case studies that should correspond to only 40 % of a full built in condition. The experimental results also pointed out the high participation of the struts to the global behaviour of the floor, particularly when well connected to the beams. Although the ceiling structure also contributes to homogenise the timber floors global behaviour, its most important effect is to decrease the vibration frequencies as result of its large participation mass.

The campaign also showed that the non-linearity of the axis of a timber beam may cause variations of its vibration frequencies above 60 %. On the other hand, the border conditions can be responsible for a vibration frequency variation higher than 100 %, depending on the degree of embedding of the beams in the walls. These two features are those that were estimated to have more influence on the vibration frequencies of a timber floor, particularly when compared to the timber class properties for which a variation of 100 % of the elasticity modulus may change the value of the first natural frequency of a beam in only around 15 %.

Acknowledgments The authors would like to thank all the support of NCREP—Consultoriaem Reabilitação do Edificado e Património, Lda, for providing the case studies under this research. The 3rd author would like to thank the FCT (Fundação para a Ciência e a Tecnologia—Foundation for Science and Technology) of Portugal for the grant funding—SFRH/BPD/88115/2012.

References

1. Ilharco T, Guedes J, Costa A, Arêde A, Paupério E (2012) Avaliação da Distribuição de Cargas em Pavimentos de Madeira através de Ensaios In-situ. Revista Mecânica Experimental, da APAET, Associação Portuguesa de Análise Experimental de Tensões, ISSN 1646-7078, vol 21, pp 1–11
2. Capuz R (2003) Métodos de ensayos no destructivos para la estimación de las propriedades físicas y mecánicas de las maderas. Ph.D. thesis, Universidad Politécnica de Valencia, Valencia, Spain
3. Teixeira JJL (2004) Descrição do sistema construtivo da casa burguesa do Porto entre os séculos XVII e XIX. Contributo para uma história da construção arquitectónica em Portugal. FAUP, Porto, Portugal

Assessment of the Mechanical Behaviour of Tabique Walls Through Experimental Tests

Andreia Rebelo, João M. Guedes, Bruno Quelhas and Tiago Ilharco

Abstract The option for substituting timber structures by other materials when renovating old constructions often leads to more invasive solutions, less compatible with the existing structure, being often the cause for more damage. On the contrary, low invasive procedures, such as preservation actions, generates more sustained and respectful interventions, closer to the recommendations expressed in international charts and documents. However, such procedure simply having a good knowledge on the materials (properties and characteristics) and behaviour of the elements (mechanical, static, dynamic, long term, short term…). The present research aims contributing to this knowledge, by assessing, experimentally, the compression mechanical behaviour of a particular type of timber walls commonly found in old masonry constructions in Portugal and referred to *tabique* walls; the lack of data on the behaviour of this type of walls is probably the main reason why interventions quite often consider their full substitution by new elements. The work consisted on constructing 4 full scale specimens that were tested at the Laboratory of Earthquake and Structural Engineering of the Faculty of Engineering of Porto University to assess the compression strength and behaviour of these elements under no initial damage conditions. The results were also compared to numerical simulations using SAP2000 software.

A. Rebelo (✉) · J.M. Guedes · B. Quelhas
Faculdade de Engenharia da Universidade do Porto (FEUP),
Rua Dr. Roberto Frias, s/n, 4200-465 Porto, Portugal
e-mail: ec10063@fe.up.pt
URL: http://www.fe.up.pt/

J.M. Guedes
e-mail: jguedes@fe.up.pt

B. Quelhas
e-mail: bruno.quelhas@ncrep.pt

T. Ilharco
NCREP—Consultoria em Reabilitação do Edificado e Património,
Lda, Praça Coronel Pacheco, 2, 4050-453 Porto, Portugal
e-mail: tiago.ilharco@ncrep.pt
URL: http://ncrep.pt/

H. Cruz et al. (eds.), *Historical Earthquake-Resistant Timber Framing in the Mediterranean Area*, Lecture Notes in Civil Engineering,
DOI 10.1007/978-3-319-39492-3_37

Keywords Timber walls · Experimental testing · Compression strength

Introduction

The rehabilitation of old buildings is becoming a priority issue in Portugal. It's a process that demands the assessment of the state of conservation and the understanding of the mechanical behaviour of the materials and structural elements that constitute these buildings. Among these elements, timber walls made of vertical and, sometimes, diagonal boards, and referred to *tabique* walls (TW), are present in most of the buildings built till the 40s–50s of the last century in Portugal [1] and represent a very important constructive and cultural element. In spite of that, these walls are often seen as secondary replaceable elements with no significant mechanical value, a belief that makes research and technological development on the assessment and rehabilitation of TW to be still quite limited. Most of the research already done consists on the assessment of the different layouts and components of TW from buildings located at the Northern part of Portugal [2]. In particular, this research involved the evaluation of the dimensions of timber boards, laths, lintels, braces and the way they are jointed (nailed) and linked to floors and ceilings, and the assessment of their materials and/or mechanical characteristics. Recent studies have simulated numerically the mechanical behaviour of TW under vertical (considering lateral buckling) and horizontal in-plane forces, acting separately [3, 4]. The results estimated a performance that goes far beyond what was initially expected for a constructive element that is commonly considered to be just a partition element with no structural functions, namely that a good in-plane (vertical and horizontal) bearing capacity may be foreseen. In particular, the results show the high contribution the vertical boards have to the walls vertical load bearing capacity, and the diagonal boards to the walls in-plane shear stiffness, i.e. that under certain conditions TW may be used as load bearing and horizontal bracing elements.

However, these numerical results still need experimental support. Until now, no compression tests have been made on TW. The only experimental campaigns found on these types of structures are reported in [5, 6]; the first presents shear in-plane tests made on full scale models involving small prototypes of a specific type of TW (with no diagonal boards), and the second flexural out-of-plane tests on TW samples retrieved from old buildings. Other experimental and numerical campaigns involved timber framed walls from *Pombalino* constructions (referred to in Portugal as *frontal* walls), a constructive element made of quadrangular type timber frames with inside crossed braces defining cells that are filled in with a combination of mortar, bricks, stones, or other materials, i.e. quite different from the previous TW and that, unlike those, are considered to have mechanical/structural functions. In particular, a *frontal* wall retrieved from an 18th century Lisbon building was tested under shear in-plane forces at LNEC, providing interesting information on their residual performance [7]. Thus, the present research aims assessing the compression behaviour of TW through experimental testing on full scale specimens, contributing also to a higher appraisal and preservation of timber walls.

Tabique Walls

TW are commonly used as interior partition walls and are often considered as non-structural and low resisting elements, having no ability to contribute to improve the overall performance of the buildings. This explains why during rehabilitation works TW are often replaced by new elements, ignoring its potential as supporting and/or bracing element. These walls, which are quite slender (height to thickness ratio usually greater than 20), are mainly made of vertical timber boards juxtaposed with a gap of around 2 cm, that may be framed and/or diagonally crossed (cutting or not the vertical boards) by thicker planks. This timber vertical structure is usually covered by a series of horizontal wooden laths nailed on the boards, a system that improves the adhesion of the finishing layer of lime mortar or earth based render. These are the so called simple TW, in opposition to the double TW, which are made of two layers of boards: a layer of diagonal boards overlaid to the layer of vertical boards. The diagonal boards can be placed parallel to each other, or in two opposite directions, making a geometrical draw. Also in the case of the double TW, the surface of the boards is usually covered with horizontal wooden laths to improve the adhesion of the finishing layer of lime mortar or earth based render.

Under this research, 2 simple and 2 double full scale TW (2.5 m long per 3.0 m high) were built by *Portilame* (www.portilame.com—the company that financed the research with the construction of the specimens) respecting the most common layout of TW in the city of Porto and Porto region, Portugal, and the average dimensions reported in [8], although higher average thicknesses have been assessed from the inspection of a large number of buildings in this region by NCREP, namely values from 3.0 to 3.5 cm. The walls were finished at the top and bottom with a 8×8 cm^2 beam that was centred at the walls axis and nailed against the boards. The specimens and the average dimensions of the elements are presented in Figs. 1 and 2, respectively. At this stage the TW were tested without mortar. Future experimental campaigns will include the test of the same walls with mortar.

Laboratory Testing Apparatus

The tests were performed at the Laboratory of Earthquake and Structural Engineering (LESE) of the Faculty of Engineering of Porto University (FEUP) and were financed by *CS Construtora* (www.csconstrutora.com). The testing apparatus was setup to represent, as close as possible, the in situ conditions of the TW at Porto old houses. Therefore, it was designed and constructed a timber basement having a central beam with a 2.5 m long longitudinal groove with 8×8 cm^2 to receive the

Fig. 1 TW specimens

Fig. 2 Average dimensions of the specimens constituents

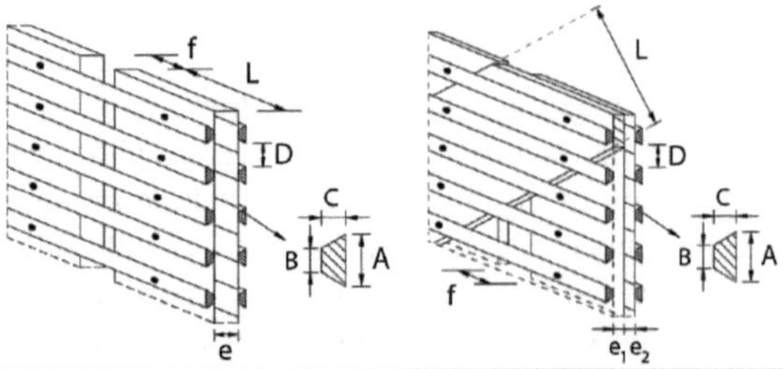

Boards (cm)			Laths (cm)			
L	*e*,e_1,e_2	*f*	*A*	*B*	*C*	*D*
26	2,5	2,0	2,5	2,5	1,3	4,0

walls bottom beam, being the whole set fixed to the laboratory floor using four steel profiles. At the top of the wall, and to receive the vertical load, it was placed a timber beam identical to the central beam of the basement. All timber elements were built by *Portilame*.

The vertical load was applied on the walls top beam to simulate, as closely as possible, a distributed load. For this, it was used a three beams setup to transfer the load of the hydraulic actuator to 4 equally distributed points, as shown in Fig. 3. This layout was determined by simulating a TW on SAP2000 software and applying different number of concentrated loads, 2, 3, 4 and 5 at the top. This simulation allowed concluding that the 4 points setup simulated, appropriately, a uniformly distributed vertical load across the wall.

The hydraulic actuator was positioned below the LESE stiff floor that was used as reaction structure. It was connected to transversal top and bottom steel beams linked together by 2 rods that transferred the load from the bottom to the top beam,

Fig. 3 Top beam and bottom basement (*left*) and blocking system of the top beam (*right*)

i.e. from the actuator to the specimens top beam. The tests were performed with displacement control and were stopped when the vertical force versus vertical displacement curves tended to a horizontal line; this force will be referred to in the text as the critical load. Collapse was avoided, since the specimens will be reused in future tests. The applied load was measured by internal and external load cells, two of them placed on the steel rods. To avoid the walls top beam moving on the horizontal direction, it was constructed a frame with three horizontal cantilever arms that blocked the top wall displacement in the out-of-plane direction, as seen in Fig. 3.

The walls were instrumented with 11 displacement transducers (LVDTs) that measured the horizontal and vertical displacements of a series of points distributed on the wall. The movements of the top beam were measured through a clinometer and 2 extra horizontal LVDTs (Fig. 4).

Experimental Results

It were performed 4 in-plane compression tests: one monotonic (ML) and one cyclic (CL) on each different TW typology: simple (S) and double (D). The tests are referred to TWS-ML and TWS-CL, for the simple TW, and TWD-ML and TWD-CL for the double TW, being this the testing sequence that was adopted. The load layout of the cyclic tests, in particular the displacements level established for each cycle, was decided after the monotonic tests. The tests were performed at a loading rate of 0.01 mm/s and an unloading rate of 0.05 mm/s. Figure 5 shows the vertical force versus vertical displacement for the simple and double TW.

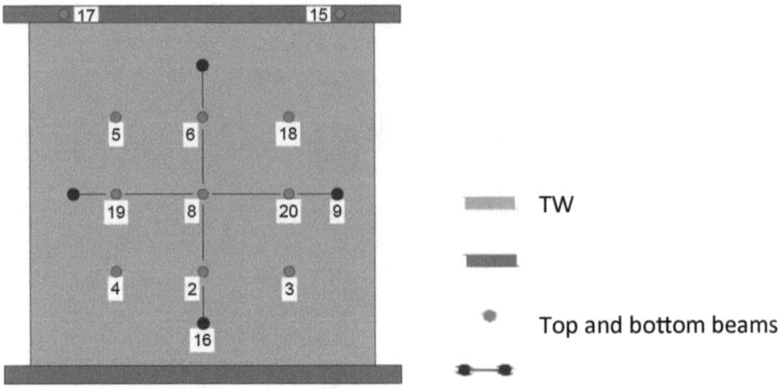

Fig. 4 Displacement monitoring setup

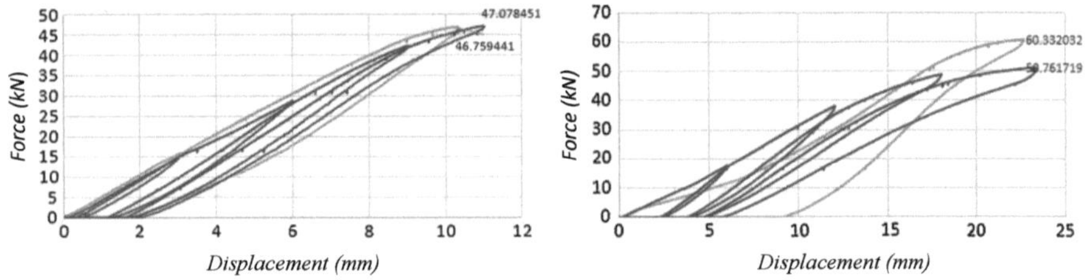

Fig. 5 Vertical force versus vertical displacement curves for the simple (*left*) and double (*right*) TW—monotonic (*blue*) and cyclic (*orange*)

The graphs in Fig. 5 show that:

- The vertical load capacity of the simple and double TW is around 18 kN/m and between 20 and 24 kN/m, respectively;
- The double TW, with double thickness compared to the simple TW, i.e. with half the slenderness, have a critical load that is only around 30 and 9 % higher than the critical load of the simple TW for the monotonic and cyclic tests, respectively. This result shows that the diagonal boards have a much lower participation to the walls vertical capacity than the vertical boards;
- The monotonic curve of the simple TW coincides with the envelop of the cyclic test. On the contrary, the monotonic test on the double TW reaches a critical load that is about 20 % higher than that of the cyclic test. This difference maybe explained by the higher expected "deterioration" of the links (nailed) between the vertical and diagonal boards during the cyclic tests. Notice that in this case the critical load of the double TW decrease to a value closer to that of the simple TW;
- The residual/inelastic displacements after unloading were higher for the double TW, namely around 18 and 26 % (using the cyclic tests as reference) for the simple and double typologies, respectively;

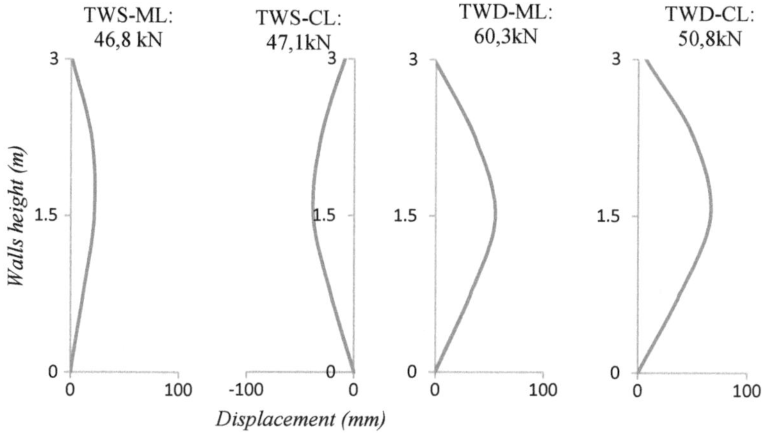

Fig. 6 Horizontal displacement profiles of the 4 TW at the maximum load

- The elastic(initial) compression stiffness for the simple and double TW, for the total height of the specimens, is around 5 and 3 MN/m, respectively. The much lower stiffness of the double TW reflects the non-symmetric layout of the boards in relation to the specimens middle plane and load position; the side of the diagonal boards is more flexible to the vertical forces and creates a disequilibrium on the vertical deformation of the two sides, promoting a higher top beam rotation, i.e. a higher flexural deformation of the wall that induces a higher vertical displacement and a(apparent) lower vertical stiffness. The higher flexural deformation of the double TW can be confirmed by the lateral profiles of the deformed TW for the maximum applied load represented in Fig. 6. Furthermore, the top beam rotations measured by the clinometer were around 1.5° and 8.5° for the maximum load applied on the cyclic tests on the simple and double TW, respectively.

Finally, the results expressed by the profiles in Fig. 6 show that the walls rotate at both extremities. The bottom basement may prevent the bottom beam to rotate, but not the boards, i.e. the connection between the boards and the bottom beam (using nails, as it happens in situ) are flexible enough not to prevent the boards from rotating at the basement. As for the top beam, not only the setup was established to allow free rotations, as the profiles in Fig. 6 indicate that it may be considered as hinged.

Numerical Versus Experimental Results

The material adopted in the simulation of the TW prototypes was pine wood (Pinuspinaster Ait.), the specimens species. Table 1 presents the average values of the longitudinal (E_{mean}) and transversal (G_{mean}) elastic modulus and density (ρ_{mean})

Table 1 Properties of pine wood—class of strength C18

E_{mean} (GPa)	G_{mean} (GPa)	ρ_{mean} (kg/m^3)
9	0.56	380

for this species, considering a C18 class of strength. The boards, laths and nails were discretized by three-dimensional "frame" type elements (SAP2000), i.e. with 6° of freedom per node.

The numerical model of the simple TW is made of 9 vertical boards, simulated by 43 frame elements each, 42 laths on each face, simulated by 10 frame elements per lath, and one beam at the top and bottom simulated by 10 elements each. Three planes were set along the walls thickness: the wooden boards are positioned at the central plane and the laths at the lateral ones, i.e. eccentrically in relation to the boards. The general dimensions and positions respected the values in Fig. 2. The connection between the laths and the boards is materialized by 3 mm square section nails (one for each lath-board connection) modelled by frame elements positioned perpendicularly to the walls. These elements were modelled assuming high axial and bending stiffness and low torsional stiffness.

The model of the double TW differs from the previous model only on the elements of the central plane that has 35 diagonal boards in addition to the 9 vertical boards of the simple TW. These two sets of boards are rigidly connected to each other and exist in the same plane. Figure 7 presents the 2.5 m wide and 3.0 m high numerical models simulated in SAP2000.

In terms of the boundary conditions, two different situations were adopted for each TW model: hinged at both extremities (top and bottom beam points prevented from having horizontal translations and top beam points prevented to move

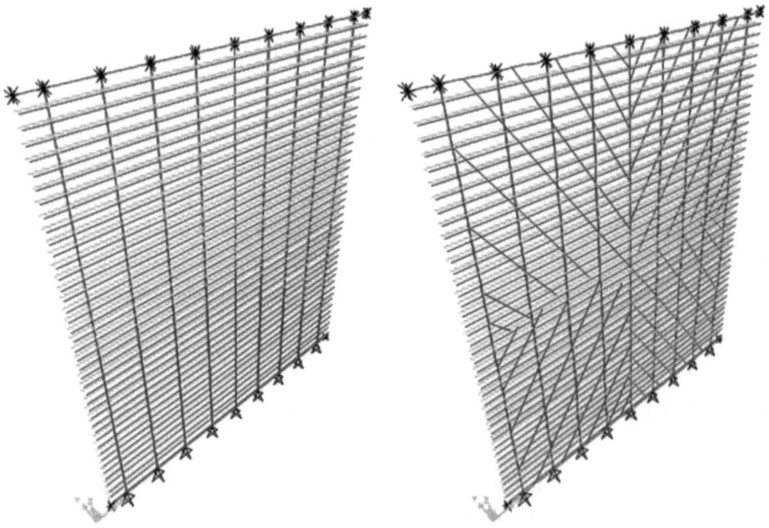

Fig. 7 The numerical model of the simple (*left*) and double (*right*) TW

Table 2 Critical vertical loads (kN/m): numerical and experimental

TW	Numerical		Experimental	
	Top and bottom hinge	Top hinge and bottom built in	Monotonic	Cyclic
Simple	12.2	25.0	18.7	18.8
Double	17.3	35.9	24.1	20.3

vertically), or hinged at the top beam and built in at the bottom beam (bottom beam points prevented from having any translation and rotation). The specimens at the laboratory were expected to respond with boundary conditions in between these two situations, since it's difficult to ensure completely free top rotations and fixed bottom conditions. The experimental and numerical results, in Table 2, show that, in fact, this is the case.

Conclusions

The present research allowed assessing, experimentally, the compression mechanical behaviour of a particular type of timber walls that is quite present in old masonry constructions in Portugal and that referred to *tabique* walls. In particular, it showed that these walls have a good mechanical performance under vertical compression forces. In particular, a simple TW (with 2.5 cm thick boards and no mortar finishing) is able to sustain 18 kN/m, a value that, considering a 2.0 safety factor, corresponds to a design applied load capacity of 9 kN/m. Moreover, this capacity can be easily multiplied by a factor of $(3.0/2.5)^3 = 1.7$ in case the thickness of the boards is 3.0 cm, which is a very common value. On the other hand, the numerical simulation was capable of forecasting the laboratory behaviour/conditions establishing the specimens limit border conditions, i.e. hinged at the top and built in or hinged at the bottom.

Finally, the double TW, with double thickness, i.e. with half the slenderness, responded with a critical load that is only 30 % higher (maximum) than the critical load of the simple TW, showing that the participation of the diagonal boards on the vertical capacity of the TW is much lower than that of the vertical boards; in the case of the cyclic tests, the critical load of the simple and double TW became quite close one to the other. Nevertheless, further investigation is still needed. In particular, this experimental campaign will include more tests on other specimens.

Acknowledgments The authors would like to thank the financial support of *Portilame* and *CS Construtora* that made this research possible. The 3rd author would like to thank the FCT (Fundação para a Ciência e a Tecnologia—Foundation for Science and Technology) of Portugal for the grant funding—SFRH/BPD/88115/2012.

References

1. Gonçalves C, Pinto J, Vieira J, Silva P, Paiva A, Ramos L, Tavares P, Fernandes L, Lousada J, Varum H (2010) Tabique construction in the Municipalities Association of the Terra Quente Transmontana. Latest trends on cultural heritage and tourism. In: Miadenov V, Bojkovic Z (eds) 3rd WSEAS international conference on cultural heritage and tourism (CUHT'10). Corfu Island, Greece. 22–24 July 2010. Published by WSEAS Press, pp 235–240. ISSN: 1792-4308. ISBN: 978-960-474-205-9
2. Pinto J, Varum H, Cepeda A, Tavares P, Lousada J, Silva P, Vieira J (2010) Study of the traditional tabique constructions in the Alto Tâmega region. In: Brebbia CA (ed) The sustainable world. WIT Press 2011, WIT eLibrary, British Library, pp 299–307. ISBN: 978-1-84564-504-5, ISSN: 1746-448X (print), ISSN: 1743-3541 (online)
3. Pires AMC (2013). Análise de paredes de tabique e de medidas de reforço estrutural; estudo numérico. Master thesis, FEUP
4. Cardoso RJS (2013) Caracterização da construção em tabique de Lamego e Alto Douro. Ph.D. thesis at UBI
5. Fonseca I, Feio AO, Machado JS (2014) Utilização de derivados de madeira no reforço de paredes de tabique. 1° Congresso Luso-Brasileiro de Materiais de Construção Sustentáveis (CLB—MCS 2014), Março, UMinho, pp 5–7
6. Ribeiro SDG (2014) Estudo do comportamento estrutural de paredes de tabique. Master thesis at UTAD
7. Tests on "Pombalina" walls (in Portuguese). Reportno. 15/97 - NCE, 21 pp. LNEC, Lisboa, 1997
8. Teixeira JJL (2004) Descrição do sistema construtivo da casa burguesa do Porto entre os séculos XVII e XIX. Contributo para uma história da construção arquitectónica em Portugal, Provas de aptidão pedagógica e capacidade científica. FAUP, Porto

Robustness Assessment of an Ancient Timber Roof Structure

Leonardo G. Rodrigues, Chrysl A. Aranha and Hélder S. Sousa

Abstract During the last decades, it has become a usual practice to replace ancient timber roof structures by concrete or steel roof structures. Despite the durability of ancient timber, the lack of knowledge about how to reinforce or retrofit timber structures may have resulted in their disuse. However, the main reason may be attributed to the difficulty in assessing the real condition of timber frames while respecting the level of conservation. Within that scope, two timber collar trusses dating back to more than a hundred years were transferred from the Laboratório Chimico in Coimbra (Portugal) and tested up to failure in the structural laboratory of the University of Minho. According to the test results using non-destructive techniques, it was possible to develop different probabilistic analyses considering both exterior and residual cross-sectional dimensions. In this work, a robustness assessment is performed considering the spatial distribution of the strength of timber elements and also the variability of the stiffness of the connections. Permanent loads, snow load and wind load variability were considered for the structural safety analysis according to the provisions of the Joint Committee on Structural Safety. The robustness index was obtained based on the reliability indices of the intact and damaged roof structure, considering the geometric variation obtained for the exterior and residual cross sections respectively. Crude Monte-Carlo simulation method was used here to estimate the probabilities of failure of the structural system. The comparison between different imposed damage scenarios covering loss of bearing capacity of connections and timber frames is finally presented.

Keywords Timber trusses · Robustness · Geometric variability · NDTs

L.G. Rodrigues (✉) · C.A. Aranha · H.S. Sousa
Department of Civil Engineering, ISISE, University of Minho,
Guimarães, Portugal
e-mail: leonardofgrodrigues@gmail.com

C.A. Aranha
e-mail: chryslaranha@gmail.com

H.S. Sousa
e-mail: sousa.hms@gmail.com

H. Cruz et al. (eds.), *Historical Earthquake-Resistant Timber Framing in the Mediterranean Area*, Lecture Notes in Civil Engineering,
DOI 10.1007/978-3-319-39492-3_38

Introduction

The definition of robustness is still not clear and allows room for different interpretations among engineers, owners and remaining involved agents. According to EN-1991-1-7 [1], robustness is the ability of a structure to withstand events like fire, explosions, impact or the consequences of human errors while not being damaged to an extent that is disproportionate to the original cause.

The design of redundant structures is one of the main recommendations given to avoid disproportionate collapse due to abnormal loads [2]. The inherent redundancy of timber structures enables load redistribution in order to avoid the propagation of damage due to local failures. Nevertheless, these alternate load paths must be composed of elements and connections with adequate reserves of strength and ductility [3].

The failure of several timber buildings in Europe under expected snow loads has shown that design and/or construction errors, unexpected deterioration, or even unexpected loading can lead to consequences far greater than the initial event. For instance, the failures of Siemens Arena and Bad Reichenhall Ice-Arena prompted a discussion about robustness design and assessment of large-span timber roofs [4]. Robustness methods have been applied to evaluate the collapse of these structures in order to find better design and construction regulations, minimizing the potential consequences of unpredictable events. A framework for robustness assessment of timber structures was proposed under the EU COST Action E55 [5]. The main indices proposed to measure robustness can be divided into three categories: deterministic, reliability-based and risk-based.

A simple robustness analysis can be performed through deterministic indicators which allow the measurement of the redundancy of structural systems and their ability to redistribute loads after a local failure. This may be achieved through a pushover analysis where the structural bearing capacity is computed for an intact and damaged structure (notional removal of an element) [6]. A reliability-based analysis is based on a probabilistic assessment and allows to evaluate how prone a structural system is to disproportionate collapse due to local failures. Here, the robustness indices are based on the probabilities of failure or reliability indices of intact and damaged structures as proposed in [7, 8]. This task involves the consideration of probabilistic distribution functions for loads and the strength and stiffness of structural elements. These distributions can be found by consulting the Probabilistic Model Code or obtained through an experimental campaign. For timber structures, it is common to use the loss of a component or reduction of strength and stiffness in elements and joints as damage scenarios. A risk-based robustness index is proposed in [9] which depends on the consideration of several hazard scenarios and their consequences. These analyses are made through a probabilistic framework, which, in turn, depends on proper modelling of the effects of the exposures on the structural systems. More generally, the risk definition is given as the mathematical expectation of the consequences of damage and failure. This analysis takes into account direct risks, associated with the consequences of components failure (damage state), and indirect risks where the consequences are disproportionate to the trigger event (e.g. collapse of the structure).

In this work a robustness analysis is conducted for an ancient timber roof structure system, based on the reliability indices mentioned in [5] and on the probabilistic models given in the Probabilistic Model Code for timber properties.

Description and NDT

The geometric assessment and visual inspection of historic timber structures is one of the most important tasks in evaluating their integrity. The inherent variability of timber and its susceptibility to decay emphasize the need to carry out a characterization of the constitutive timber elements. The information on the geometry and the state of degradation can be measured by means of nondestructive tests (NDT), mechanical tests, and other means [10, 11].

In the case of the historic timber trusses considered in the current study, a geometrical assessment was performed at intervals of 40 cm for each member. The assessed sections were marked and identified so that the dimensions of these same sections could be measured using non-destructive tests in order to obtain the effective cross section. Information on the density, the presence of voids and their location and the modulus of elasticity was obtained using impact penetration, drilling resistance and ultrasounds tests. In Fig. 1 the structure of the collar truss is seen with at least one of the measured sections of each member mapped on the diagram.

The collar truss was tested at the Structural Laboratory of the University of Minho, Portugal. The truss was subjected to downward displacement at the rate of 0.05 mm/s until failure, as seen in the setup presented in Fig. 2. Three main failures were found during the mechanical test, namely, failure of both rafters in the sections

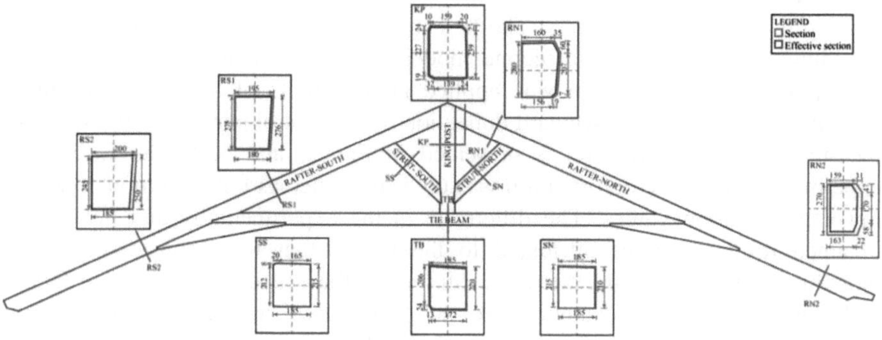

Fig. 1 The collar truss with representative sections obtained from the geometrical survey mapped

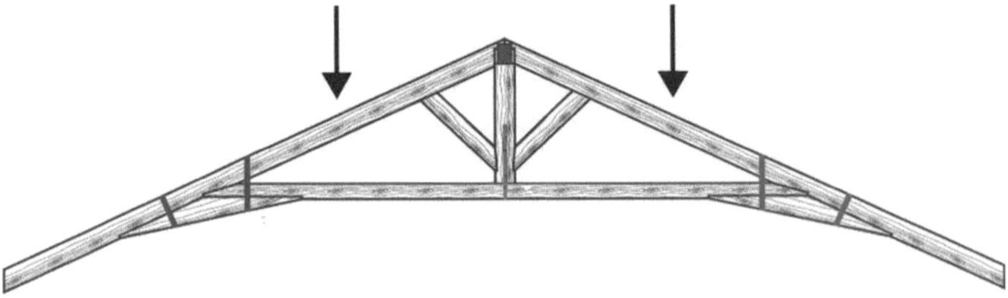

Fig. 2 The test set-up showing the points of application of the displacement

near the connection with the tie-beam and failure near the connection of one of the rafters with the post. Those failures coincided with sections having lower visual grades due to the presence of significant knots. The reaction at the points of loading and the deflection of the tie beam and rafters were monitored using actuators and strategically placed LVDTs.

In most cases, the seismic damage of historic timber roofs is due to the inadequate behaviour of the connections and the timber members themselves remain elastic [12]. To determine the effect of the connections between the rafters and the tie beam as well as the connection between the post and the tie beam on the behaviour of the truss [13], two models of the truss were created using SAP2000. Both models had the same geometrical and material properties but differed in the definition of the joints. By averaging the dimensions obtained from the geometrical assessment of each timber frame element, a uniform rectangular cross-section was assigned to each of them. Since the measured values of the modulus of elasticity of the rafters were much lower than in all the other frame members, the average value of the modulus of elasticity of the rafters was calculated and an average value for the rest of the members was calculated separately, according to bending tests made to individual elements of a truss retrieved from the same building. The modulus of elasticity of the rafters was defined as 4680 MPa while for the rest of the members, it was taken as 9600 MPa. The shear modulus and density were set to 555 MPa and 575 kg/m^3 for all the members. The first model consisted of frame members with pin-jointed ends while in the second model, the frame members had rigid joints. A displacement of 90.7 mm was applied at the points of application as in the experimental tests. The downward deflection of the central part of the tie beam was measured in both cases and is presented in Figs. 3 and 4. In the case of the experimental tests, the deformation was 63.5 mm whereas both the models yielded lower deformation values. The difference in the experimentally and numerically obtained results may be attributed to the values that were considered for the modulus of elasticity of the elements. Moreover, the existence of localized failures and out-of-plane displacement were not considered in the numerical model, although they occurred during the experimental tests.

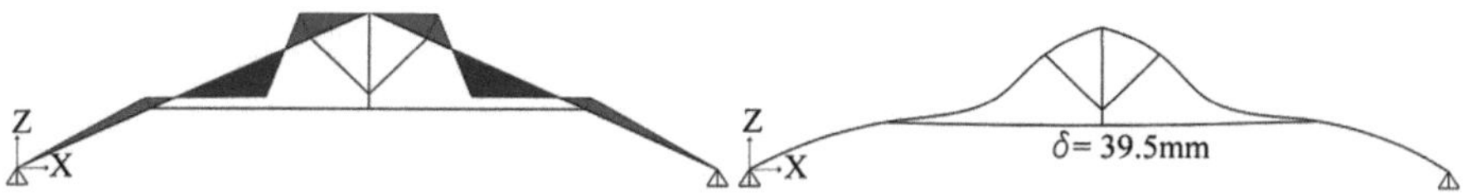

Fig. 3 Bending moment diagram (*left*) and deformed shape (*right*) of the truss model with hinged joints

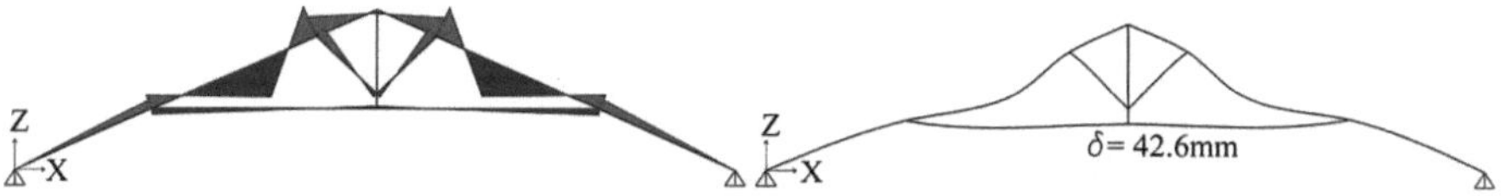

Fig. 4 Bending moment diagram (*left*) and deformed shape (*right*) of the truss model with rigid joints

Probabilistic Model

For the structural analysis, a linear finite element model was developed within the Opensees framework. The truss is modelled using linear elastic frame elements connected with zero-length elements as seen in Fig. 5. The truss schematization considered rigid joints as given in Fig. 4.

Each structural element is divided in frame segments of in 40 cm length as per the locations of the non-destructive tests that were previously carried out. From the results of the experimental campaign, it was possible to assume deterministic values for the cross-section of the truss. The joints between the truss elements, as well as the connection of the purlins, are assumed to be rigid. This feature allows to consider load redistribution to the undamaged parts of the structure in case of a local failure. The supports of the truss on the walls were modelled by hinges. The considered failure modes are related with the ultimate limit state verifications for bending and combined axial and shear forces in the timber elements according to Eurocode 5 [14], as follows:

$$\left(\frac{\sigma_{m,d}}{f_{m,d}}\right) + \left(\frac{\sigma_{c,0,d}}{k_{c,z} \cdot f_{c,0,d}}\right)^2 \leq 1.0 \quad (1)$$

$$\frac{\sigma_{t,0,d}}{f_{t,0,d}} + k_m \frac{\sigma_{m,d}}{f_{m,d}} \leq 1.0 \quad (2)$$

$$\frac{\tau_{v,d}}{f_{v,d}} \leq 1.0 \quad (3)$$

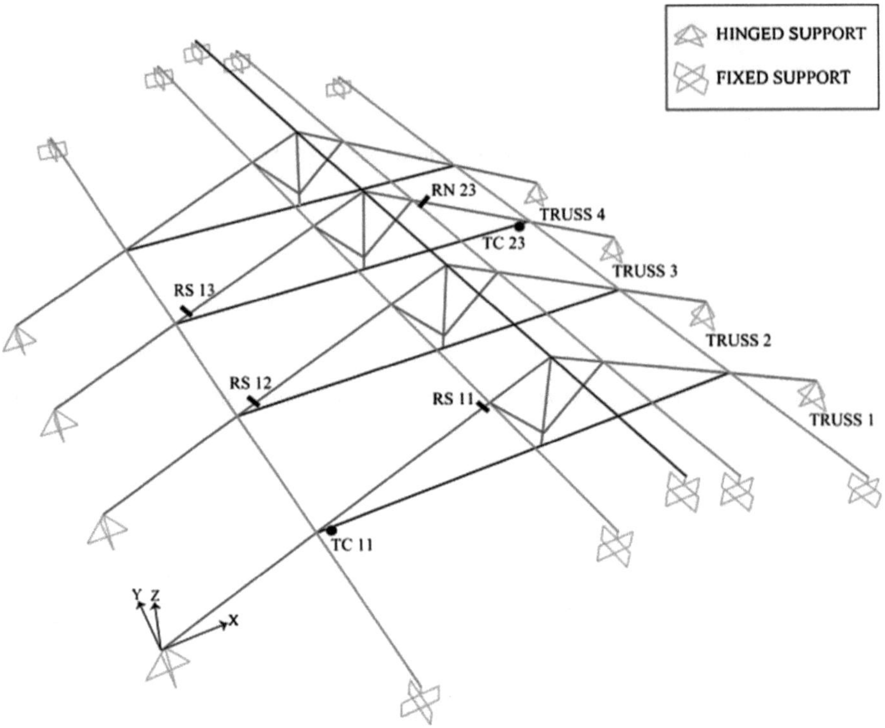

Fig. 5 Finite element model of the roof structure

The reliability index is based on limit state functions (Eq. 1–3), where σ_m, σ_c, σ_t and τ_v correspond to the imposed stresses in bending, compression, tension and shear, respectively. Consequently, the Monte-Carlo method was used instead of First Order Reliability Methods. The probability of failure is then given by the ratio between the number of failures (z_0) and the number of realizations (z).

$$p_f \approx \frac{z_0}{z} \tag{4}$$

Given the probability of failure (p_f), it is possible to reach the structural reliability index (β) through the inverse cumulative distribution function (Φ^{-1}) of standard normal distribution.

$$\beta = \Phi^{-1}(1 - p_f) \tag{5}$$

The number of Monte Carlo simulations is strictly dependent on the coefficient of variation for failure events (V_{pf}) and on the expected order of magnitude postulated for probability of failure.

$$V_{p_f} = \frac{1}{\sqrt{z} \cdot p_f} \quad (6)$$

A coefficient of variation (CoV) of 10 % and a probability of failure of 10^{-4} were assumed, resulting in one million realizations z. Thus, it is necessary to generate one million length samples for the mechanical characteristics of each element of the truss. These samples follow the distribution types recommended in the Probabilistic Model Code [15] for timber properties. The reference mechanical properties for structural timber are bending strength, bending modulus of elasticity and density. The relations between the reference properties and the remaining strength properties are given by the Probabilistic Model Code and presented in the Table 1.

In this study, permanent loads (*G*) due to self-weight and two variable actions namely wind (*W*) and snow (*S*) were considered. The permanent loads (weight of the trusses, roof tiles and sheeting) were modelled with a normal distribution with an expected value of 1.5 kN/m^2 and a coefficient of variation of 10 %. The wind load was assumed to be a Gumbel distribution having a mean value of 0.32 kN/m^2 for downwind and 0.23 kN/m^2 for upwind exterior wind pressure. These values were computed according to Eurocode 1 [16] with a CoV = 35 %. The snow load was calculated in agreement with Eurocode 1 [17], resulting in a mean value of 0.18 kN/m^2 and a CoV = 35 %. Experimental tests were performed to achieve the stochastic models for the reference parameters.

To take into account the spatial distribution of strength and stiffness, the components are considered as a set of correlated random variables. In order to generate the samples for each analysis, it was necessary to consider the classification of correlation coefficients proposed in [16]. These values are presented in Table 2. According to the visual inspection and considering the traditional methods of selection that were used in the construction period of the tested trusses, the elements with similar length had higher correlation coefficients. In this structure, the rafters

Table 1 Probabilistic models for reference properties of truss elements

Property, X	Distribution	E[X]	CoV[X]
Bending strength f_m	LN	15.0 MPa	0.25
Bending MOE E_m	LN	8880 MPa	0.30
Density ρ_{den}	N	575 kg/m^3	0.1
Tension strength $f_{t,0}$	LN	$E[f_{t,0}] = 0.6E[f_m] = 9.0$ MPa	$COV[f_{t,0}] = 1.2COV[f_m] = 0.3$
Compression strength $f_{c,0}$	LN	$E[f_{c,0}] = 5E[f_m]^{0.45} = 16.9$ MPa	$COV[f_{c,0}] = 0.8COV[f_m] = 0.2$
Shear modulus G_v	LN	$E[G_v] = E[E_m]/16 = 555$ MPa	$COV[G_v] = COV[E_m] = 0.3$
Shear strength f_v	LN	$E[f_v] = 0.2E[f_m]^{0.8} = 2.8$ MPa	$COV[f_v] = COV[f_m] = 0.25$

Table 2 Correlation coefficients

	E_m	ρ_{den}	$f_{t,0}$	$f_{c,0}$	G_v	f_v
f_m	0.8	0.6	0.8	0.8	0.4	0.4
E_m	1	0.6	0.6	0.6	0.6	0.4
ρ_{den}		1	0.4	0.8	0.6	0.6
$f_{t,0}$			1	0.5	0.4	0.6
$f_{c,0}$				1	0.4	0.4
G_v					1	0.6

Where f_m = Bending strength, E_m = Bending MOE, ρ_{den} = Density, $f_{t,0}$ = Tension strength//grain, $f_{c,0}$ = Compression strength//grain, G_v = Shear modulus, f_v = Shear strength

and the ties have almost the same length, consequently their correlation is high and a coefficient of 0.8 is assumed. On the other hand, the correlation between a rafter and a post is low since their lengths are significantly different and a coefficient of 0.4 is assumed.

To apply the Monte-Carlo method, it was assumed that failure occurred when the inequalities of Eqs. (1–3) were not satisfied. Given that the Opensees finite element framework allows user—defined computations, it was possible to assess if the elements reached their load bearing capacity through procedures developed in tcl/tk [18]. For each simulation, the most critical element was identified and saved.

Reliability Analysis

The main goal is to assess the condition of the actual structural system comparing the reliability indices for original and residual cross sections. Three combinations of loads were considered to perform this study. In COMB 1, only permanent loads were considered; in COMB 2, permanent and wind loads were applied and in COMB 3, permanent and snow loads were taken into account. The modification factor, k_{mod}, was determined by considering solid timber and service class 1 resulting in a value of 0.6 for COMB 1 and 0.9 for COMB 2 and COMB 3.

After performing the predefined number of simulations, it was possible to understand that no failure occurs for any of the combinations. Nevertheless, with the values obtained from Eqs. (1–3), it is possible to estimate the parameters of a lognormal distribution for the safety margin and then evaluate the probability of failure. The results of the simulations revealed that the probabilities of failure were very low, allowing to conclude that this structure was initially over-designed. From the results presented in Table 3, it was also inferred that the reliability indices obtained are higher than the minimum value recommended in Eurocode 1 for the consequence classes RC3 for an operational life time of 1 and 50 years (β = 5.2 and β = 4.3).

Table 3 Reliability indices and probabilities of failure (in brackets) for intact and damaged structures

	Original cross-section	Residual cross-section
COMB 1	>10 (−)	>10 (−)
COMB 2	6.35 (1.08×10^{-10})	4.83 (6.66×10^{-7})
COMB 3	>10 (−)	>10 (−)

Robustness Analysis of the Roof

The robustness of the system is assessed through reliability-based robustness indices:

$$\beta_R = \frac{\beta_{(\text{int}act)}}{\beta_{(\text{int}act)} - \beta_{(damaged)}}, \quad \forall\beta_{(\text{int}act)} \geq 0, \ \forall\beta_{(damaged)} \geq 0 \tag{7}$$

$$I_{Rob} = \frac{\beta_{(damaged)}}{\beta_{(\text{int}act)}}, \quad \forall\beta_{(\text{int}act)} \geq 0, \ \forall\beta_{(damaged)} \geq 0 \tag{8}$$

These robustness indices are related to a damage scenario that involves the loss of a component or reduction of strength and stiffness in elements and joints. In Eqs. 7 and 8, β_{intact} is the reliability index of the undamaged structure and $\beta_{damaged}$ is the reliability index of the damaged structure. High robustness indices indicate high ability of the structure to redistribute forces and withstand local failures. COMB 2 was considered to perform this robustness analysis as lower reliability levels were found in the previous analysis. The intact, or undamaged, system coincides with the model that considers the actual condition (effective cross sections) of the structure and a reliability index $\beta_{intact} = 4.83$. From the reliability analysis, it was also possible to conclude that the cross sections of rafter-south next to the connection of the tie in truss 2 and 3 are the critical ones (RS 12 and RS 13 in Fig. 5). Thus, two of the damage scenarios that were considered resulted in the loss of capacity of these cross sections. To evaluate the influence of the connections between rafters and ties, one of the damage scenarios indicated a reduction in their rotational stiffness. Another damage scenario consists in the loss of bearing capacity of two of these connections at a given time (TC11 and TC23 in Fig. 5). The results are presented in Table 4.

For the initial damage scenario, failure occurred due to loss of bearing capacity of the rafter-south of truss number 3 (RS 13 in Fig. 5). When the analysis simulated the loss of capacity of cross section of the rafter-south of truss 3, this resulted in the failure of section RS 12, which was the opposite of what happened previously. The reliability indices decreased to 35 and 36 % of the intact value. Nevertheless, it is possible to state that the system is able to withstand these failure events. From Table 4, it is possible to understand that the effect of the reduction on the rotational stiffness of the rafter-tie is negligible. When a local failure occurs in the rafter-tie

Table 4 Reliability indices, probabilities of failure (in brackets) and respective robustness indices for different damaged scenarios

Damage scenario	$\beta_{damaged}$ (P_f)	β_R	I_{Rob}
1. Fail rafter-south of truss 2 (RS 12)	1.70 (4.4×10^{-2})	1.54	0.35
2. Fail rafter-south of truss 3 (RS 13)	1.73 (4.2×10^{-2})	1.56	0.36
3. Reduced stiffness of rafter-tie connections	4.78 (8.62×10^{-7})	93.5	0.99
4. Loss of bearing capacity of joint 1 (TC 11)	2.05 (2.03×10^{-2})	1.74	0.42
5. Loss of bearing capacity of joint 2 (TC 23)	1.98 (2.39×10^{-2})	1.69	0.41

connections, the most critical section obtained through the finite element model is located on the correspondent rafter next to the connection with the stud. It is important to stress that the reliability indices for the intact structure are high which leads to the conclusion that the system is robust for this damage scenario.

Conclusions

A robustness assessment is presented for an ancient timber collar roof. The evaluation is based on reliability indices obtained through Monte Carlo simulations for the intact and damaged roof structure modelled with finite elements. Permanent loads, snow load and wind load variability were considered along with the spatial distribution of strength of timber elements.

The results of the reliability analysis confirm that this structure was overdesigned and is able to remain in service. The robustness indices obtained demonstrate that the structure has the ability to withstand the local failure assumed. In the future, more failure modes and damage scenarios may be considered e.g. reduction of bearing capacity of timber to wall joints. The capabilities of non-linear finite element models can be exploited to study the propagation of failure up to collapse. Nevertheless, these results lead to the conclusion that if the secondary system allows load redistribution, the structure can also be considered as robust.

Acknowledgments The information obtained through training school: Assessment and reinforcement of timber elements and structures (Guimarães, Portugal) within scope of FP1101 and RILEM TC 245 RTE is acknowledged. The support of the Structural Lab from University of Minho is acknowledged. Financial support of the Portuguese Science Foundation (Fundacão de Ciência e Tecnologia), through PhD grant PD/BI/52583/2014 included in InfraRisk-PhD program, is gratefully acknowledged.

References

1. CEN, EN1991-1-7:2006, Actions on structures. P. 1-7: general actions—accidental actions
2. Ellingwood BR, Dusenberry DO (2005) Building design for abnormal loads and progressive collapse. Comput Civ Infrastruct Eng 20–3:194–205

3. Knoll F, Vogel T (2009) Design for robustness, vol 11. IABSE
4. Munch-andersen J (2009) Robustness considerations from failures in two large-span timber roof structures, pp 1–8
5. Sorensen JD (2011) Framework for robustness assessment of timber structures. Eng Struct 33–11:3087–3092
6. Cavaco ES, Casas JR, Neves LA, Huespe AE (2010) Robustness of corroded reinforced concrete structures—a structural performance approach. Struct Infrastruct Eng 9–1:1–17
7. Frangopol DM (1987) Effects of damage and redundancy on structural reliability. J Struct Eng 113–7:1533–1549
8. Fu G, Frangopol DM (1990) Balancing weight, system reliability and redundancy in a multiobjective optimization framework. Struct Saf 7(2–4):165–175
9. Baker JW, Schubert M, Faber MH (2008) On the assessment of robustness. Struct Saf 30–3:253–267
10. Faggiano B, Grippa MR, Marzo A, Mazzolani F (2009) Combined non-destructive and destructive tests for the mechanical characterization of old structural timber elements. In: 3rd international conference on advances in experimental structural engineering
11. Cruz H, Yeomans D, Tsakanika E, Macchioni N, Jorissen A, Touza M, Mannucci M, Lourenço PB (2013) Guidelines for the on-site assessment of historic timber structures. Int J Archit Herit. doi:10.1080/15583058.2013.774070
12. Parisi MA, Piazza M (2002) Seismic behaviour and retrofitting of joints in traditional timber roof structures. J Soil Dyn Earthq Eng 22:1183–1191
13. Branco JM, Cruz PJS, Piazza M, Varum H (2006) Behaviour of traditional Portuguese timber roof structures. In: Proceedings of the 9th WCTE, Portland, Oregon, USA
14. CEN, EN 1995-1-1:2005, Design of timber structures. P. 1-1: general—common rules and rules for building
15. Köhler J, Sørensen JD, Faber MH (2007) Probabilistic modeling of timber structures. Struct Saf 29–4:255–267
16. CEN, EN 1991-1-4:2005. Actions on structures. P. 1-4: general actions—wind actions
17. CEN, EN 1991-1-3:2003. Actions on structures. P. 1-3: general actions—snow actions
18. Welch BB (1995) Practical programming in Tcl and Tk, vol 391. Prentice Hall PTR NJ, USA

FP1101 and RILEM TC 245 RTE Training School on Assessment and Reinforcement of Timber Elements and Structures: The Role of the Academic Community in Disseminating Knowledge

Jorge M. Branco, Hélder S. Sousa, Eleftheria Tsakanika, Marco Jorge and Paulo B. Lourenço

Abstract In the past decades, timber has been considered as a secondary construction material owing to the emergence of other materials such as concrete and steel, and as a consequence a large amount of knowledge on how to use timber has been forgotten or even lost. This has led to badly designed interventions on timber structures, which have misinterpreted the past structural solution and led to significant structural performance problems. However, the European research community has made a significant effort towards the dissemination of knowledge regarding the design and assessment of timber structures. Within that scope, COST FP1101 had the objective of increasing the acceptance of timber in both the design of new structures, as well as in the repair of old buildings by developing and disseminating methods to assess, reinforce and monitor them. To that aim, FP1101 and RILEM TC 245 RTE in collaboration with the University of Minho (Portugal) organized a Training School (TS) on the assessment and reinforcement of timber elements and structures during 11–14 May, 2015. The training school had the objective of disseminating knowledge to young researchers, as well as to others interested in wood and timber construction and the rehabilitation market. During the TS, several experts provided information on the assessment of existing timber

J.M. Branco (✉) · H.S. Sousa · M. Jorge · P.B. Lourenço
ISISE—Department of Civil Engineering, University of Minho, Guimarães, Portugal
e-mail: jbranco@civil.uminho.pt

H.S. Sousa
e-mail: sousa.hms@gmail.com

M. Jorge
e-mail: marco@civil.uminho.pt

P.B. Lourenço
e-mail: pbl@civil.uminho.pt

E. Tsakanika
School of Architecture, National Technical University of Athens, Athens, Greece
e-mail: etsakanika@arch.ntua.gr

H. Cruz et al. (eds.), *Historical Earthquake-Resistant Timber Framing in the Mediterranean Area*, Lecture Notes in Civil Engineering,
DOI 10.1007/978-3-319-39492-3_39

structures, visual grading, interventions on heritage buildings and on analysis, repair and reinforcement of traditional and dowelled timber connections and structures. Also, full-scale tests on an old timber trusses and on connections were made examining both the original conditions and post failure reinforcement. This work presents the different areas of knowledge that were scope of that training school and its main achievements on the dissemination of knowledge to young researchers and other interested groups.

Keywords Assessment · Reinforcement · Timber trusses · FP1101

Introduction

Timber was, for many centuries, one of the most important and used widely structural materials. Its durability, if properly designed, combined with high ratio of strength to weight, allowed its use in several building types. The most common types of timber structures are found in roofs, from the simplest (e.g. wooden beams and purlins) to the most complex (e.g. three-dimensional trusses), however many types of timber walls and floors are also found across Europe (e.g. half-timbered walls).

Nevertheless, in the past decades, timber has been considered as a secondary construction material owing to the emergence of other materials such as concrete and steel, and as a consequence a large amount of knowledge on how to use timber has been forgotten or even lost. This has led to badly designed interventions on timber structures, which misinterpreted the past structural solution and led to significant structural performance and durability problems. However, the European research community had now made a significant effort towards the dissemination of knowledge regarding the design and assessment of timber structures. COST is an intergovernmental framework for European Cooperation in Science and Technology, allowing the coordination of nationally-funded research on a European level. It contributes to reducing the fragmentation in European research investments and opening the European Research Area to cooperation worldwide. One of its areas of interest is the management of Forests, their Products and Services.

In the memorandum of understanding of the European COST Action FP1101: Assessment, Reinforcement and Monitoring of Timber Structures [1], is clearly stated that the objective of that Action is to increase the acceptance of timber in the design of new structures and in the repair of old buildings by developing and disseminating methods to assess, reinforce and monitor. The assessment, reinforcement and monitoring of timber structures may be needed for several reasons, such as the loss of materials properties due to aging or degradation, exceptional incidents, or even change of use. To reach the objectives of the Action a multidisciplinary approach was used with contributions of involved parties from different areas of expertise. This fomented synergies between the various contributors and provided an effective way of discussing and disseminating the results from ongoing projects. The final result of that Action is to ensure an increase in the confidence of designers, authorities and end-users in the safe, durable and efficient use of timber and consequently increase its use in construction.

Within the scope of advancing scientific knowledge related to construction materials, systems and structures, and to encourage transfer and application of this knowledge world-wide, RILEM (International union of laboratories and experts in construction materials, systems and structures) also has a research activity entitled TC 245-RTE: Reinforcement of Timber Elements in Existing Structures, in which a specific milestone was to organize a workshop on state-of-the-art reinforcement methods.

One of the tools that these actions had for dissemination of knowledge was the training school, producing synergies between institutions aiming at reaching a larger audience. For those activities, students, lecturers, and practitioners were invited to participate. In the case of FP1101, its members had the teaching obligation to pass on knowledge to current and future users with the emphasis on the opportunities and not the limitations of timber as a structural material.

To that aim, FP1101 and RILEM TC 245 RTE, in collaboration with the University of Minho, (Guimarães, Portugal) organized a Training School (TS) on the assessment and reinforcement of timber elements and structures during 11–14 May, 2015. Here it presented the different areas of knowledge that were included in that training school and its main achievements in the dissemination of knowledge to young researchers and other interested groups.

Training Schools

A training school dealing with assessment and reinforcement of timber elements and structures took place at the University of Minho (Guimarães, Portugal) from the 11th to the 14th of May 2015. The training school had the contribution of FP1101 and RILEM TC 245 RTE actions. Members of those actions were invited to attend, priority of choice being given to young researchers (PhD students or early doctorate), as the objective of the training school was to disseminate information among the future generation of wood and timber specialists. People from other backgrounds, such as master students and industry professionals were also invited to attend and to share their experience.

This training school was the fifth one organized within the scope of FP1101. The first occurred in Greece, on September 2013, dealing with assessment. Shortly after, in December of the same year, a training school was organized in Belgium dealing with assessment and reinforcement. Following, on June 2014, a training school about monitoring took place in France. Finally, on December 2014, a training school in Ireland discussed issues related to reinforcement.

Objective and Framework. The training school, to which this papers refers, had the objective of disseminating knowledge to young researchers, as well as to

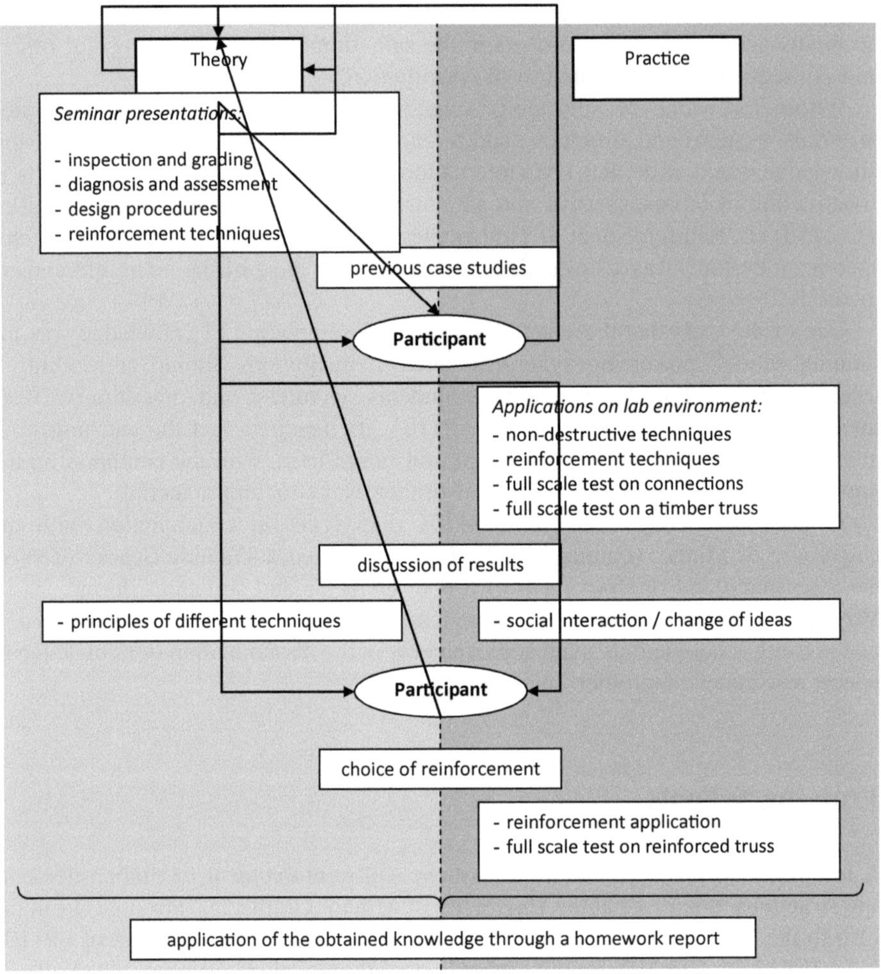

Fig. 1 Layout and interaction of participants in the training school activities

others of the wood and timber construction and rehabilitation market. The global layout of the training school is presented in Fig. 1.

During the training school, several experts provided information on the assessment of existing timber structures, visual grading, intervention on heritage buildings and on analysis, repair and reinforcement of traditional and dowelled timber connections and structures. The theoretical framework was provided by describing the methods and guidelines (standardization and design rules) through presentations and by sharing the experience of the different lecturers in previous case studies in which they were involved.

After the initial theoretical framework, the participants were invited to participate in practical applications on a laboratory premises (Structural Laboratory of

University of Minho). The diagnosis made of a timber collar truss was presented to the participants and onsite applications of non-destructive tests, namely penetration impact, drilling resistance and ultrasound tests, were shown. Different strengthening and repairing techniques with composite materials and their application were also presented. The main laboratory activities comprised the full scale tests of traditional carpentry and dowelled connections and finally of the full scale test of an old collar timber truss with span of over 13 m taken to failure.

Following that, the participants had the possibility of choosing two different repair techniques to be applied to the elements of the truss that were damaged during the test. To help in their choice, the theoretical design considerations of commonly used techniques were provided. Moreover, the interaction between participants was encouraged so that the solution could fit different ideologies. The only restriction was related to the availability of materials and time needed to for application.

Finally, the participants could see the application of the repair techniques that they chose and then observe the behaviour of the truss in a new full scale test.

Lectures and Participants. The lecturers were chosen regarding their past experience on the each relevant topic of the training school. Lecturers from different institutions were present with the intention of presenting a multi-perspective approach to each topic. Besides lecturers from the organizing institution (University of Minho), also lecturers from the University of Coimbra (Portugal) and National Technical University of Athens (Greece) were present. Although all came from the academic and research fields, their experience is also vast in practice application and was taken into consideration when giving the lectures. The lecturers comprised experts from Full Professor positions to young doctorates.

The participants, although all working on the topic of wood and timber, had very distinct technical backgrounds and experience, coming from several institutions from all over Europe, however with strong percentage of participants coming from countries of the Mediterranean Region. Fourteen different countries were represented among the participants, with 55 % coming from institutions of the Mediterranean Region (Fig. 2).

Lectures. Experts discussed important topics related to interventions on old timber structures such as inspection, diagnosis, non-destructive methods, joints analysis and strengthening techniques. The different areas of knowledge, although very specific, were complementary and were presented in such a way and order that resembled a common intervention. Therefore, inspection and definition of existing timber buildings, followed by grading of materials and structural assessment were considered. Finally the basis of design and case studies were discussed.

During the training school the importance of preserving timber structures was emphasised, mentioning that one must ensure that the responsibility for cultural heritage and its management belongs, in the first place, to the cultural community that has generated it, and subsequently to that which values it.

It was also made clear that conservation is difficult and requires a different approach and skills from those employed in designing new construction. This

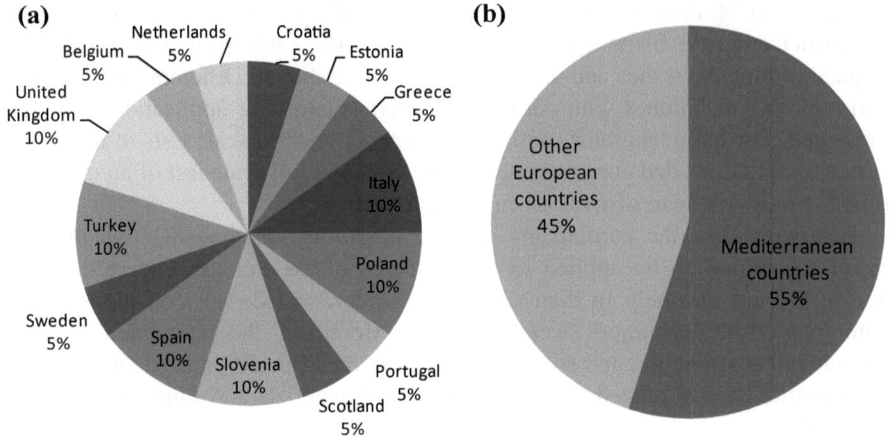

Fig. 2 Distribution of participants: **a** by countries; **b** by region

premise was also taken into consideration when mentioning to the participants the different perspectives of the different groups involved.

The main message that was passed to the participants was that although conservation, assessment and intervention may have simple theoretical concepts they may be very difficult to apply in practice. To sum up, durability, compatibility, reversibility and minimum intervention are priorities, however there should be no bargaining on safety. The increase in knowledge allows the production of better results with lower costs. To that matter, traditional and novel intervention techniques may be acceptable, as well as change of use (e.g. a living cultural heritage). Conservation is an art, requiring specific education and experience but some assets can hardly survive without significant remedial measures.

Experimental Demonstrations. During the training school three main practical applications were considered in laboratory conditions. The first comprised demonstrations of repair and strengthening techniques, as well as non-destructive techniques. The participants were invited to actively help and try each one of those techniques (Fig. 3).

After that full-scale tests on timber connections were shown to the participants. The timber connections comprised traditional carpentry and dowelled connections (in original and reinforced conditions) (Fig. 4). The connections with and without dowels were tested under compression and tension. The load bearing capacity was measured and the specimen deformed plastically in the contact area, and local shear failure (on one side only) was observed due to a knot very near to the notched area. Shear failure of the notched connection caused brittle failure of the connection, not allowing a large plastic yield deformation.

The last practical application, comprised the full-scale tests on a timber truss in both original and repaired conditions. The setup of the tests and layout of the measuring equipment (LVDTs) were explained to the participants and while the test was made the level of load and displacement was recorded (Fig. 5).

(a) **(b)** **(c)**

Fig. 3 Hands-on applications: **a** repair techniques for timber elements; **b** strengthening techniques using composite materials; **c** non-destructive testing of existing elements

(a) **(b)** **(c)**

Fig. 4 Full scale tests on timber connections: **a** test setup for a dowelled connection; **b** failure of a carpentry connection; **c** load-displacement curve

(a) **(c)**

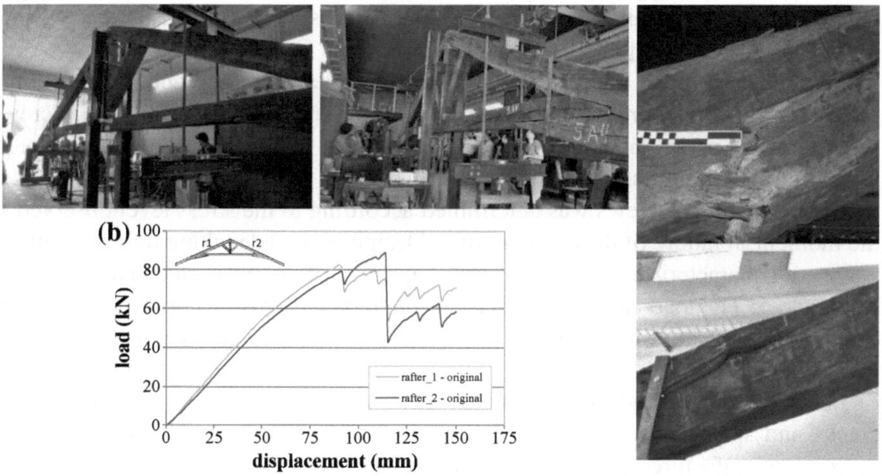

Fig. 5 Full scale tests a timber truss: **a** test setup; **b** load-displacement curves; **c** failures on the rafters

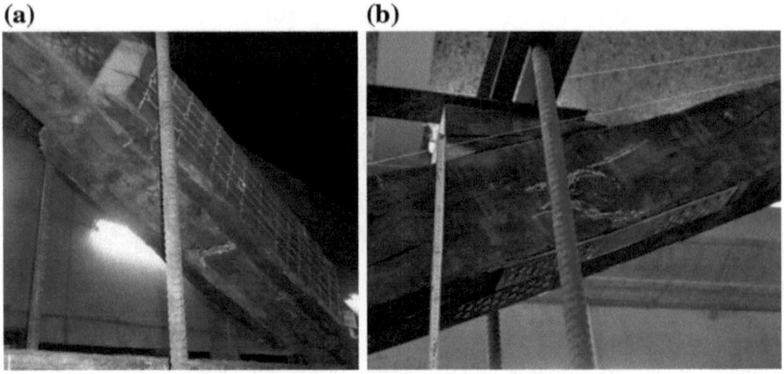

Fig. 6 Repair techniques: **a** screwed timber elements; **b** screwed metal plates

After seeing the test made to the full-scale timber truss, the participants had the final decision on the type of repair technique. Two different methods for repair were considered: either using timber element or with metal plates screwed to the original rafters. By the use of these two simple repair methods it was possible to obtain approximately 80 % of the original load bearing capacity of the truss (Fig. 6).

In the first repair method, the cross-section was retrofitted with timber elements placed on the lateral sides of the cross-section by means of screws. The cross-section of the added timber elements had half the width of the original cross-section and the same height, therefore the area and the inertia of the repaired cross-section is equal to the original one if no contribution of the original cross-section is considered. The number of screws used in the connection between timber elements was calculated accordingly to the appropriate code [2] taking into consideration that the screws before and after the failed cross-section had to be sufficient to resist the existing normal stress on the rafter.

In the second repair method, metal plates were placed on the bottom face of the cross-section, corresponding to the tension side where failure was initiated. Screws were inserted perpendicular to the timber element fixing it to the metal plates. Owing to the configuration of the cracks, two metal plates were used. The number and diameter of the screws was determined according to the stress level observed at the moment of failure of the original truss. Design was made taking into account the value of characteristic load carrying capacity of screws calculated through Eurocode 5. The load carrying capacity of the repaired cross-section was estimated considering the simplified analysis of mechanically jointed beams provided also in Eurocode 5 [2].

In both repair methods, layout of the screws was defined attending to the existing cracks and spacing was determined according to Eurocode 5 [2] and to the connectors' manufacturer ETA (European Technical Assessment), aiming at maximizing the number of effective connectors.

Fig. 7 Group photo with participants and lecturers of the training school in the laboratory premises

Conclusions and the Role of the Academic Community

The objective of the training school was to disseminate information among the future generation of wood and timber specialists on the preservation of existing timber structures and to encourage the construction of new timber buildings. A holistic approach was used to transmit the main guidelines and procedures for a correct and full assessment of existing timber structures and following interventions. Nevertheless, the role of the training school was not limited to its time duration. After the training school, the participants prepared different homework reports where they were able to use the information provided and even to go deeper on the topics that were mentioned in the training school. The homework reports may be found in [3–5].

Through this procedure, it is clear that this training school served as an initial step for further investigation and also allowed a younger generation to obtain the necessary tools to better understand, assess and intervene in existing timber structures (Fig. 7).

This paper mentioned the different areas of knowledge that were scope of that training school and how they interacted as to provide a sustainable framework for knowledge transference.

Acknowledgments The information obtained through training school: Assessment and reinforcement of timber elements and structures (Guimarães, Portugal) within scope of FP1101 and RILEM TC 245 RTE is acknowledged. The support of the Structural Lab from University of Minho is also acknowledged. The assistance of the companies Rotho Blaas and Augusto de Oliveira Ferreira e Companhia Lda (AOF) are greatly acknowledged.

References

1. European Cooperation in Science and Technology—COST Action FP1101 (2011) Memorandum of understanding for the implementation of a European concerted research action designated as COST action FP1101: assessment, reinforcement and monitoring of timber structures, 24 p
2. CEN (2004) EN 1995-1-1:2009—Eurocode 5—design of timber structures—part 1-1: general —common rules and rules for buildings. National specifications, national comments and national supplements concerning ÖNORM EN 1995-1-1. European Committee for Standardization (CEN), Brussels (EN 1995-1-1:2009)
3. Cibecchini D, Bjorngrim N, Sousa HS, Jorge M, Branco JM (2015) Repairing of a timber truss through two different techniques using timber elements and screwed metal plates. HEaRT2015, Lisbon, Portugal
4. Šobra K, Avez C, Aktaş YD, Rijk R, Burawska I (2015) Load-bearing capacity of traditional dovetail carpentry joints with and without dowels: comparison of experimental and analytical results. HEaRT2015, Lisbon, Portugal
5. Šobra K, Rijk R, Aktaş YD, Avez C, Burawska I (2015) Experimental and analytical assessment of the capacity of traditional single notch joints and impact of retrofitting by self-tapping screws. HEaRT2015, Lisbon, Portugal

MIX
Papier aus verantwortungsvollen Quellen
Paper from responsible sources
FSC® C105338

If you have any concerns about our products,
you can contact us on
ProductSafety@springernature.com

In case Publisher is established outside the EU,
the EU authorized representative is:
Springer Nature Customer Service Center GmbH
Europaplatz 3, 69115 Heidelberg, Germany

Printed by Libri Plureos GmbH
in Hamburg, Germany